CLINICAL ANATOMY AND PATHOPHYSIOLOGY

for the Health Professional

by Joseph V. Stewart, MD

MedMaster, Inc., Miami

ISBN#0-940780-06-2

Made in the United States of America

Published by
MedMaster, Inc.
P.O. Box 640028
Miami, Fla. 33164

*formerly entitled Clinical Anatomy and Physiology for the Frustrated and Angry Health Professional.

DEDICATION

For my mother, Lelia Stewart, and my father, Joseph Stewart, Sr.,
for their great love and attention in early years

and

To the student, who **likes** a short book from which to learn much

ACKNOWLEDGMENTS

I wish to thank Alexander Lane, whose enthusiasm for teaching and criticism of tradition helped inspire the book. Thanks to Nancy Gorman, who typed the first bumpy draft of the manuscript, to Judi (my wife, and a nurse) for her critique of the book, to Ken Smith for his splendid art work, to Joyce Romz who typed the final draft, and to Sixten Netzler who prepared the cover.

TABLE OF CONTENTS:

CHAPTER

PREFACE

What do I absolutely need to know in anatomy and physiology? What is important? Am I wasting time learning too much detail (other more poignant terms are often used)? Does the instructor really know what I need? Questions like this from students are endless. The writing of this book attempts to answer questions one and two. The answer to the third question is yes, usually time is wasted learning a voluminous amount of trivia. Does the instructor really know what I absolutely need to know? Often not.

It is my firm belief that the beginning student—and I suspect most others—does not learn much from large texts, including many labeled "essential", "basic", "manual" and such. Frequently they are merely textbooks in disguise, rarely containing fewer than 400 pages. In medical school I learned little from big books unless I already knew much of the material. I did waste a lot of precious time roaming around medical bookstores in hopes of coming across a new small paperback that would give me only the basics of a particular subject. I really did not want to know more than that, and it always offended me when I encountered the same old disguised text. Occasionally, however, I was rewarded, and used that little book from which to learn.

This book originated from 15 years of teaching anatomy, physiology and pathophysiology to college students, mostly in nursing, many with little background in science. Most books in the nursing field are written by those with sparse exposure to medicine. As a result, the student is overwhelmed with facts. Frequently, important material is not stressed or is omitted completely.

A significant amount of faulty information has been perpetuated by some large textbook companies, (accountable to no-one) that have dominated the nursing market for several years, mainly because of colorful illustrations. I recently thumbed through two best sellers. I was curious to see how the authors handled the thorny problem of a simple explanation of the relationship of thyroid disease to the immune system. In the first book, nothing was said about thyroid disease. In fact, hypothyroidism and hyperthyroidism were not mentioned at all—only a vague phrase about the hormone, thyroxin, and the "basal metabolic rate", which has not been used to evaluate thyroid function in twenty years. In the second book, the hypo- and hyper- conditions were mentioned, but the diagnostic test described, the "protein-bound iodine" was also discarded in the 1960's for the more accurate thyroxin level. No mention was made in either book of immune mechanisms and thyroid problems. The authors were unaware, evidently, that any exist.

I have been in the unique position of being able to evaluate graduates from a wide variety of programs and schools as they come to work in the hospital. Commonly, the student is deficient not only in sophisticated concepts, but in minimal basic anatomy and physiology needed to function on the ward. This is not the fault of the student, but of the instructor and textbook.

A fundamental, provocative question inevitably arises at some point in any anatomy and physiology course, asked by the angry and frustrated student: "Why do I have to know this?" The response from the irritated instructor is often, "Because it is important". Important to whom? The practicing neurosurgeon? The biochemist? It is my belief that the instructor must justify everything he teaches as to relevance and importance.

This outline presents an up-to-date summary of the important aspects of human anatomy and physiology, with pertinent pathophysiology for the beginner. Clinical correlation is stressed, since students seldom study anatomy and physiology for a general fund of knowledge; they learn with a goal in mind — someday the material will be useful. Consequently, disorders relevant to the anatomy and physiology just studied are presented at the end of the chapter. Only common or important problems are included.

This book was created not only for students, but for instructors as well. Disorders at the end of the chapter may be selectively excluded, according to needs.

An alternative title for this book could be A SHORT COURSE IN PRACTICAL ANATOMY AND PHYSIOLOGY— practical because it by-passes items which people seldom use. For instance, the names of the foramina of the skull and many hand muscles are omitted, as the need for this information rarely occurs. Physics is completely excluded and chemistry is downplayed (which should be a refreshing change for the beginner). In my opinion, physics, general chemistry and the intricacies of cell structure and function have no place in an introductory book.

The last chapter consists of regional anatomy, including surface anatomy. Students must be aware of which structures (nerves, arteries, etc.) are in various locations of the body because of vulnerability to local trauma. Also, the insertions of intravenous lines and the drawing of blood require knowledge of certain landmarks. Trying to find structures in the antecubital fossa, for example, using the traditional "systems" approach, is an exercise in futility.

Basic, or core, material is highlighted in italics to facilitate learning and to downplay less important material for review.

Joseph V. Stewart M.D. —Department of Emergency Medicine
Silver Cross Hospital
Joliet, Illinois
—Former Instructor, Anatomy and Physiology,
Triton College, Rivergrove, Ill.

-ONE-

ORGANIZATION OF THE BODY

I. *Surface, regional and systemic anatomy.*
   A. *Surface anatomy* is the identification of structures that can be seen or palpated. An ideal way to study surface anatomy would be to have an artist's model available at all times, so that one could observe, palpate and identify important anatomical features whenever one wished! Since this is impractical, we are forced to rely on pictures or drawings.
   B. *Regional anatomy* is the identification of structures found in a region of the body (e.g. elbow, neck).
   C. *Systemic anatomy* is the study of individual organ systems (e.g. endocrine system, respiratory system). In this book, the systemic approach is used in each chapter except the last, which discusses important regions, including surface anatomy.

II. *Body planes (sections)* (Fig. 1-1):
   A. *Sagittal:*
      1. *Midsagittal:* divides the body into equal left and right parts.
      2. *Parasagittal:* divides the body into unequal left and right parts.
   B. *Frontal (coronal):* divides the body into anterior and posterior parts.
   C. *Transverse (horizontal, cross, X-section):* divides the body into upper and lower parts.
   D. *The "anatomical position"* of man is with feet together, arms to the sides and palms facing anteriorly.

III. *Direction* (Fig. 1-2):
   A. *Superior* (cranial, cephalic): up
   B. *Inferior* (caudal): down

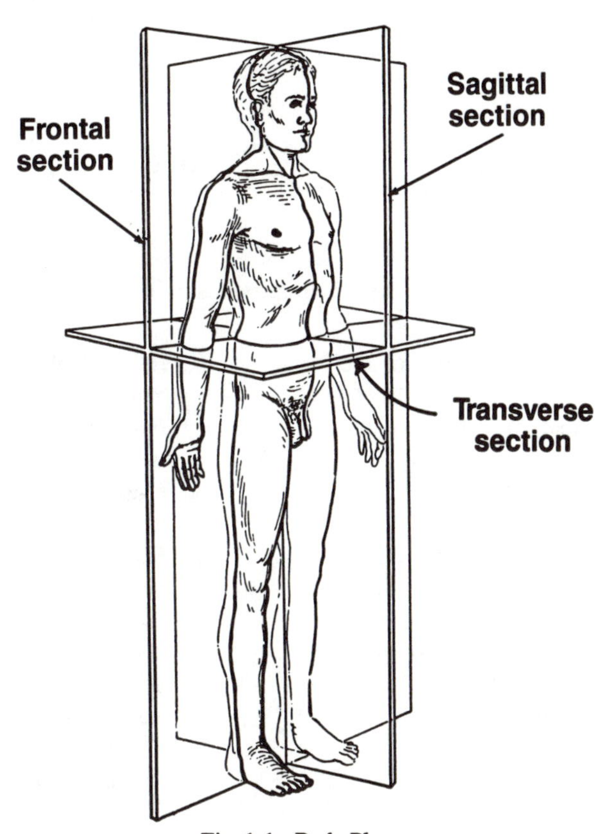

Fig. 1-1 Body Planes.

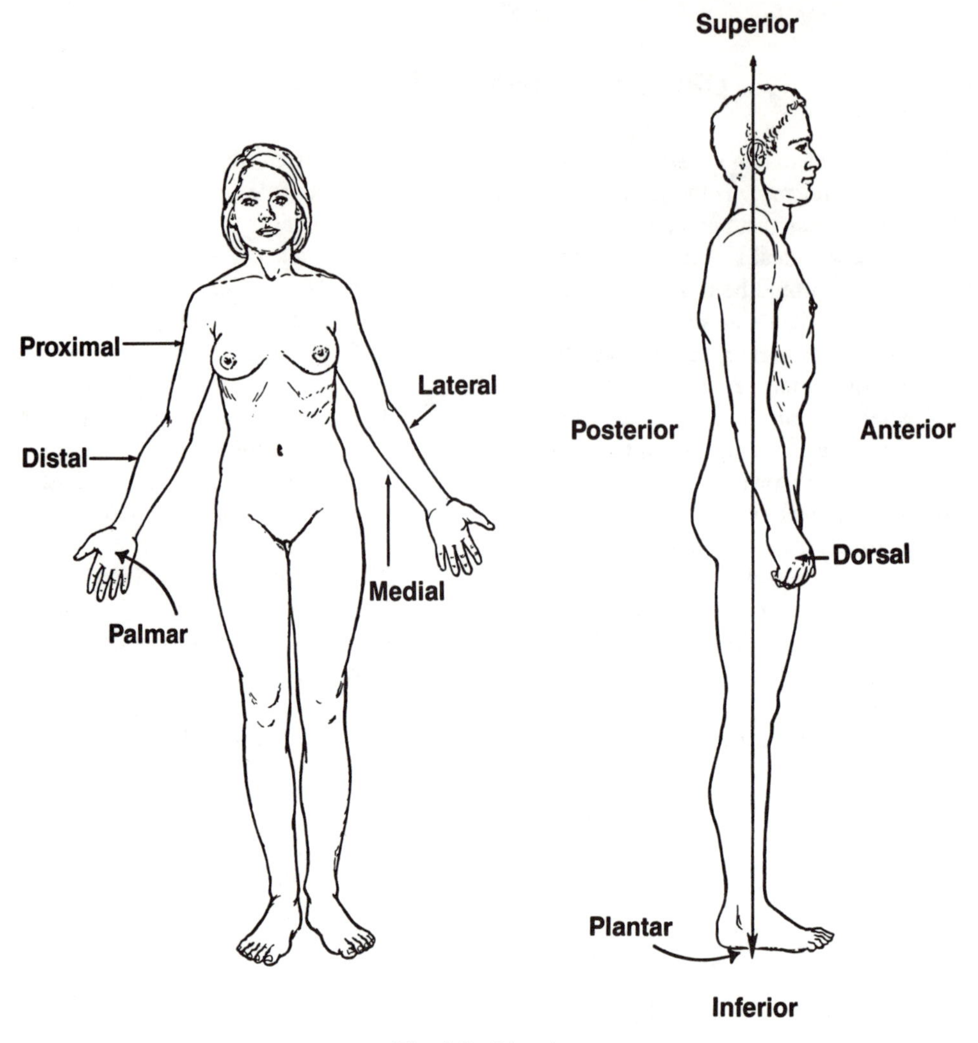

Fig. 1-2 Direction.

C. *Anterior* (ventral): the front of the body
D. *Posterior* (dorsal): the back of the body
E. *Medial:* the midline of the body
F. *Lateral:* away from the midline of the body
G. *Proximal:* close to the torso (used mainly for extremities)
H. *Distal:* away from the torso (used mainly for extremities)
I. *Superficial* (external): outside the body
J. *Deep* (internal): inside the body
K. *Palmar* (volar): the palm surface of the hand
L. *Plantar:* the sole of the foot

IV. *Movements* (Fig. 1-3):
A. *Flexion:* decreasing the angle between two bones
B. *Extension:* increasing the angle between two bones
C. *Abduction:* away from the midline (used mainly for extremities)
D. *Adduction:* toward the midline (used mainly for extremities)
E. *Rotation:* turning around a central axis
F. *Circumduction:* describing a circle

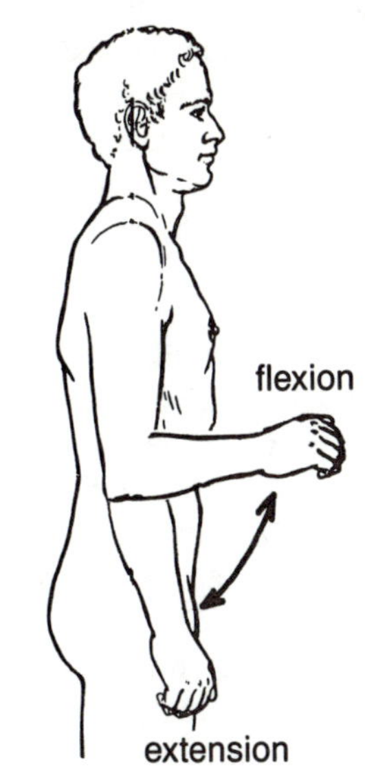

a. Flexion/extension

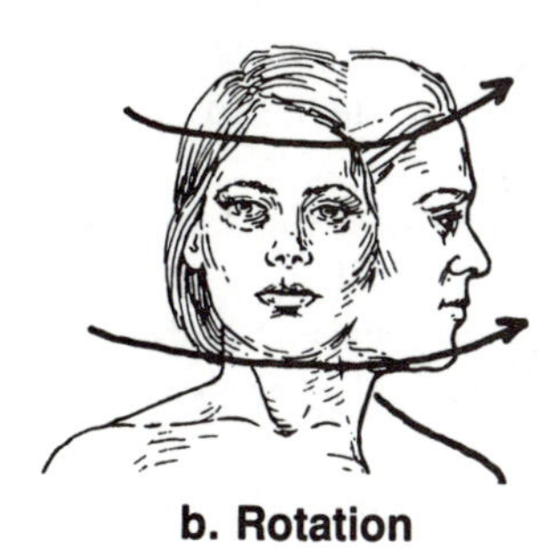

b. Rotation

c. Circumduction

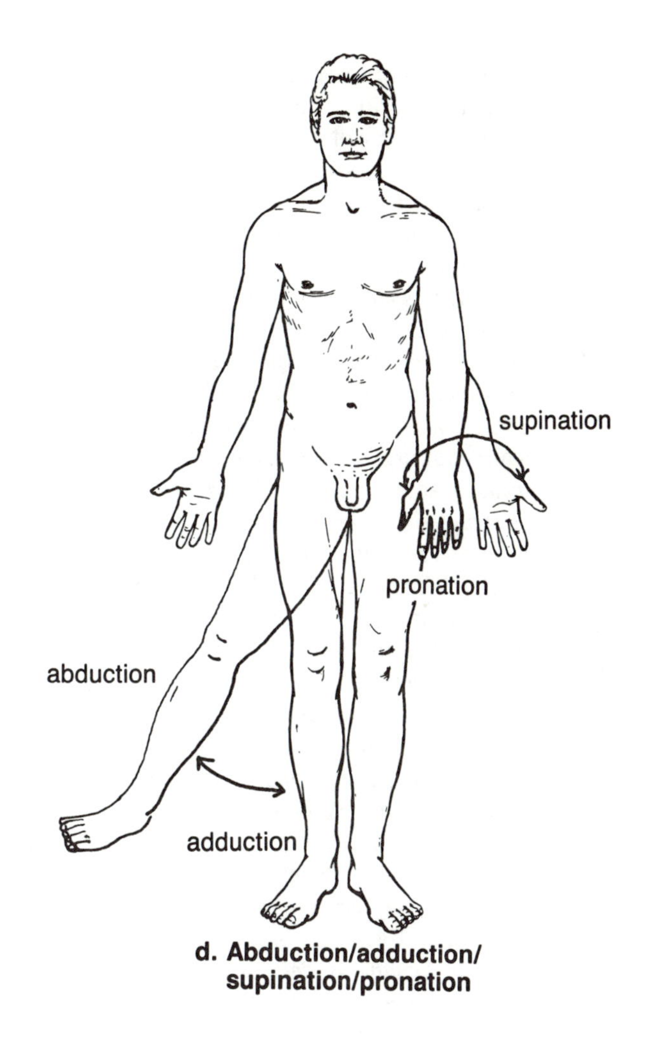

d. Abduction/adduction/ supination/pronation

eversion

inversion

e. Eversion/inversion

Fig. 1-3 Movements.

G. *Supination:* movement of hand with palm anterior ("soup"ination, as if holding a cup of soup)

H. *Pronation:* movement of hand with palm posterior

I. *Inversion:* movement of sole of foot inward

J. *Eversion:* movement of sole of foot outward

V. *Tissues, organs and organ systems.*
Many cells, the smallest living functional units, make up a tissue, various tissues make up an organ, and a number of organs make up an organ system. The *four types of tissues are epithelial, connective, muscular* and *nervous.* An organ such as the heart or kidney is composed of several tissues contributing to a common function. Each chapter in this book covers one, or occasionally two, organ systems.

A. *Epithelial tissue* (epithelium) is *lining tissue*. Its functions are *protection* (e.g., skin), *absorption* (e.g., digestive tract) or *secretion* (e.g., glands).

1. *Epithelial cells* may be *one or more layers thick* (simple or stratified) and may assume *various shapes* (Fig. 1-4):

a. *Squamous cells* are flattened types: e.g., simple squamous epithelium is found lining the interior of the lungs, heart, and blood vessels (endothelium). It is the outer layer of peritoneum surrounding the abdominal organs. Skin is composed of stratified squamous epithelium.

a. Simple Squamous

b. Cuboidal

c. Columnar

d. Strafified Squamous

Fig. 1-4 Types of Epithelial Tissue.

b. *Cuboidal cells* are cube-shaped: e.g., glands are composed of simple cuboidal epithelium.

c. *Columnar cells* are cylinder-shaped: e.g., the digestive tract is composed of simple columnar epithelium.

2. *Moist membranes* are *double serous* (serous: serum-like) *structures* consisting of simple squamous epithelium and connective tissue which contain blood and lymph vessels. A small amount of fluid filters from the capillaries through the epithelium into the spaces between the membranes, preventing friction. The membrane lying on the organ itself is the *visceral* (viscera: organ) portion; the membrane surrounding the body cavity is the *parietal* (cavity wall) portion (Fig. 1-5). Examples are:

a. The *pericardium* surrounding the heart.

b. The *pleura* on the lungs and inner chest wall.

c. The *peritoneum* on the abdominal organs and inner abdominal cavity.

d. The single *synovial membrane* of a joint also secretes a similar type of fluid, although in this case the lining cells are connective tissue.

e. Sometimes, disease results in increased amounts of fluid collecting in these spaces, which must be evacuated (e.g. pleural effusion from congestive heart failure or cancer).

B. *Connective tissue* forms the structure of the body (Fig. 1-6).

1. *Classification*:

a. *Connective tissue proper* is the framework for most organs.

b. *Fat* (adipose) functions for energy storage and metabolism.

c. *A tendon* connects muscle to bone (e.g. Achilles tendon of

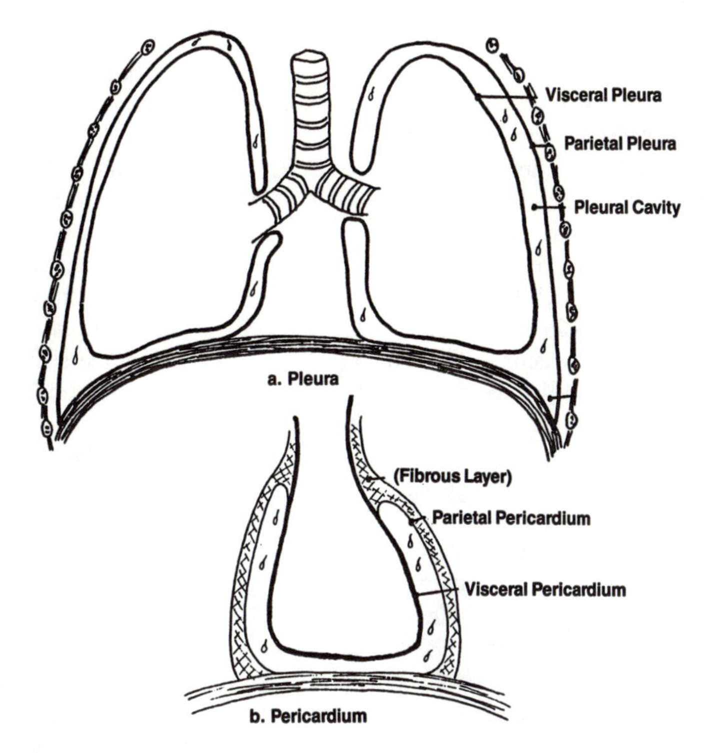

Fig. 1-5 Moist Membranes. (Modified, with permission, from Goldberg, Clinical Anatomy Made Ridiculously Simple, MedMaster, 1984.)

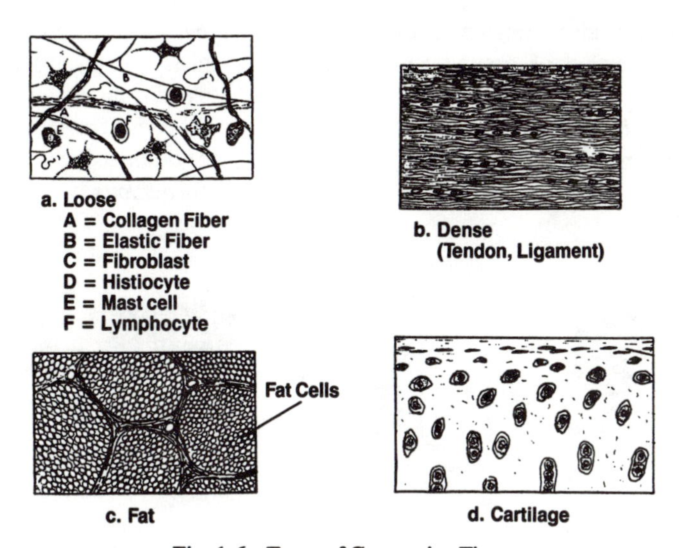

Fig. 1-6 Types of Connective Tissue.

heel). An *aponeurosis* is a flattened, sheetlike tendon.

d. *A ligament* connects bone to bone (e.g. a joint capsule).

e. *Cartilage* is firm connective tissue (e.g. nose, ear)

f. *Bone* — (see chapter 4).

g. *Blood and lymph* — (see chapter 10).

2. *Constituents*:
   Connective tissue is categorized as *loose* or *dense* and consists of cells, fibers and ground substance.
   a. *Cells*. The *fibroblast* manufactures fibers and ground substance and is responsible for repair. Other cells found in connective tissue are macrophages, mast cells, plasma cells and white blood cells, described in chapters 2 and 10.
   b. *Fibers*. Various types of fibers are found — collagen, elastic and reticular. *Collagen* is the more important, and makes up the majority of total body protein. It is a large polypeptide and is made by the fibroblasts. When *vitamin C* is lacking, fibroblasts make defective collagen.
   c. *Ground substance*, manufactured by the fibroblasts, fills the spaces between cells and fibers. Collagen also constitutes the main portion of ground substance.
3. *Functions of connective tissue*.
   a. *Support*: mostly by fibers
   b. *Defense*: by white blood cells
   c. *Repair*: by fibroblasts
   d. *Transport*: nutrients move through the ground substance
   e. *Structure*: derivatives such as cartilage, bone

C. *Muscular tissue* is designed for contraction, thus the movement of other tissues and organs (Fig. 1-7). The three types of muscular tissue are:
   1. *Skeletal (voluntary, striated)*: muscles of the skeleton, composed of fibers and smaller fibrils, containing actin and myosin filaments. *Cross-striations* are present.
   2. *Smooth (involuntary)*: composed of long thin cells arranged in sheet-like fashion, forming the walls of internal organs.
   3. *Heart* (cardiac): also contains *cross-striations*, resembling skeletal muscle. However, when one area of the heart is stimulated, the entire heart contracts, rather than contracting in an isolated fashion like skeletal muscle.

D. *Nervous tissue* is composed of *neurons*, specialized cells for the transmission of *electro-chemical impulses* to and from various parts of the body. The *central nervous system* consists of the *brain* and *spinal cord*; the *peripheral nervous system* comprises the *spi-*

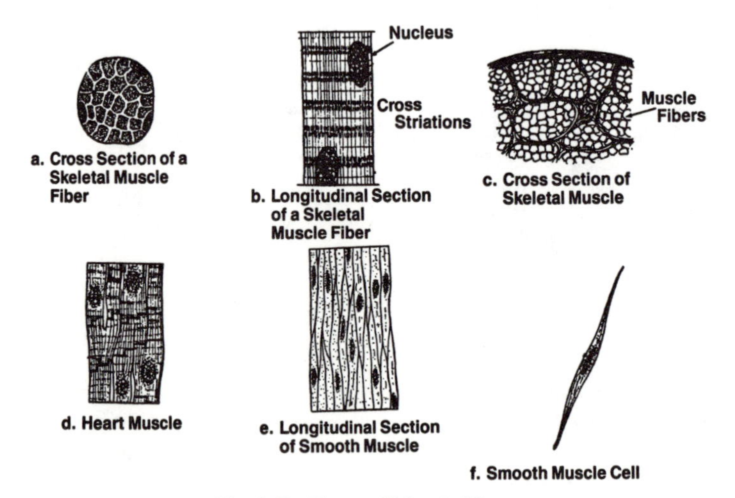

Fig. 1-7 Types of Muscle Tissue.

nal nerves, *cranial nerves* and the *autonomic nervous system* (Fig. 1-8).

1. *Structure of a neuron* (Fig. 1-9):
   a. *Cell body*
   b. *Axon*
   c. *Dendrite*, or dendrites
2. The term "*nerve fiber*" means an axon or dendrite. Most fibers are axons surrounded by a lipid-protein layer called *myelin*, responsible for the *rapid conduction* of *the nerve impulse.*
3. *Synapses* (gaps) exist between neurons, and between a neuron and a muscle (discussed in chapter 5). The nerve impulse is conducted from the axon of one cell across the synapse by a *synaptic transmitter* (*neurotransmitter*) to the cell body or dendrite(s) of a second neuron.
4. The *neurotransmitter* is released by the axon, bridges the synapse for a moment, then dissipates and is reformed in the axon. Some examples of neurotransmitters are:
   a. *Acetylcholine* is at the synapse between nerve and muscle (*motor end-plate*) and in the *autonomic* and *central nervous systems.*
   b. *Norepinephrine* is in the *autonomic* and *central nervous systems*. Neurons that secrete norepinephrine are *adrenergic*; those that secrete acetylcholine are *cholinergic*. The effects of autonomic adrenergic stimulation are usually the opposite of cholinergic stimulation.
   c. Due to similarities in chemical structure and action, the neurotransmitters *epinephrine (Adrenalin), norepinephrine* and *dopamine* are called the *catecholamines* (see chapter 7).
   d. Twenty-five other *neurotransmitter substances,* including serotonin, glycine, gamma amino butyric acid (GABA), glutamic acid, Substance P, enkephalins, endorphins, histamine and prostaglandins, have been identified in various parts of the nervous system.
5. *Sensory* (afferent) fibers receive impulses from the skin, muscles,

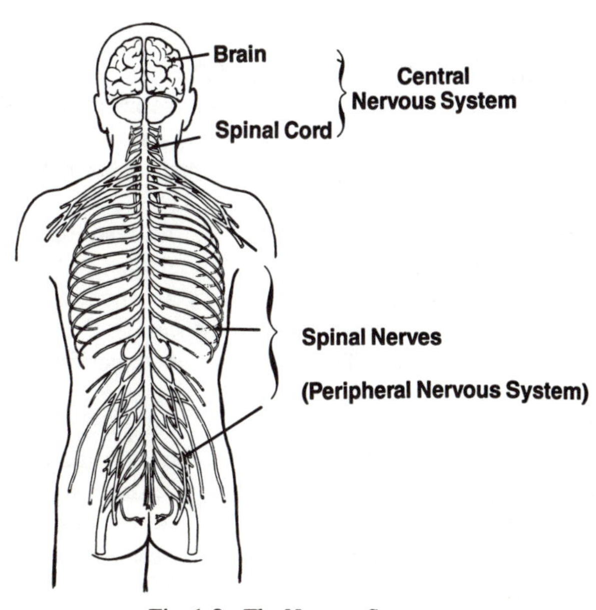

Fig. 1-8 The Nervous System.

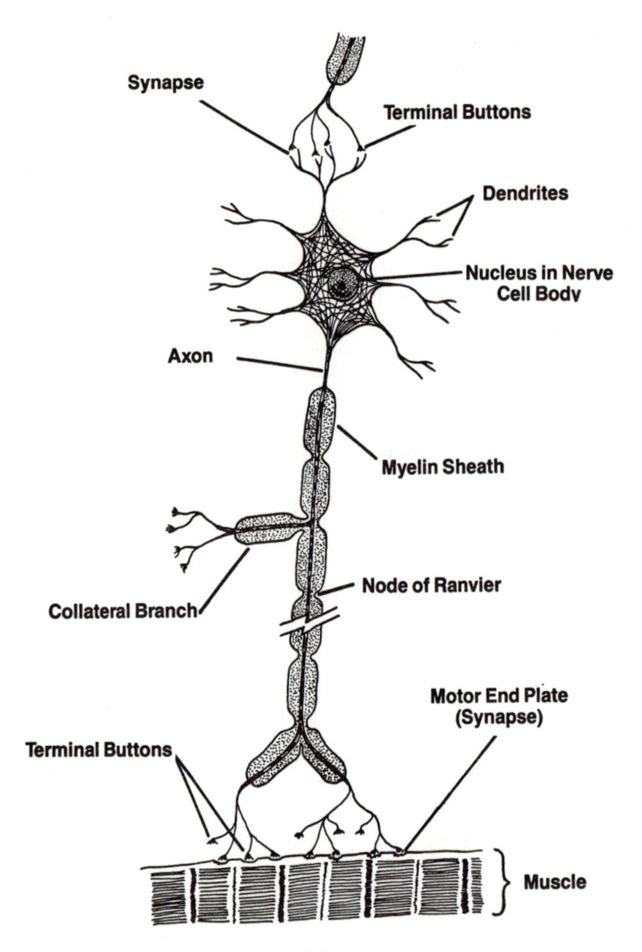

Fig. 1-9 A Motor Neuron and Synapses.

tendons or internal organs and relay them to the brain or spinal cord. A loss of sensation is *anesthesia*. A partial loss is *hypesthesia*. An abnormal or "tingling" feeling is *paresthesia*.

6. *Motor* (efferent) fibers react to sensory impulses. They originate in the brain or spinal cord and terminate on muscles or glands. A loss of motor function is *paralysis*. A partial loss is *paresis*.
7. *A "nerve"* means *many nerve fibers*. Most peripheral nerves (e.g. radial, sciatic) are *mixed* nerves — that is, they contain both *sensory and motor fibers* (Fig. 1-10).

VI. *The nerve and muscle impulses*.

Nerve and muscle cells are specialized for the conduction of electro-chemical impulses down the length of the cell (Fig. 1-11).

A. At rest, there is an abundance of *sodium* on the *outside* of the cell, and an abundance of *potassium* on the *inside*. If the charges of all substances, both positive and negative, are added up on each side of the nerve or muscle cell, there are more positively charged ones on the outside. Thus, a cell in the resting state is *considered to be positive on the outside* and *negative on the inside*. This is the *resting potential*.

B. When the cell is stimulated, the impulse proceeds down the cell in

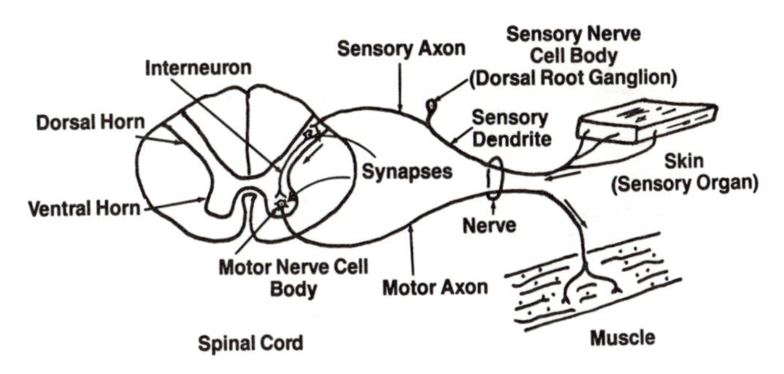

Fig. 1-10 Sensory and Motor Neurons.

fuse-like fashion. This is the *action potential.* The cell becomes permeable to sodium, and *sodium flows into the cell*, making the inside more positive. This is *depolarization*.

C. *Potassium then flows out of the cell*, restoring electrochemical balance. This is *repolarization*. The amount of sodium and potassium exchanged during an impulse is small. However, after a certain number of impulses, significant amounts of sodium and potassium are on the wrong sides of the membrane.

D. In order to *restore ionic balance* (move sodium back outside and potassium back inside the cell), energy is required, supplied by adenosine triphosphate (ATP). Sodium is pumped out and potassium in. This is the *sodium-potassium pump*.

E. *Calcium* plays a role in the nerve impulse, and an important role in muscle depolarization and contraction. This is described in chapters 5 and 9. The electrocardiogram (EKG) is a recording of these events in heart muscle.

F. *Summary of the nerve/muscle impulse*:

1. *Resting state*: cell positive on the outside, more sodium on the outside.
2. *Depolarization*: sodium moves into the cell, making the inside positive and the outside negative (in muscle, after depolarization, contraction takes place).
3. *Repolarization*: potassium moves outside, making the outside positive again.
4. *Ion restoration*: using ATP (energy), sodium is pumped out and potassium is pumped in.

VII. *The cell* is the *smallest living functional unit* (Fig. 1-12).

A. *Using the light microscope*, one may observe:

1. *The cell membrane*, a semipermeable structure composed of proteins and fats.
2. *The cytoplasm* contains many substances and particles, and appears pink with eosin stain. The older term, *protoplasm*, means "living material" and is used to denote nucleus and cytoplasm together.
3. *The nucleus* appears purplish with hematoxylin stain.
   a. *The nuclear membrane* is semipermeable, permitting the passage of large molecules between the nucleus and cytoplasm.
   b. *The purple chromatin (chromosomes)* contains genetic material. Each chromosome consists of protein and the compound *deoxyribo-nucleic acid (DNA). Genes are specific regions along the double-coiled DNA molecule (double-helix)* (Fig. 1-13). There are 46 chromosomes in all cells of man except the egg and the sperm, which contain

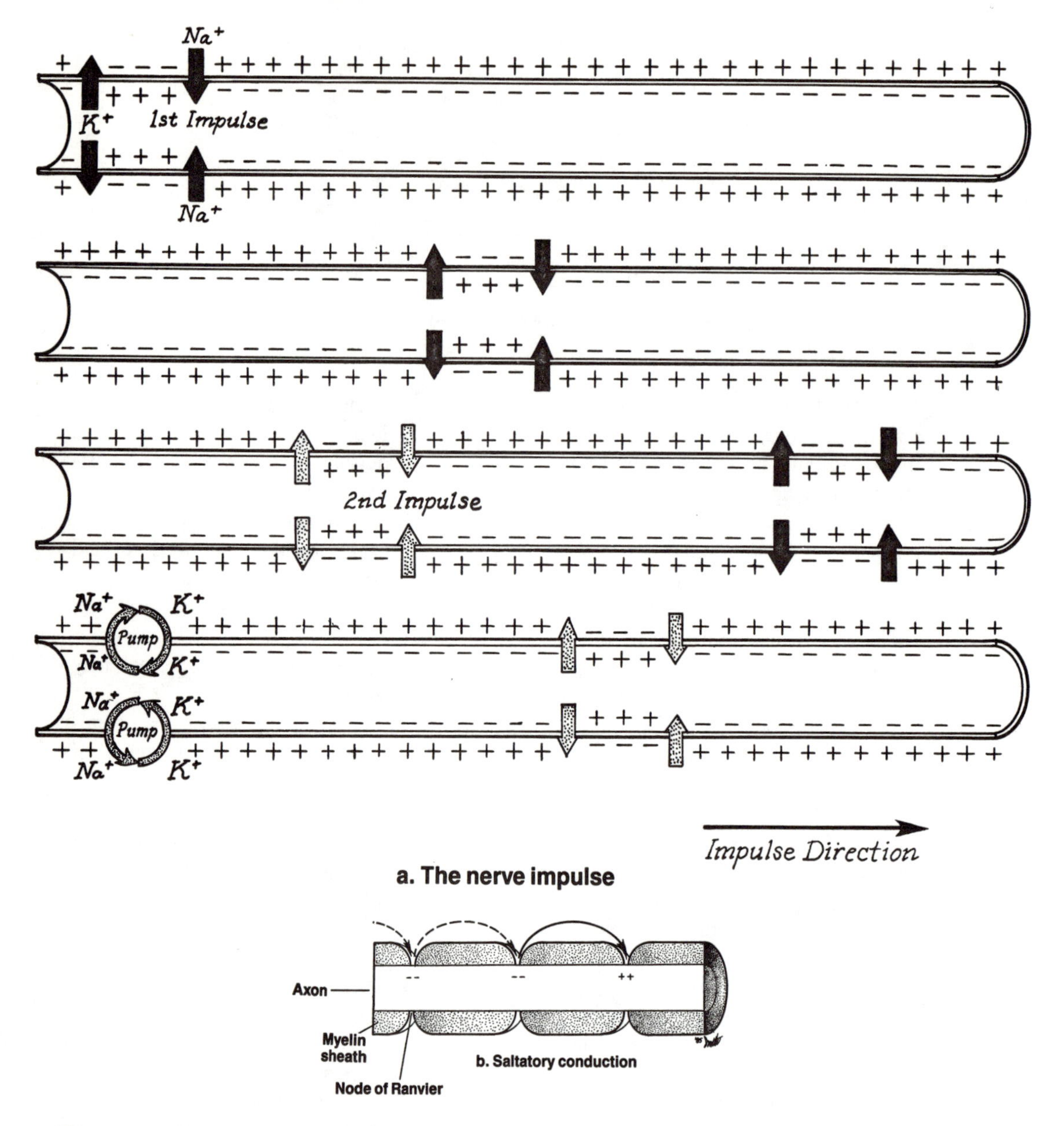

Fig. 1-11 In most myelinated neurons, the impulse jumps from node to node. This type of conduction (saltatory conduction) speeds up transmission, since only a small part of the fiber is depolarized and repolarized.

23. Genes determine enzyme activity, and such diverse things as the color of hair, skin, body stature, susceptibility to certain diseases, behavioral patterns, etc.

c. *The nucleolus, or nucleoli*, are one or more dark purple bodies inside the nucleus, containing *ribo-nucleic acid (RNA)*.

B. With the *electron microscope*, usually not accessible to the student, one can further observe:

1. *Mitochondria*, threadlike or oval-shaped structures, providing high-energy ATP.

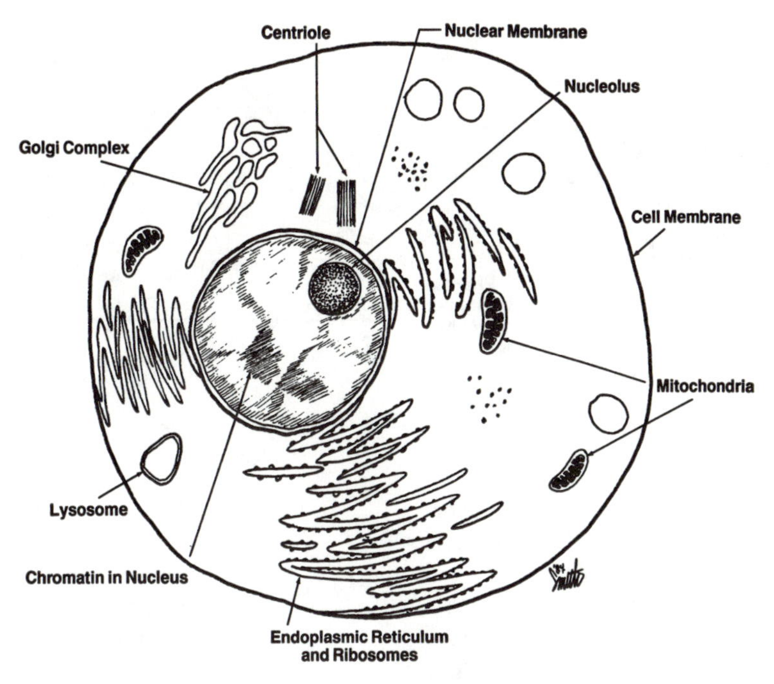

Fig. 1-12 The Animal Cell.

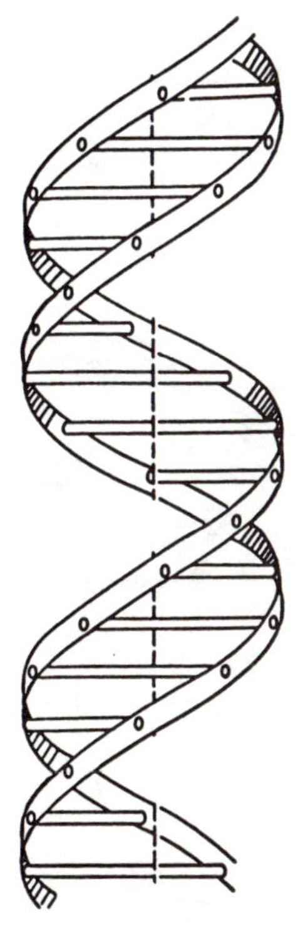

Fig. 1-13 A Portion of the DNA Double-Helix. (Modified, with permission, from Watson and Crick, "The DNA Molecule", Nature, 1953).

2. *The endoplasmic reticulum*, a complex series of tubules in the cytoplasm.
3. *Golgi complex,* a collection of tubes and vesicles that produces hormones, enzymes and lysosomes.
4. *Lysosomes*, vesicles containing enzymes that digest foreign and other material.
5. *Ribosomes*, small granular structures rich in RNA, that are the sites of protein formation.

C. *When cells divide, chromosomes divide* (Fig. 1-14). Coiled DNA strands separate from one another, each forms a complementary strand of DNA, and restores the double helix. The daughter cells formed are identical to the original. This process is cell division, or *mitosis*. As seen in Fig. 1-14, mitosis is a short part of the *cell cycle.* The longest phase is *interphase* (33 hours) where the synthesis and duplication of DNA take place.

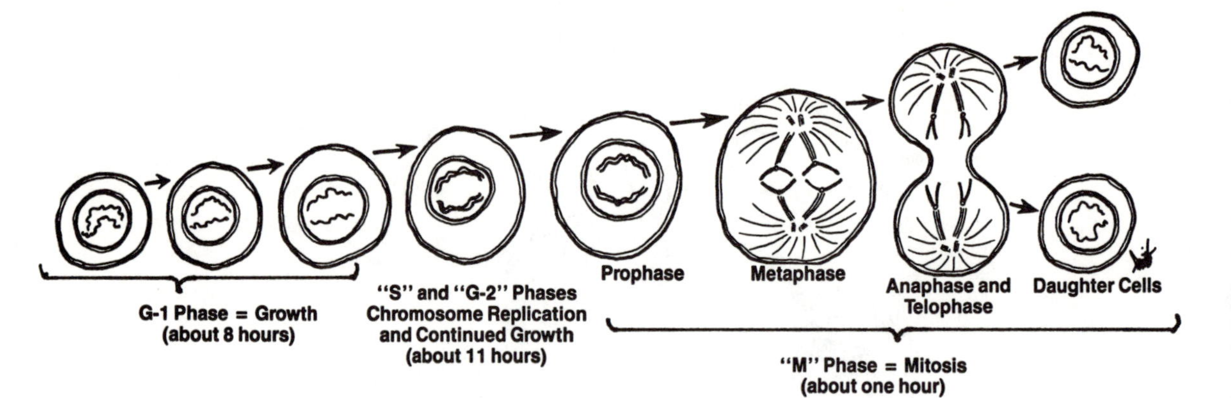

Fig. 1-14 The Cell Cycle.

D. The *synthesis of proteins* is an important part of cell functioning. The cell is composed of carbohydrates, fats and proteins in varying amounts. Reactions of these substances require *enzymes*, which are *protein catalysts* (a catalyst is a substance that accelerates, or sometimes decelerates, a chemical reaction, and is not directly involved in the reaction). A substantial portion of cell metabolism is involved with the manufacture of enzymes and other proteins, such as combinations with carbohydrates, fats, DNA and RNA (glycoproteins, lipoproteins, nucleproteins), many hormones, actin and myosin in muscle, plasma proteins (albumin, globulin), heme and numerous other substances (see chapter 13). DNA manufactures proteins as follows (Fig. 1-15):

1. An uncoiled strand of DNA acts as the template for the formation of messenger-RNA (m-RNA), a single-stranded molecule. This is

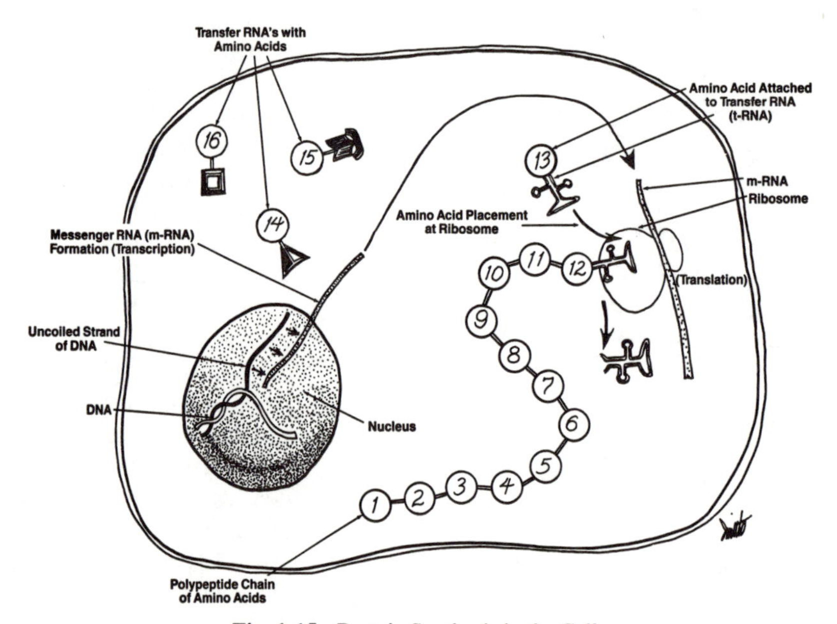

Fig. 1-15 Protein Synthesis in the Cell.

transcription.

2. m-RNA migrates from the nucleus to the cytoplasm, and attaches to a ribosome.
3. Small double-coiled molecules of transfer-RNA (t-RNA) place amino-acids (the structural units of proteins) sequentially on the m-RNA at the ribosome according to a code, and form a chain. Each amino-acid has its individual t-RNA. This is *translation*.
4. When the chain of amino-acids is *completed*, synthesis stops. A protein (polypeptide) has been formed.

-TWO-

BASIC PATHOLOGY AND THE IMMUNE SYSTEM

Pathology is the study of disease. The immune system involves foreign particles (antigens) and substances produced to combat them (antibodies).

I. *Inflammation — reaction to injury*. Signs of inflammation are an increase in temperature, pain, redness, swelling, and a disorder of function in the area. The body reacts to injury by an immediate vessel (vascular) and a slower cell (cellular) response:

A. *The vascular response*. Tissue injury releases vaso-active (vaso: vessel) substances, such as *histamine* from tissue cells, the *mast cells*. Histamine causes smooth muscle relaxation in vessels (vasodilation), smooth muscle contraction in organs, and an increased permeability to blood in the small vessels (arterioles, capillaries and venules). Blood passes into the tissue spaces as an "exudate", causing the increased temperature, nerve irritation, redness and swelling in the area.

B. *The cellular response*. *White blood cells* are cells of inflammation. After the vascular response, certain white cells, *phagocytes,* leave the vessels and migrate to the area of inflammation. They are attracted by bacteria, or by substances released from injured tissue cells. This is *chemotaxis*. In *phagocytosis*, small particles such as bacteria, cell fragments and foreign matter are engulfed by certain white blood cells. Neutrophils and macrophages are the common phagocytes (Fig. 2-1). *Infection*, or invasion of the body by microorganisms such as bacteria and viruses, is a common cause of inflammation.

C. *Drugs used to treat inflammation*.

1. *Antihistamines* are frequently used to *block the toxic effects of histamine* and other vasoactive substances, the kinins. Antihistamines such as diphenhydramine (Benadryl) and chlorpheniramine (Chlor-Trimeton) counteract the redness and itching of *hives* (urticaria) and the irritation of *hay fever* (allergic rhinitis).
2. *Epinephrine (adrenalin)*, the *physiologic antagonist of histamine*, is frequently used to counteract severe reactions to histamine such as *asthma* and *anaphylaxis*.
3. The *synthetic adrenocorticosteroids* (corticosteroids, steroids) are hormones patterned after those of the adrenal cortex; they *reduce inflammation and allergic reactions by limiting capillary dilatation and the increased permeability* that follows. They prevent the release of the vasoactive kinins. Examples such as *dexamethasone* (Decadron) and *prednisone* are widely used to suppress inflammation in a variety of conditions and diseases (see Chapter 12).
4. *Non-steroidal anti-inflammatory agents* were originally developed to replace steroids in the treatment of diseases such as arthritis, as steroids often have undesirable side-

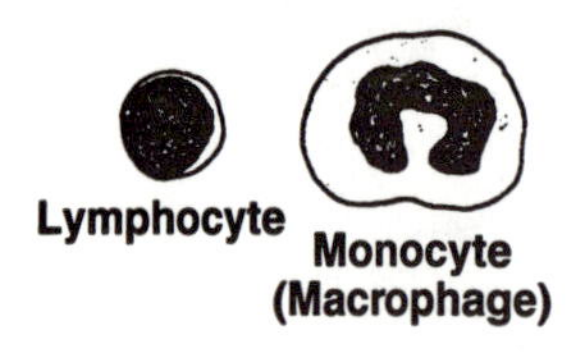

Fig. 2-1 White Blood Cells.

effects. Now they are widely used in the treatment of many types of inflammatory conditions. These drugs *block prostaglandin synthesis*. The group includes *aspirin* (the oldest member of the group), indomethacin (Indocin), ibuprofen (Motrin), naproxen (Naprosyn, Anaprox), fenoprofen (Nalfon) and sulindac (Clinoril) (see Chapter 12).

II. *Regeneration and repair.*

Healing involves two stages: the reabsorption of dead tissue and debris (*resolution*), followed by *regeneration* (replacement of destroyed cells by new ones) or *repair* (scarring, or the replacement of original cells by connective tissue).

- A. *Regeneration* takes place at different degrees and rates in various tissues:
  1. *Labile cells* regenerate easily and quickly. These are cells of the *lymphatic system*, *epidermis*, *bone marrow* and *gastrointestinal tract*.
  2. *Stable cells* (most common) regenerate at slower rates. These are the *parenchymal cells of glands* (the parenchyma is the epithelial portion of an organ), as well as *connective tissue cells* and derivatives. For example, intestinal cells regenerate in 1-2 days, liver cells in 3-5 days and cells of the kidney tubules in 7-14 days.
  3. *Permanent cells* do not regenerate, or regenerate minimally. These are *nerve and muscle cells*. Regeneration may take place if cut nerve axons are well approximated. Muscle cells may also minimally regenerate.
- B. *Types of repair, or scarring*:
  1. *Primary union (1st intention)*. The wound margins are closed surgically. Healing is rapid and scar size is minimal.
  2. *Secondary union (2nd intention)*. The margins of a wound are *poorly* approximated (e.g. a large wound healing without suturing). Connective tissue and inflammatory cells migrate to and remodel the area (*granulation tissue*). Healing is slow. Scar size is large with possible complications such as proud flesh (much connective tissue) and keloids (a *keloid* is an excess amount of collagen in an area of scarring).
  3. *Collagen* is the chief constituent of scar tissue. *Wound healing* correlates with the amount of collagen synthesized. Factors that promote healing are Vitamin C, youth, warmth and good nutrition (particularly protein).

III. *Infectious diseases.*

An infection is an *invasion of the body by parasites*: viruses, bacteria, one-celled animals, fungi, worms or insects. The invasion causes inflammation, characterized by fever, headache, muscle aches, itching, cough, etc. Factors determining the outcome of the infection are the numbers of microorganisms, their toxicity, and the host's defenses.

- A. *Bacteria* are *one-celled microorganisms* causing illness either by direct invasion or the elaboration of a *toxin*. A toxin (exotoxin) is a protein poison excreted from the bacterial cell. Common bacterial infections are pneumonia, urinary tract infections, skin infections, throat infections, ear infections, venereal disease, bronchitis and tuberculosis. Toxins cause diptheria, tetanus, botulism, other kinds of food poisoning, cholera and scarlet fever. A bacterial infection is treated with an *antibiotic*.
- B. *Viruses* are the *smallest infectious microorganisms*, consisting of either RNA or DNA in a protein shell. In contrast to other infectious agents, *viruses only grow in living cells*. DNA or RNA from the virus becomes incorporated into the host cell's RNA or DNA, causing disease by altering the host cell. Common viral infections are influenza (flu), herpes infection,

mononucleosis, upper respiratory infection, warts, mumps, measles, chickenpox, polio and gastroenteritis. Antibiotics are not effective against viruses. In some cases, an *anti-viral agent* may be used. Some invaded cells produce *interferon,* a low-molecular-weight protein that inactivates the virus. It is released into the body fluids and prevents viral synthesis inside similar cells. Interferon is presently being investigated as a possible treatment for some viral diseases and cancer. The *immune system* plays an important role in the defense against viral disease.

C. *Fungi* are small *plants*, some of which infect man. They frequently cause skin infections. Yeasts are oval fungi. Common fungal infections are yeast vaginitis, jock-itch, ringworm, athlete's foot, and thrush. Treatment is with an *anti-fungal medicine*.

D. *Protozoa* are *one-celled animals*. Protozoan diseases are rare in this country, except for trichomoniasis, a common vaginal infection. Usually occuring in tropical climates, other protozoal infections are amebic dysentery, sleeping sickness and malaria. Treatment is with an *anti-protozoal drug*.

E. *Helminths* are *worms*. These infections are also rare in this country, except for pinworms, and perhaps trichinosis from poorly cooked pork. Worms thrive in wet, warm areas. Other examples are elephantiasis, fluke and tapeworm infections. Treatment is with an *anti-helminthic drug*.

F. *Insects* often infect man. Ticks may carry rocky mountain spotted fever. Scabies is caused by a small mite. Lice may infect the head, body and groin (crabs). Treatment is with an *anti-parasitic drug*.

IV. *The immune system* (immunis: safe) involves antigens and antibodies. *Antigens* are foreign substances, such as bacteria and viruses. *Antibodies* are proteins (gamma or immune globulins) formed in response to the antigens. Lymphocyte derivatives (*plasma cells*) produce antibodies in tissue and in blood. Antibodies bind to the antigens that stimulated their production, and prevent toxic effects (Fig. 2-2).

A. *Origins of lymphocytes*:
During embryonic life, a *stem cell* from the fetal liver and other organs gives rise to all blood cells, including lymphocytes. Some lymphocytes migrate through the *thymus gland* and become *T-lymphocytes (T-cells)*. Others pass through the bone marrow or the fetal liver and become *B-lymphocytes (B-cells)*. In these areas (thymus, bone marrow and fetal liver), lymphocytes are programmed to recognize the body's tissues, or to distinguish *"self" from "non-self"*. Most recognition is done by the T-cells. If this system fails, as in autoimmune diseases, antibodies are produced to one's own tissues. T and B-cells migrate to lymph nodes and bone marrow after birth. In the adult, they are found in the spleen, tonsils, blood and connective tissue (particularly in the digestive and respiratory tracts).

B. *Types of immune responses*:
When antigens enter the body, *macrophages are activated*, and present the antigens to receptors on the B or T cell membrane. This stimulates the cell to change and divide.

1. *B-cells are responsible for antibody-mediated (humoral) immunity*, the major defense against most *bacterial infections*. After the antigen binds to a B-cell receptor, a family (*clone*) of identical *plasma cells* manufactures *antibodies* specific for the antigen. A *complement system* (blood enzymes required for most humoral activity) combines with antigen and antibody, forming an *immune complex*. Clumping (agglutination), precipi-

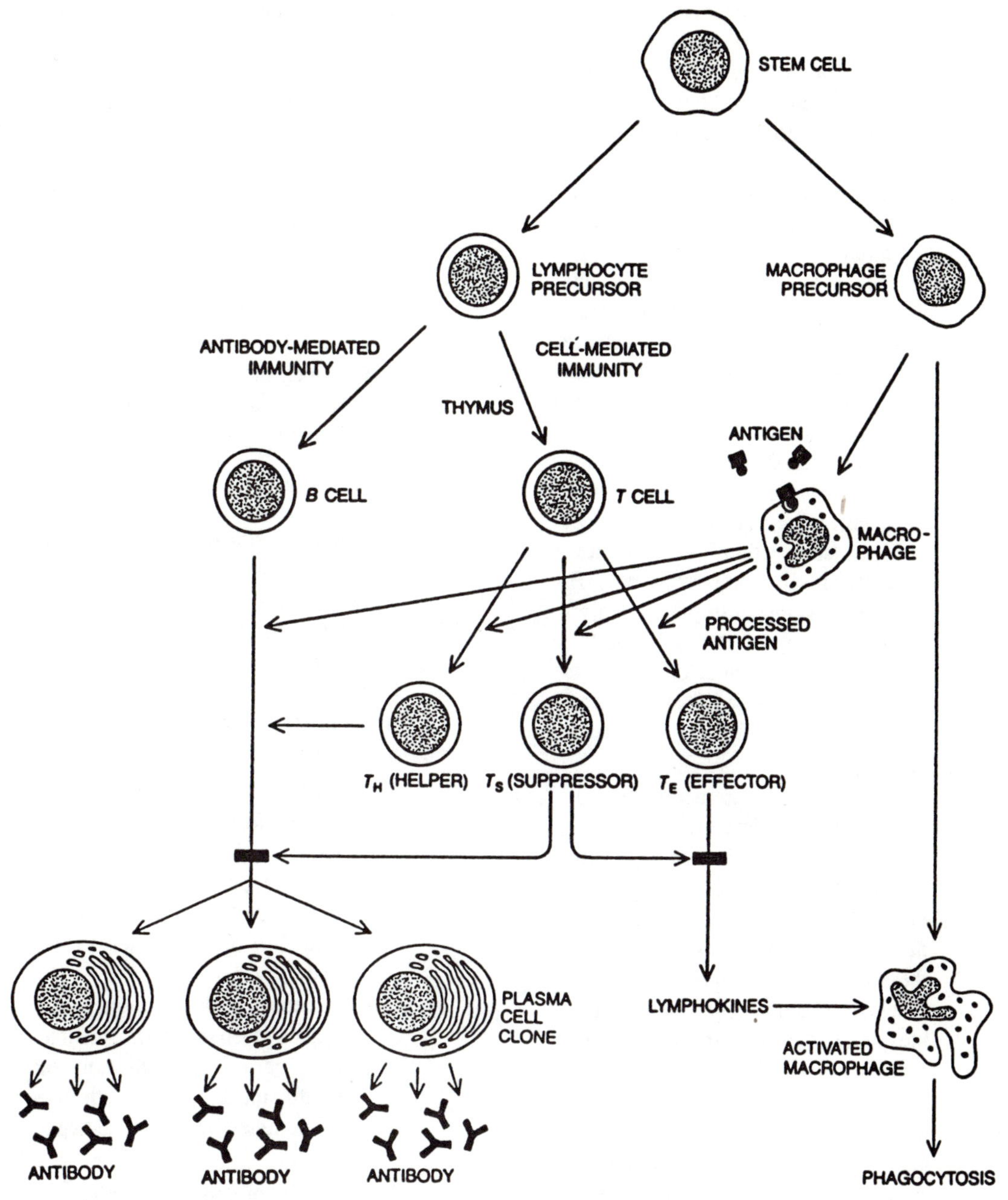

Fig. 2-2 The Immune Response. (Modified, with permission, from Rose, "Autoimmune Diseases", Scientific American, 1981).

tation, neutralization, or rupture of the cell membrane (lysis) may take place. The immune complexes are *phagocytized by neutrophils or macrophages*. Some B-cells do not form plasma cells, but remember the antigen. A second or subsequent exposure triggers a quicker and greater response because of these *memory cells*. This is why booster shots of a vaccine are important.

2. *T-cells are responsible for cell-mediated (cellular) immunity,* the

major defense against most *viral infections, cancer, and is the major cause of graft rejection.* When the antigen enters the body, a clone of T-cells is produced, similar to the plasma cell clone. T-cells differentiate into sub-populations with various functions (e.g., *helper cells, suppressor cells, effector cells*). Helper cells cooperate with B-cells to produce plasma cells and antibodies. T-cells also release substances (*lymphokines*) that mobilize, stabilize and assist with phagocytosis. One lymphokine (interleukin-2) causes proliferation of T-cells, and is presently under investigation as an anti-cancer drug (see section V-C). Memory cells are also present as with B-cells. T-cells constitute the majority of lymphocytes (70%) and live for months or years, in contrast to the B-cells whose life span, except for memory cells, is days or weeks.

C. *Hypersensitivity (allergy)*:
Antigen/antibody interactions sometimes form complexes that stimulate cells to release a variety of toxic substances. In other words, the *antigen-antibody reaction causes inflammation*. Antigens producing these reactions are *allergens*. In most cases, the reaction begins within minutes and last only hours. *B-cells and antibody-mediated immunity* are involved. Examples are allergies to pollens, foods and animal danders, causing hay fever, asthma or hives. Allergies to antibiotics, particularly penicillin, also cause a reaction. An important example is *anaphylaxis* (anaphylactic shock), a life-threatening condition occurring seconds to minutes after the ingestion of a drug such as *penicillin*, or after an insect sting (usually *bee or wasp*). The patient must have previously been exposed to the antigen, although he may not have realized it. The following events occur:

1. Certain antibodies attached to the surfaces of *mast cells* react with the antigens and trigger the release of *granules* containing *histamine* and similar substances.
2. The reaction is characterized by a sudden flush, shortness of breath, itching, nausea and wheezing, occasionally followed by collapse and shock.
3. Primary treatment consists of epinephrine (to antagonize histamine and open the airway), intravenous fluids, and steroids (to decrease the inflammatory response).

D. *Immunization*:
Protection from disease takes place by innoculation with antibodies that *neutralize* bacterial or viral antigens (*passive immunity*), or by the injection of a small amount of antigen to stimulate the formation of the person's own antibodies (*active immunity*). Passive immunity is temporary. Active immunity is long-term.

1. *Examples of passive immunity*:
   a. Toxins formed by clostridium tetani or clostridium botulinum (tetanus or botulism) may be neutralized by an immediate injection of antibodies to the toxin (*antitoxin*).
   b. In certain viral infections, antibodies given during the incubation period may prevent or modify the disease (e.g., the use of immune globin in viral hepatitis).
2. *Examples of active immunity:*
   Vaccination consists of injecting live or killed microorganisms, their antigens or toxins, into the body. *Booster shots* of the antigen at subsequent intervals increase the length of the immunity as a result of *memory cells.* Diptheria and tetanus toxoids (a toxoid is a toxin of low toxicity), pertussis (whooping cough) bacterial antigens, as well as live polio, mea-

sles, mumps and rubella viruses are administered routinely during childhood.

E. *Impaired immunity:*
Impaired immunity is a continuum involving immunosuppression on the one hand in which the bone marrow produces less blood cells (i.e. cancer, chemotherapy, radiation therapy), auto-immune diseases in which the body reacts with its tissues, and viral destruction of lymphocytes (AIDS) on the other hand.

1. *Autoimmunity:*
An autoimmune disease is one in which the body fails to recognize its tissues as its own and produces antibodies against them (antibody-mediated) or reacts with them (cell-mediated). Causes include tissue damage, drugs, a genetic deficiency, nutrient deficiencies and infectious agents, particularly viruses. Autoimmune diseases are more common in females and in old age. Examples include hyperthyroidism, hypothyroidism, adrenal insufficiency, myasthenia gravis, multiple sclerosis, rheumatoid arthritis, lupus erythematosis and some viral infections. In the latter case, viral RNA or DNA is incorporated into host genes and transmitted as part of the genetic information of the cells. Viruses create antigenicity on the host cell surface and in other ways injure T-cells. Misprogrammed T-cells then inappropriately stimulate B-cell antibodies. Autoimmune diseases in which viruses have been implicated include systemic lupus erythematosus (oncornavirus) and multiple sclerosis (paramyxovirus). *Two common autoimmune diseases:*

a. *Lupus erythematosus is a multiple-system disease,* primarily affecting young females. T-cell function is defective, resulting in an overactivity of B-cells, the production of multiple auto-antibodies and the trapping of antigen-antibody complexes in capillaries. Joints, skin, kidneys and the central nervous system are affected. Numerous auto-antibodies are formed: anti-nuclear, anti-DNA, anti-immunoglobulin, anti-lymphocytic and anti-red cell. Steroids improve survival rates.

b. *Rheumatoid arthritis is a chronic inflammatory disorder of the synovial membranes of joints,* again mostly affecting young females. Antigen-antibody complexes are deposited in joint spaces. Rheumatoid factor (an anti-immunoglobulin antibody) is present in 90% of cases, as are anti-nuclear antibodies (see Chapter 4).

2. *Acquired immune deficiency* means that lymphocytes and/or macrophages are compromised in function. Commonly seen in leukemia and other cancers or in patients receiving chemotherapy or radiation, an emerging entity is the *acquired immune deficiency syndrome (AIDS). AIDS is a constellation of rare viral, fungal, bacterial and protozoal infections appearing years after invasion of the body by the AIDS virus (human immunodeficiency virus, HIV). The virus destroys helper T-cells, impairs B-cells and macrophages, and compromises the immune system so badly that the body becomes susceptible to opportunistic infections.* Originating in Africa where it continues to wreak havoc, it has spread to most parts of the world. The primary mode of transmission in Africa is heterosexual, whereas in this country it is by way of

needle-sharing drug-abusers and homosexual anal intercourse. The virus is carried in most bodily fluids, including semen. Target cells of the virus, in addition to the helper T-cell, are the macrophage and possibly cells of the brain, spinal cord, colon and rectum. *The virus binds to the T-cell at a protein receptor site on the membrane (CD4 antigen, or binding site) and the viral material is integrated into the DNA of the host cell.*

a. *Signs and symptoms:* about a month after invasion by the virus, vague complaints such as fever, sweating, muscle and joint pains appear lasting a few days to two weeks. Antibody formation (seroconversion) occurs one to two months later. The AIDS test detects this antibody. A positive test means likely progression to opportunistic infections. The time from antibody formation to infection is about 2 to 6 years. During this time the person is usually asymptomatic. Early symptoms of AIDS are weight loss, fatigue, diarrhea and respiratory problems. Later clinical findings include pneumonia (the most common infection), skin lesions, enlarged lymph nodes, oral lesions and neurologic impairment.

b. *Treatment:* most cases end fatally in less than 5 years in spite of treatment. However, some opportunistic infections respond to new therapies. Research is threefold: finding a successful vaccine to the virus, blocking viral replication, and blocking viral attachment to cells. Three anti-viral drugs blocking viral replication have been used experimentally, the most successful of which is zidovudine (azidothymidine, AZT). AZT is able to reduce the number and severity of some infections. A new synthetic protein having similarities to the CD4 T-cell binding-site protein appears to block viral attachment to cells. When it is introduced into the host, the virus mistakenly binds to it, and the T-cell is bypassed, or spared. A toxin is then released, killing the virus. *Prevention* is with a safe partner, or a condom.

F. *Grafts (transplants).*

1. *Transplants* of organs or tissues are of four types:
   a. *Autograft — from different sites on the same person (tolerated).* An example is the transplantation of skin from one part of the body to another (e.g. in burns).
   b. *Isograft — between identical twins (tolerated).*
   c. *Allograft (homograft) — between members of the same species (variably tolerated).* An important consideration is a close genetic matchup (brother, mother) as well as the vascularity of the tissue. A transplant with few or no vessels (thus few white blood cells) such as the cornea, is usually successful.
   d. *Xenograft (heterograft) — between members of different species (rejected).*
2. The main problem in transplants is *rejection*. The transplanted organ is the antigen and the body produces antibodies to it. *Cell-mediated immunity* is triggered. *T-cells* penetrate the transplant and destroy the cells. Inflammation, infiltration by neutrophils and monocytes, antibody production, edema, hemorrhage, thrombosis, blood clotting and tissue death (necrosis) occur. If donor and

recipient are well matched (close relatives or others having a compatible genetic make-up), long-term survival of a transplanted organ or tissue is enhanced.

3. Certain drugs such as *cyclosporin* suppress rejection by destroying or arresting the growth of lymphocytes. These are *immunosuppressive agents*. They are used in autoimmune diseases, organ transplants and in cancer chemotherapy, since they *inhibit or retard cell growth*. Unfortunately, since lymphocytes and macrophages are important constituents against infection, the person then becomes *increasingly susceptible to infection*.

V. A *tumor* is an abnormal growth. When it is localized and non-invasive, it is *benign*. If it invades surrounding tissue or spreads by way of blood or lymph to a distant site (metastasize), it is *malignant*, or *cancerous*. Benign tumors are more common than malignant ones, and are either not treated (e.g. warts, moles) or are removed surgically. Malignant tumors are treated by combinations of *surgery*, *chemotherapy* and *radiation*.

A. *Mutant cells*:

When normal cells divide (mitosis), a small number do not replicate correctly. This abnormal replication, or alteration, is a *mutation*. Mutations are a common occurrence. A mutation may be an alteration of a gene, genes, a segment of DNA, or a part of, or an entire, chromosome. Many mutant cells have limited survival capacity and die. Those that survive are usually destroyed by T-lymphocytes and are phagocytosed by macrophages, since they are recognized as "non-self". About 1 in 100,000 genes is a mutant. If the immune system is not capable of proper "*surveillance*", mutant cells may take hold and grow. The result is a tumor. *Many cancers involve a weakened immune system*.

B. *Incidence of cancer in the U.S.*:

Cancer is the *second leading cause of death* (the first is heart disease). The leading causes of cancer deaths in the United States are:

	Male	*Female*
1. Most common	Lung	Breast
2. Second most common	Colon	Colon
3. Third most common	Prostate	Lung

C. *Characteristics of cancer cells*:

1. Do not follow normal mitotic patterns and show bizarre cell-cycle patterns such as early and multiple cell-divisions and abnormal numbers of chromosomes per cell.
2. More fragile than other cells, and have a tendency to break off and travel through blood and lymph to other organs, where they implant at distant sites (metastasis).
3. Compete with normal cells for nutrients. The metabolism of cancer cells is high, and normal tissues gradually die. The person loses weight.
4. Many cancers are caused by environmental factors (e.g., chemicals, viruses, x-rays, ultra-violet light), and some are hereditary (e.g., some colon and breast cancers).
5. Some cancers are *caused by* radiation and chemotherapeutic agents; some cancers are *successfully treated by* radiation and chemotherapeutic agents.
6. The above agents (in #5) interfere with and/or stop normal mitosis, preventing growth of the fast-growing cancer, but kill normal cells. Thus, the exact identification and dosage for a particular cancer is important.
7. Viruses may induce cancer by combining with the DNA of cells.
8. Cancer cells have surface antigens, thus stimulating both antibody-mediated and cell-mediated immune responses.
9. Generally speaking, the *labile* cells (lymphatic system, epidermis,

FIVE LEADING CAUSES OF DEATH UNITED STATES, 1981		
Frequency	Cause of Death	Number of Deaths
1.	Heart Diseases	753,788
2.	Cancer	422,094
3.	Cerebrovascular Diseases	163,504
4.	Accidents	100,704
5.	Chronic Obstructive Lung Disease	58,832
Source: Vital Statistics of the United States, 1981		

Table 2-1: Leading Causes of Death

bone marrow and gastrointestinal tract) which regenerate easily and quickly, and the *stable* cells (glands) are more *prone to cancer* than the permanent cells (nerve and muscle), because of the high turnover (thus higher mutation) rates.

10. Recently, interest has focused on the lymphokine *interleukin-2* (IL2) as a possible anti-cancer agent. IL2 dramatically increases T-cell subpopulations (helper T-cells, etc.), shoring up the deficient immune system.

E. *Summary of events in cancer formation*:
   1. Prolonged exposure to a *carcinogen* (e.g., tar in cigarette smoke) causes an increased number of cell mutants (tar is a *mutagen*).
   2. An increase in the number of cell mutants results in increased activity of the immune system (T and B-cells, plasma cells, macrophages) attempting to destroy the antigenic tumor cells.
   3. If the immune system is weak, mutants may take hold and grow, forming a cancer.

F. *Definitions:*
   1. *Neoplasm*: a tumor (benign or malignant)
   2. *Cancer*: a malignant tumor
   3. *Carcinoma*: a cancer derived from epithelial tissue
   4. *Sarcoma*: a cancer derived from connective tissue
   5. *Metastasize*: spread of cancer cells to distant sites.

G. *Nomenclature of tumors*:
Benign tumors usually have the suffix *"-oma"* attached to the name of the tissue (e.g. a tumor of lipid cells is a lipoma, of bone cells an osteoma). In *cancer*, the name of the tissue often precedes the term (e.g. squamous cell carcinoma, basal cell carcinoma, osteosarcoma).

ABSCESS: a local accumulation of pus

ANEURYSM: local dilation (from weakness) of a vessel wall

ATROPHY: a decrease in cell number and size

BOIL (FURUNCLE): abscess of skin involving a hair follicle

CARBUNCLE: more deeply located abscess with communicating tracts (sinuses)

CELLULITIS: widespread inflammation of connective tissue (usually from infection)

CONGESTION: engorgement of a vessel with blood

CONTUSION: bruise

EDEMA: increase of fluid in tissue spaces

EMBOLISM: clot breaking off and traveling in the bloodstream

EXUDATE:	fluid and cells that pass from the bloodstream to the tissue spaces during acute inflammation
FISTULA:	an abnormal communication between two structures
GANGRENE:	death of tissue associated with bacterial invasion and putrefaction
HEMATOMA:	blood clot
HYPERPLASIA:	an increase in cell number
INFARCT:	an area of necrosis caused by obstruction of blood supply
INFECTION:	invasion by parasites, e.g. viruses, bacteria
INFLAMMATION:	the body's reaction to injury
ISCHEMIA:	deficiency of blood due to obstruction of blood supply
"-ITIS":	means inflammation; often implies infection
NECROSIS:	death of tissue
PURULENT:	pus-containing
PUS:	dead neutrophils, tissue breakdown products
PYOGENIC:	pus-producing
SIGNS:	physical examination (phenomena observed and recorded by examining physician)
SUPPURATION:	purulent exudate
SYMPTOMS:	medical history (complaints of the patient)
THROMBOSIS:	intravascular clotting
ULCERATION:	an excavation in an epithelial surface, resulting in localized inflammation and necros'

-THREE-

SKIN

Skin is about 1/2 inch thick and consists of a thin outer layer of epithelial cells, the *epidermis*, and a thick connective tissue layer, the *dermis*, containing blood vessels, nerve endings, and other structures. Deep to the dermis is a thick layer of loose connective tissue, the subcutaneous tissue (*subcutis*) containing fat, as well as nerves, glands, vessels and hair follicles. Fats are mobilized here. Below the subcutis are muscles, tendons, ligaments and bone (Fig. 3-1).

I. *Layers*:

A. *Epidermis*.

1. *General*: On most of the body, the epidermis is about the width of tissue-paper, although it is 1 mm or more on the palms and soles. Much of the epidermis is made up of keratin-containing cells of various shapes. *Keratin is a hard protein.* As the lower layers form new cells, they are pushed upward. The outermost layer consists of dead cells, matted together, which have lost their nuclei and are filled with keratin. When the dead cells are 12 to 25 cells thick, the outer part is shed.
2. *Cells*.
   a. The basal layer is made up mostly of *keratinocytes*.
   b. Sandwiched between keratinocytes are *melanocytes*, cells responsible for skin color. They produce the brown pigment melanin (also found in the iris and choroid of the eye), which protects the body from excessive ultraviolet radiation. The darker races have more melanin granules per cell than the lighter races. Faulty synthesis of melanin causes *albinism*. Sunlight

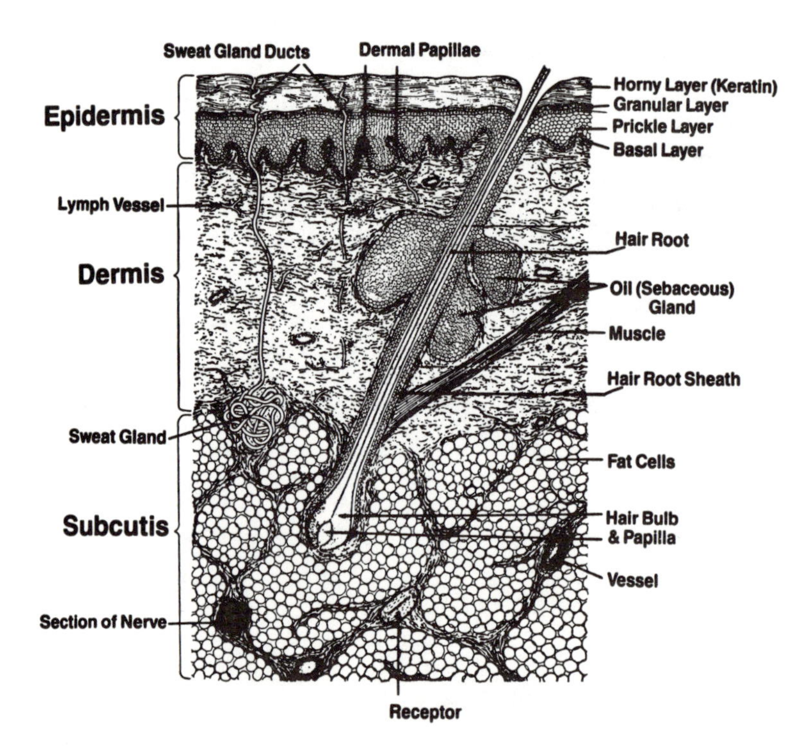

Fig. 3-1 Skin. (Modified, with permission, from "Section of Skin (Human)", General Biological, Inc., 1947).

darkens melanin and increases its rate of production. *Freckles* are an example, in fair-skinned people, of a partial response of melanocytes to sunlight. The *nevus cell*, found in moles and in the malignant melanoma, is derived from the melanocyte.

B. *Dermis*, or true skin, is thick, tough and flexible. It contains:
   1. Connective tissue cells and elastic fibers
   2. Blood vessels
   3. Lymph vessels
   4. Occasional smooth muscle cells
   5. Sensory nerve endings (receptors)
   6. *Skin appendages* (epithelial derivatives embedded in the dermis):
      a. Hair follicles
      b. Nails
      c. Sweat and oil glands

   (fingerprints are specific protrusions of the dermis up into the epidermis)

II. *Functions*:

A. *Epidermis*:
   1. The *main function* of the epidermis is *protection from trauma, bacteria and drying* by the actions of keratin and oil.
      a. *Sebum* from oil glands retards water evaporation and *inhibits the growth of bacteria and fungi* on the skin surface.
      b. The combintation of *sweat and oil* is acid which also helps *kill bacteria*.
   2. *Regeneration of epithelial cells* takes about 7 days; thus, superficial wounds heal readily.
   3. As there are no blood vessels in the epidermis, *nutrition* takes place by diffusion up from the dermis.

B. *Dermis*:
   1. *Sensory nerve endings (receptors)*
      a. *Pain and temperature*
      b. *Touch*
      c. *Pressure*
   2. *Glandular secretion*:
      a. *Sebaceous (oil) glands* surrounding hair follicles secrete oil (sebum).
      b. *Sweat glands* secrete salt and water. Under normal conditions about 1/2 liter of fluid is lost in this way per day (*insensible water loss*). Several liters of water and NaC1 may be lost per day under strenuous conditions.
   3. *Storage of pro-vitamin D*. Sunlight converts the pro-vitamin to the active form of vitamin D (see chapter 13).
   4. *Temperature regulation*. The temperature is lower in the morning, and increases during the day.
      a. Average *oral* range: 97.6 to 99.6 °F (36.4 to 37.6 °C).
      b. Average *rectal* temperature: 99.5 °F (37.5 °C).
      c. Average *axillary* temperature: 98 °F (36.7 °C).
      d. Temperature varies with environmental conditions:
         1) Cold causes vessels in the dermis to *constrict*, conserving heat. In addition, piloerection ("goose-flesh") takes place: the small muscle attached to the hair-follicle contracts, the hair "stands on end", and heat is trapped. Impulses from the brain cause increased muscle tone resulting in shaking, or shivering, further increasing body heat.
         2) Heat causes vessels in the dermis to *dilate*, sweat glands to be activated, and heat is lost by perspiring and evaporation.
      e. A temperature below 95 or above 109 F is usually not compatible with life for very long (some exceptions).
      f. Temperature is one of the *4 vital signs* (the others are pulse rate, respiratory rate and blood pressure).

III. *Fever* is a *defense against infection*. The mechanics of fever are as follows:

A. *Toxins, bacteria and viruses act upon certain white cells,* causing them to release a fever-inducing substance called endogenous pyrogen.

B. *Endogenous pyrogen is carried by the blood to the temperature (thermoregulatory) center in the brain.* The above events reset the center to a higher degree.

C. Since the temperature of the blood is now lower than in the brain, *responses to conserve heat* take place: vessels constrict, and shivering takes place, resulting in "chills" until the brain temperature is reached.

D. When the infection is resolving, the opposite effect takes place: the temperature center is reset to a lower degree, and *heat-dissipating mechanisms* such as vasodilation and sweating cool the blood.

E. *Anti-pyretic* (anti-heat) drugs such as aspirin and acetominophen (Tylenol) lower the brain thermostat, and thus fever.

IV. *Heat stroke (sun stroke)* primarily affects *older people* and *alcoholics* on hot days. The person may be indoors or outdoors — the main problem is heat, not sun. There is a *failure of the heat dissipating mechanisms*. Sweating ceases. This is a life-threatening emergency.

A. *Signs and symptoms*:
1. Dizziness, headache, lethargy
2. Skin *hot, dry* and usually red (may be ashen)
3. Rapid heart rate (tachycardia)

B. *Primary treatment*:
1. Transfer to cool area, and lay person on his back.
2. Loosen clothing and apply anything cool to body, preferably cool water. *Cooling with cold or ice water is sometimes necessary.*

V. *Heat exhaustion (heat prostration)* is caused by prolonged heat exposure and *loss of salt and water* from excessive sweating.

A. *Signs and symptoms*:
1. Weakness, dizziness, faintness
2. *Skin cool and clammy*, with normal temperature
3. Tachycardia, with weak pulse
4. Sometimes muscle cramps (heat cramps)

B. *Primary treatment*:
1. Transfer to cool area and lay person on his back
2. *Give salted fruit drinks, or salt in water*. Often intravenous fluids are given.

VI *Common skin disorders*.

A. *Terms*:
1. *Bulla*: large fluid-filled elevation, e.g. bullous impetigo
2. *Crust*: dried exudate, e.g. impetigo
3. *Macule*: a flat lesion, e.g., freckle
4. *Papule*: a raised lesion, e.g., wart
5. *Scale*: flake of dried epidermis, e.g., psoriasis, dandruff
6. *Tumor*: solid mass often extending into dermis, e.g., basal cell carcinoma
7. *Vesicle*: small elevated lesion containing fluid, e.g., chickenpox lesion
8. *Wheal*: red, itchy, fluid-containing papule, e.g., hives

B. *Disorders*:
1. *Acne* (acne vulgaris) is a disorder in which the *oil glands are overactive*, then plugged (blackhead). Irritation by the blackhead can cause inflammation and infection (pimple). Oily skin is hereditary. Certain foods, lack of sleep, and nervous tension aggravate the condition. Primary treatment consists of increased personal hygiene, an "acne diet", and an antibiotic.
2. A *boil* (furuncle) is *an abscess of a hair follicle or sebaceous gland.* Commonly seen on the buttocks or back of the neck, it is caused by the staphylococcus bacterium. Primary treatment is local incision and drainage.
3. *Burns* are caused by fire, hot

solids, liquids, gases, chemicals, electricity and radiation. Classification (Fig. 3-2):

a. *First degree*: involves only the *epidermis*. The epithelium regenerates from skin appendages (sweat and sebaceous glands). No scar on healing. Example: mild sunburn.

b. *Second degree* (partial thickness): involves *both epidermis and upper dermis*, with blistering. Epithelium regenerates from appendages. No scar on healing. Example: severe sunburn.

c. *Third degree* (full thickness): involves the *entire epidermis and dermis, with damage extending into the subcutaneous tissue*. No epithelial regeneration is possible because the appendages are lost. *Charring and anesthesia are present*. Small areas heal with scar formation. Larger areas may require skin grafting.

d. *Fourth degree* burns are the same as third degree burns except that deep structures such as muscle, tendon and bone are involved.

e. *Primary treatment* depends on the severity of the burn. Minor burns may be treated with pain medication, a topical antibiotic such as sulfadiazine (Silvadene), sterile dressings and tetanus prophylaxis. Major burns involving over 15% of body surface, having a third degree component, involving the face, or involving explosions or electrical or radiation injury, require hospitalization. Several types of *skin grafts* are used. Autografts are common. Pig (porcine) skin is a xenograft employed as a temporary measure. Recently, cultured cadaver skin allografts

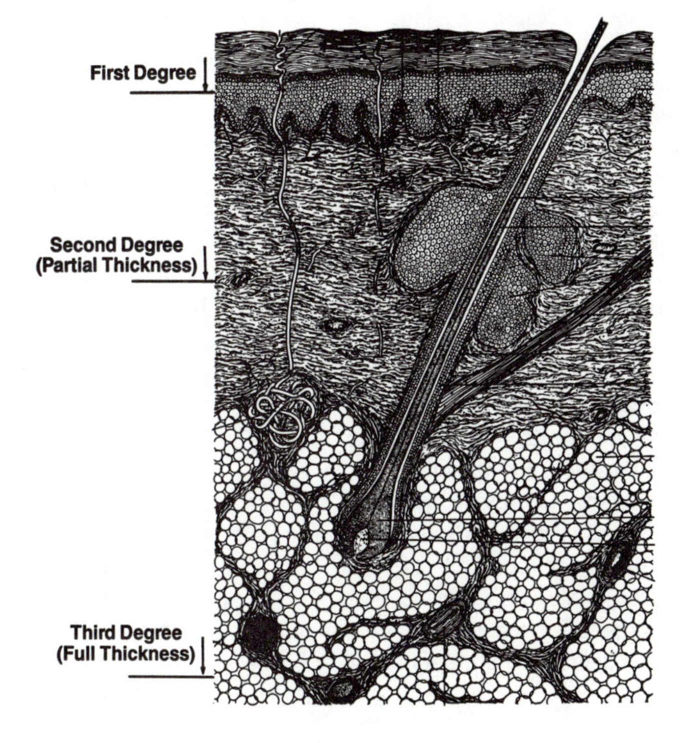

Fig. 3-2 Burns.

have been successfully used. The epidermal layers induce minimal rejection.

4. *Contact dermatitis* is a common *inflammation of the skin* from such things as soaps, eye-shadow and *poison ivy*. Primary treatment is removal of the offending substance and protection from irritant. Wet lesions respond to cool compresses. A steroid cream and oral prednisone are useful. An antihistamine such as diphenhydramine (Benadryl) or hydroxyzine (Atarax) may be needed for itching.
5. *Dandruff* (seborrheic dermatitis) affects about half of the population. *The epidermis shows thickening and scaling*. Although obvious on the scalp, scaling may also be seen on the ears, face, chest and groin. It is related to acne (acne-seborrhea complex), and may be kept under control with selenium (Selsun) or a tar shampoo.
6. *Herpes simplex* (fever blister, cold sore) is a *viral infection* causing vesicle formation around the mouth and lips, or the genital area. The lesions are painful and self-limiting. They often follow minor infections, trauma, stress or exposure to the sun. No ideal treatment is available, although the anti-viral drug acyclovir (Zovirax) offers promise (see chapter 15 for a discussion of genital herpes).
7. *Hives* (urticaria) is a common *allergic condition* characterized by red wheals and welts over large portions of the body, accompanied by intense itching. Causes include foods and drugs, as well as anxiety. Primary treatment consists of removing the cause. Epinephrine and an antihistamine are helpful. Steroids may decrease the inflammation.
8. *Scabies* is caused by a *female mite burrowing into the epidermis* and depositing eggs and feces, causing severe itching, particularly at night. The mite may persist for months and even years in an unclean environment ("seven-year-itch"). Isolated red excoriations are usually seen on the lower abdomen, arms and legs. Primary treatment: an anti-parasitic lotion, such as gamma benzene hexachloride (Kwell).
9. *Skin tumors*.
   a. *Benign tumors* include *warts* (verrucae) and *moles* (nevi). Warts are small viral tumors, which may come and go for no apparent reason. Sometimes suggestion therapy, particularly in children, may bring good results. Electrosurgery or liquid nitrogen treatment is occasionally effective. *Moles* are benign (e.g. brown, black, purple) skin growths which may contain hairs. They are thought to originate from melanocytes. Primary treatment is surgical excision. A "*suspicious*" *nevus*, such as one with a recent color change, is removed and biopsied, since it may undergo transformation into a *malignant melanoma*.
   b. *Malignant tumors* include the basal-cell carcinoma, squamous-cell carcinoma and malignant melanoma. The basal-cell and squamous-cell carcinomas are the *most common of all cancers*, but are rarely fatal. The basal-cell carcinoma is derived from the basal-cell layer of the epidermis, and the squamous-cell carcinoma from the superficial epidermis. These growths usu-

ally appear as nodules on the face (particularly the lower lip in the case of the squamous-cell carcinoma). Primary treatment is *surgical excision*. Radiation may be used with the squamous-cell carcinoma. The cure rate is high in both cases.

c. The *malignant melanoma* is a fortunately rare, but *highly malignant*, skin tumor. *Over half of melanomas arise from nevi which have changed color and consistency* (often to black or purple nodules). Current thought is that nevi and melanomas represent stages along a continuum. Primary treatment consists of surgical excision and lymph node dissection. Chemotherapy may be necessary. The mortality rate is high, as lymph node involvement and metastasis are common.

-FOUR-

BONES AND JOINTS

Bones should be studied on a skeleton. It is impractical to identify structures on an isolated bone, since there are no isolated bones in real life. The reader is asked to remember only those bony prominences (and muscles, in the next chapter) that are commonly encountered (Fig. 4-1).

I. *Structure and function of bones and joints*:
   A. Bones are composed mostly of calcium phosphate crystals. Important parts of a long bone are (Fig. 4-2):
      1. *Shaft*
      2. *Epiphyses*: ends of bone
      3. Cartilage *growth plates* in children between ends of bone and shaft
      4. *Marrow cavity*: inner space filled with marrow. Yellow or fatty marrow is present in long bones; red or bloodforming marrow is present in bones of thorax and pelvis.
      5. *Joint cartilage*: cartilage covering bone ends
      6. *Periosteum*: highly vascular membrane surrounding bone. It manufactures new bone.
   B. *Functions of bone*:
      1. *Support*
      2. *Movement*
      3. *Protection* (ribs, vertebral column, cranium)

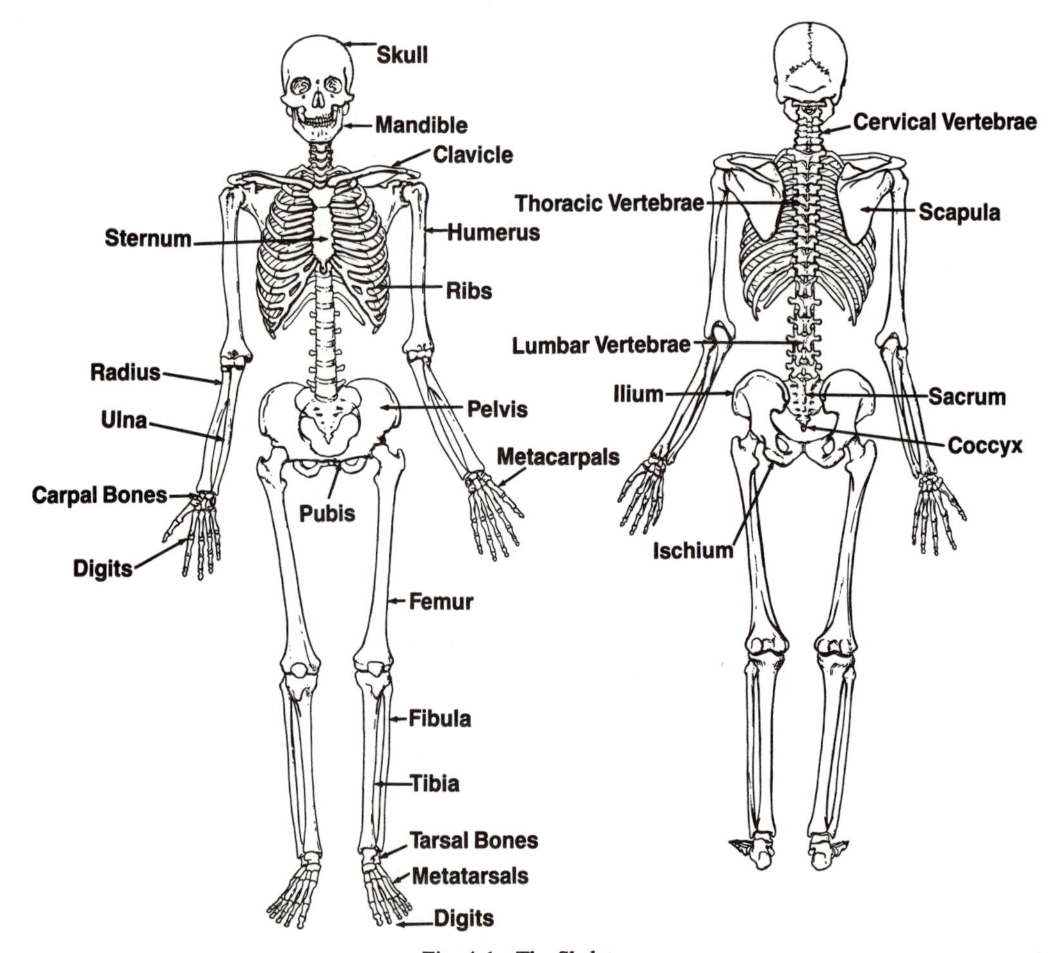

Fig. 4-1 The Skeleton.

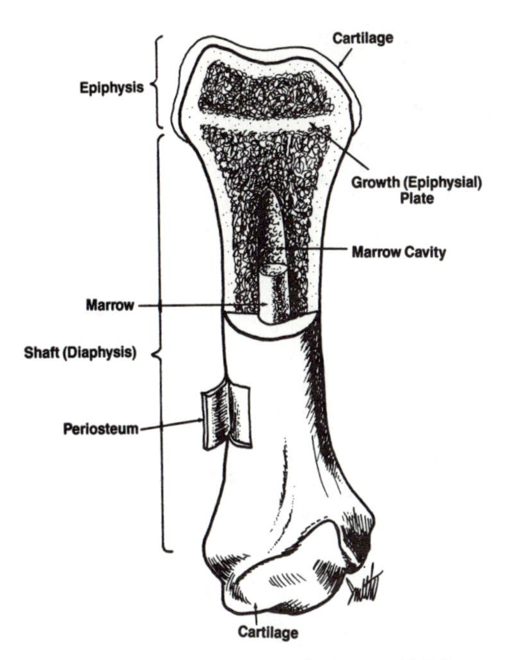

Fig. 4-2 Structure of a Long Bone (Child).

4. *Stores and releases calcium* to bloodstream
5. *Manufactures blood cells* (ribs, vertebral column, sternum, clavicle, ilium)

C. *Kinds of bony unions.*
1. *Synarthrosis*: immovable (e.g. sutures of the skull)
2. *Amphiarthrosis*: slightly movable (e.g. interpubic joint ((symphysis pubis)), intervertebral joints)
3. *Diarthrosis*: freely movable, or true *joint*. Parts of a joint are (Fig. 4-3):
   a. *Capsule* (ligament)
   b. *Synovial membrane*
   c. *Synovial fluid*
   d. *Cartilage*

D. *Types of joints* (See Fig. 1-3):
1. *Gliding*: sliding movement (e.g. clavicle joints)
2. *Hinge*: flexion and extension (e.g. elbow) (Fig. 1-3a)
3. *Pivot*: rotation (e.g. neck joint) (Fig. 1-3b)
4. *Ball and socket*: all movements (shoulder and hip joints)
5. *Condyloid*: circumduction (e.g. wrist) (Fig. 1-3c)

E. *Bursae* are small *connective tissue sacs* of *synovial fluid* lying near joints, providing lubrication between ligaments, tendons, bones, and muscles where friction normally develops.

II. *Bones and joints of the head and neck.*

A. *Skull bones* (the number of each bone is in parenthesis) (Fig. 4-5):
1. *Cranial bones*
   a. *Frontal* (1) contains frontal sinuses

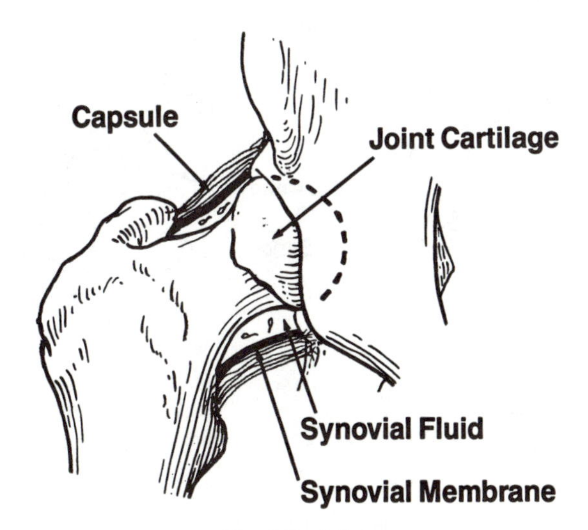

Fig. 4-3 A Joint.

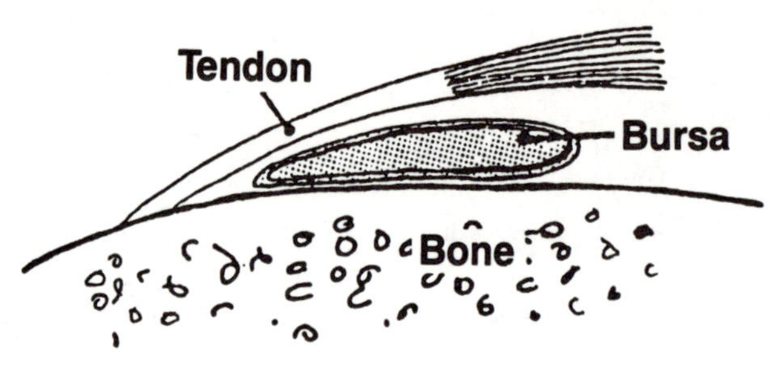

Fig. 4-4 A Bursa. (Reproduced, with permission, from Goldberg, Clinical Anatomy Made Ridiculously Simple, MedMaster, 1984).

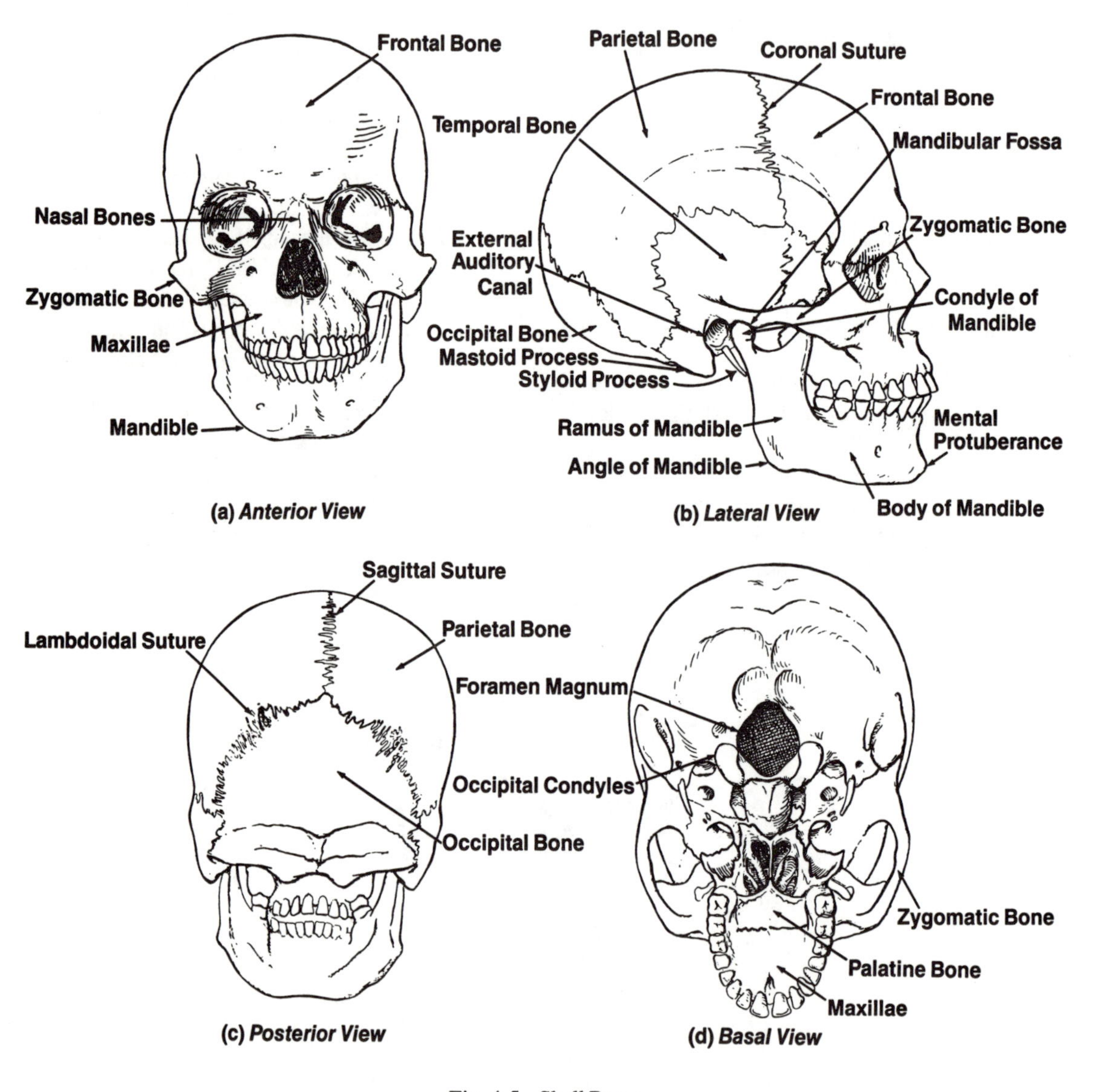

Fig. 4-5 Skull Bones.

b. *Parietal* (2) (parietal: wall)

c. *Temporal* (2) (temporal = lateral head) Important parts include the mastoid (mastoid: breastlike) process, external auditory canal, styloid (stylus: a pointed marker) process, and mandibular fossa (fossa: shallow depression).

d. *Occipital* (1) contains the occipital condyles (condyle: round process), and foramen magnum ("large hole"). The spinal cord lies in the foramen magnum.

e. *Sphenoid* (1) (sphenoid: wedge-shaped) contains the sphenoidal sinuses and the sella turcica ("turkish saddle"), which surrounds the pituitary gland.

f. *Ethmoid* (1) (ethmoid: sieve-like) The olfactory nerves (cranial nerve I) pass through the cribriform plate of the ethmoid bone. A fracture here may result in cerebro-spinal fluid leakage into the nasal cavity. Thus, a runny nose after a head injury may indicate a serious condition.

2. *Facial bones*.
   a. *Nasal* (2)
   b. *Maxilla* (2) contains the palatine process (hard palate) and the maxillary sinuses.
   c. *Lacrimal* (2)
   d. *Palatine* (2) forms a part of the hard palate.
   e. *Zygomatic* (2) is the cheek bone.
3. The *mandible* is composed of a *body*, containing the mental protuberance (chin), alveolar borders (for teeth), and a right and left *ramus* (ramus: branch) which includes an angle and condyle. A fractured mandible often requires the lower and upper teeth to be wired together to immobilize the jaw.
4. *Interior of the skull* (Fig. 4-6):
   a. *Anterior cranial fossa* (note cribriform plate of ethmoid) houses the *frontal lobes* of the brain.
   b. *Middle cranial fossa* houses the *temporal lobes* of the brain.
   c. *Posterior cranial fossa* houses the *cerebellum*, *pons* and *medulla*.
5. *Sutures* are junctions of the skull bones. (Fig. 4-5b,c) The main ones are:
   a. *Sagittal* (between the parietal bones)
   b. *Lambdoidal* (between the parietal bones and the occipital bone)
   c. *Coronal* (between the parietal bones and the frontal bone)
6. *Fontanelles* are unclosed junction

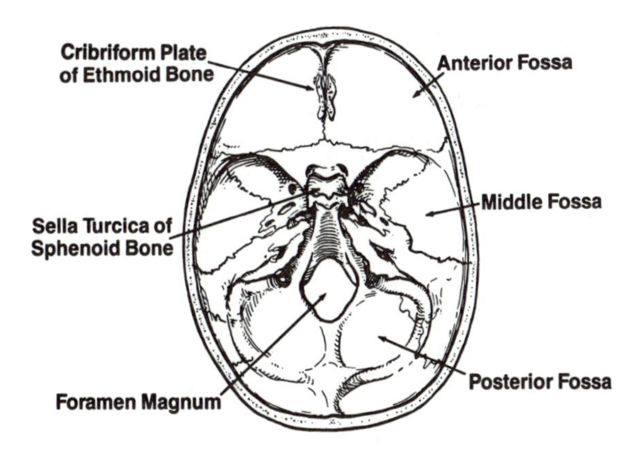

Fig. 4-6 Interior of Skull.

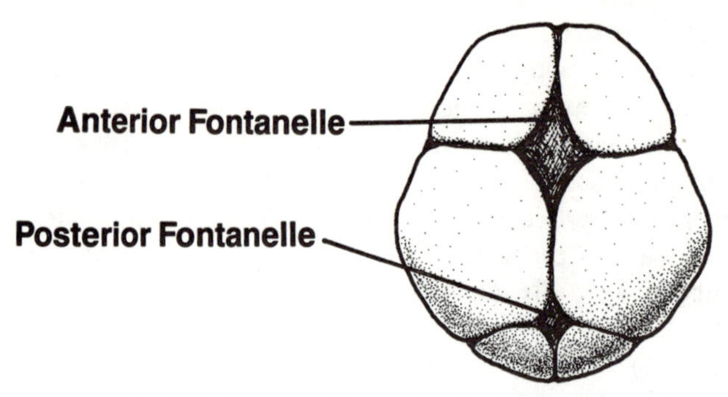

Fig. 4-7 Important Fontanelles in an Infant.

areas in the newborn skull (Fig. 4-7). They are the "soft spots" on an infant's head. The more important are the:

a. *Posterior fontanelle* (between sagittal and lambdoidal sutures) which closes at approximately three months.
b. *Anterior fontanelle* (between the coronal and sagittal sutures) which closes at the middle of the second year and is the larger of the two.
c. *Importance*:
    1) The fontanelles may close late in *hydrocephalus*, a condition in which the brain swells because of the accumulation of fluid.
    2) The anterior fontanelle may bulge out in *meningitis*.

7. *Paranasal air sinuses* (see chapter 11).

B. *Temporomandibular (TM) joint* (Fig. 4-5b):

1. *Bones*: condyle of *mandible* and mandibular fossa of *temporal* bone.
2. *Type*: combination hinge and gliding.
3. *Importance*: The jaw is usually dislocated in the forward position (a dislocation is the displacement of the bones of a joint; a subluxation is a partial dislocation); this is reduced by placing the thumbs on the mandible and pressing down and back — but quickly removing thumbs!

C. *Structure of a typical vertebra* (Fig. 4-8):

1. *Arch* surrounding spinal cord
2. *Body*, and facets for ribs
3. *Vertebral (spinal) foramen* for spinal cord
4. *Intervertebral disc* acts as a cushion for vertebral bodies

D. *Unique features of cervical vertebrae* (Fig. 4-9):

1. Small size
2. *Transverse foramina* containing the vertebral artery
3. The first cervical vertebra, the *atlas*, is circular and has no body.
4. The second cervical vertebra, the *axis*, has a bony protrusion, the *odontoid process*, that lies superiorly.
5. *Importance*:
    a. Because of their small size, the cervical vertebrae are dislocated and fractured more often than other vertebrae.
    b. Diving into shallow water and striking the head is a common cause of a cervical fracture and dislocation. This type of accident often causes spinal cord injury with paralysis of both legs (paraplegia) or both arms and legs (quadriplegia). It may be fatal if the odontoid process is driven into the spinal cord.

E. *Neck joints* (Figs. 1-3b, 4-9, 4-10b):

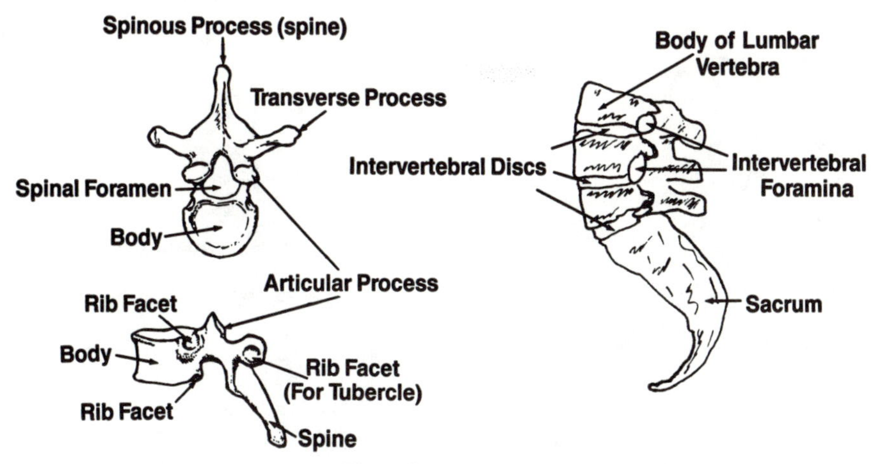

Fig. 4-8 A Typical Vertebra.

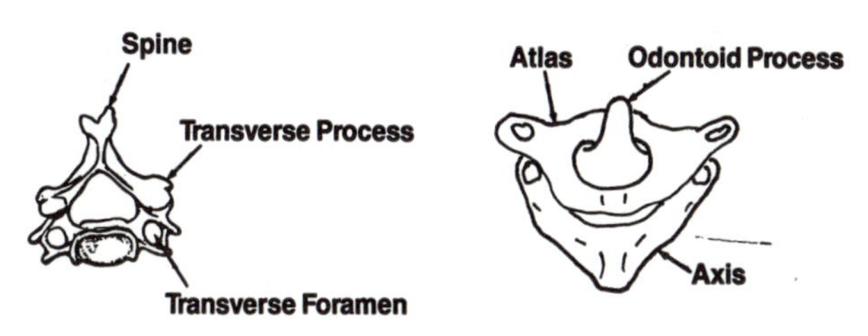

Fig. 4-9 Cervical Vertebrae.

1. *Atlanto-occipital joint*:
   a. Bones: *occipital* condyles and *atlas*
   b. Type: condyloid
   c. *Importance*: may be involved in whiplash injury and arthritis.
2. *Atlanto-axial joint*:
   a. Bones: atlas and axis
   b. Type: pivot
   c. *Importance*: (see 1)

III. *Bones and joints of the thorax, lumbar vertebrae, sacrum and coccyx* (Fig. 4-10):

A. *Bones of the chest (thorax)*:

1. *Sternum* (breastbone). Parts = manubrium (manubrium: handle-shaped), body, sternal angle and xiphoid process.
2. *Ribs*:
   a. *Number*: rib pairs 1-7 attach to sternum and vertebrae ("true ribs"); 5 lower rib pairs are "false ribs": 8, 9 and 10 attach to 7, and 11 and 12 have no anterior attachment ("floating"). The *costal margins* are the left and right arches formed by the inferior ribs on each side.
   b. *Importance*:
      1) The *intercostal spaces (interspaces)* are the areas between the ribs. Numbering begins with the first interspace below the first rib.
      2) Rib pain from bruised ribs or a hairline fracture during inspiration may be helped by splinting the thorax with a "rib belt".
      3) If more than 3 to 4 ribs are fractured in different locations, breathing may be impaired ("flail chest"). Otherwise rib fractures heal with little problem.
3. *Thoracic vertebrae*:
   a. *Parts*: larger than cervical, with oblique spinous processes.
   b. *Importance*: pathologic fractures are common in the elderly and as a result of bone cancer.

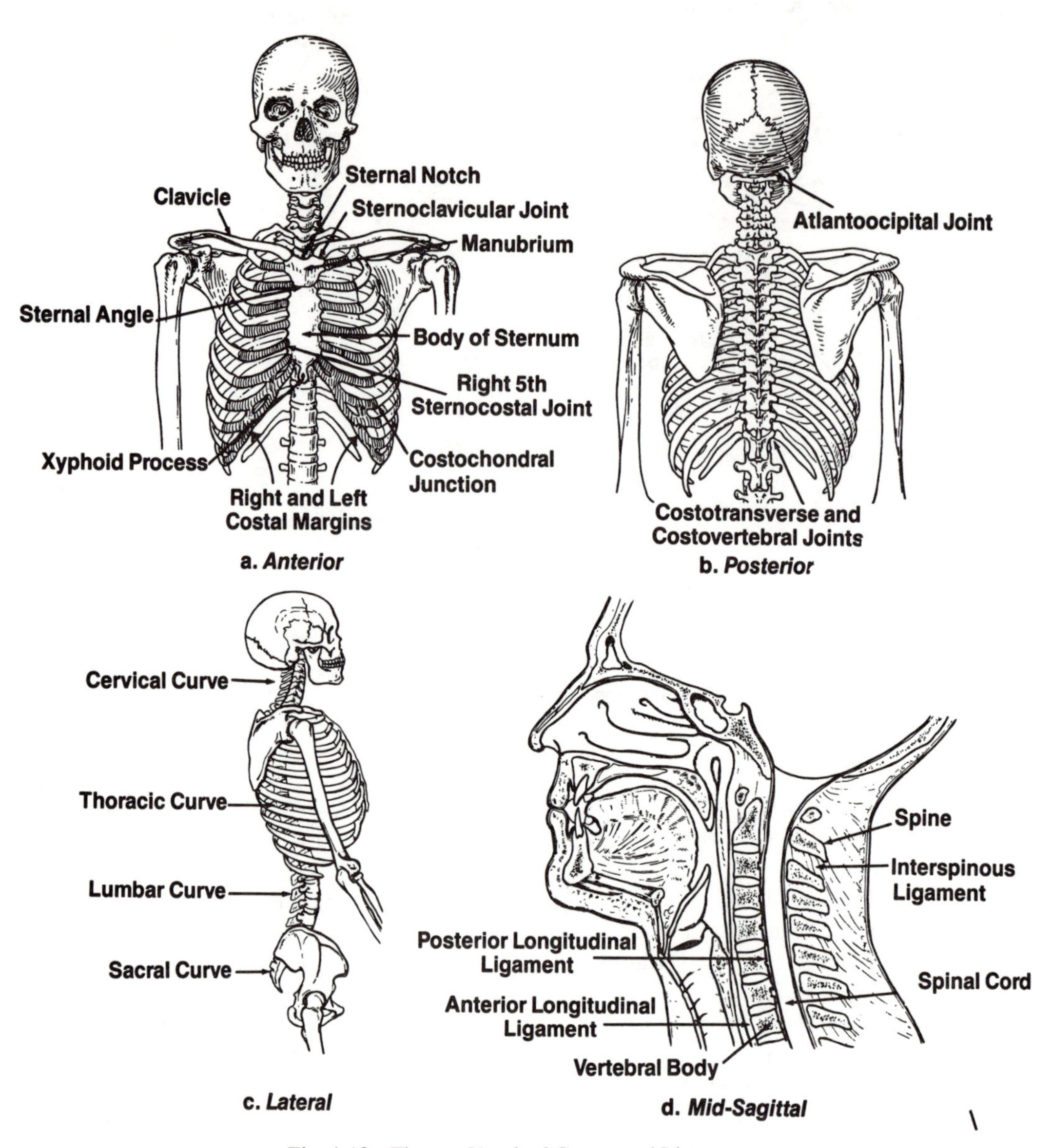

Fig. 4-10 Thorax, Vertebral Curves and Ligaments.

B. *Lumbar vertebrae*:
   1. *Parts*: larger than thoracic, with thick spinous processes.
   2. *Importance*: pathologic fractures as above.

C. *Joints of the thorax*:
   All are gliding joints involving the ribs: *sternocostal*, *costovertebral* (head of rib and vertebral body) and *costotransverse* (tubercle of rib and transverse process of vertebra). These joints are responsible for movement of the thoracic cage during respiration.

D. The *sacrum and coccyx (tailbone)* form the lower portion of the vertebral column (Fig. 4-11). The sacrum forms the posterior part of the *pelvis*. Sacral nerves to the leg run through the *sacral foramina*. The *promontory* is an important landmark for pelvic measurements in childbirth.

E. *The vertebral column as a whole*:

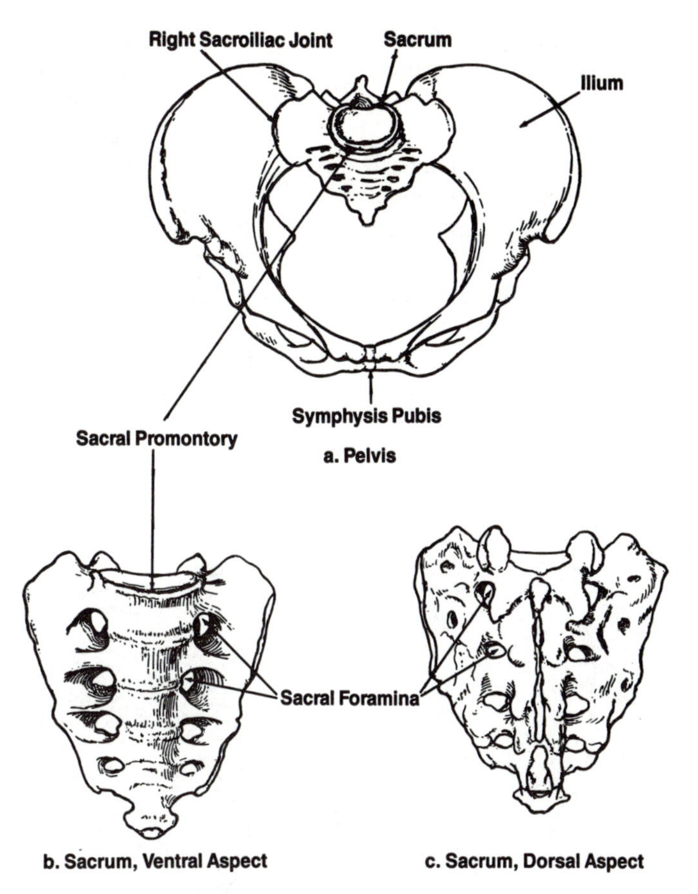

Fig. 4-11 Pelvis and Sacrum.

The vertebral column exhibits several normal curves, as illustrated (Fig. 4-10c). Each vertebra joins a successive vertebra, forming an *amphiarthrosis*. The amphiarthroses function additively, however, accounting for the flexibility of the vertebral column: flexion, extension, lateral movements, as well as some rotation.

1. The main ligaments of the vertebral column are the *anterior and posterior longitudinal ligaments*, and the *interspinous ligaments* (Fig. 4-10d).
2. *Importance* of the vertebral column:
   a. The *spinal cord* lies in the *vertebral canal*.
   b. Spinal nerves exit through the *intervertebral foramina*, formed by each successive vertebra (Fig. 4-8).
   c. If developing arches fail to fuse into the spinous processes, *spina bifida* results: skin, muscle and meninges instead of bone cover part of the dorsal spinal cord. This condition may range from a mild case to spinal cord compromise and paraplegia. The most common site is the *lumbo-sacral* region.
   d. The anterior longitudinal ligament and disc are sometimes injured in *whiplash*, caused by sudden hyperextension of the head (usually from a rear-end auto accident). Usually, it involves only muscle strain. In the rare severe case, the ante-

rior longitudinal ligament may tear, and a disc subluxation may occur ("*acute cervical sprain*" (see also p. 68).

e. A *slipped disc* in the lumbar region causes back pain because it impinges on a nerve root (see chapter 7).

f. Accentuation of the thoracic curve results in a *hunchback* (kyphosis). This condition is seen often in elderly people.

g. *Scoliosis*, lateral curvature, is common, usually in the thoracic region. Curvatures of 30 degrees or greater in children often require bracing. In severe cases, breathing and growth are affected because the twisted ribs impinge on the lung.

IV. *Bones of the upper extremity*:

A. *Clavicle*: one of the most frequently fractured and dislocated bones, particularly in children (Fig. 4-12):

B. *Scapula* (Fig. 4-12):

1. *Parts*: spine, acromion (acromion = shoulder) process, coracoid (coracoid = crow's beak) process,

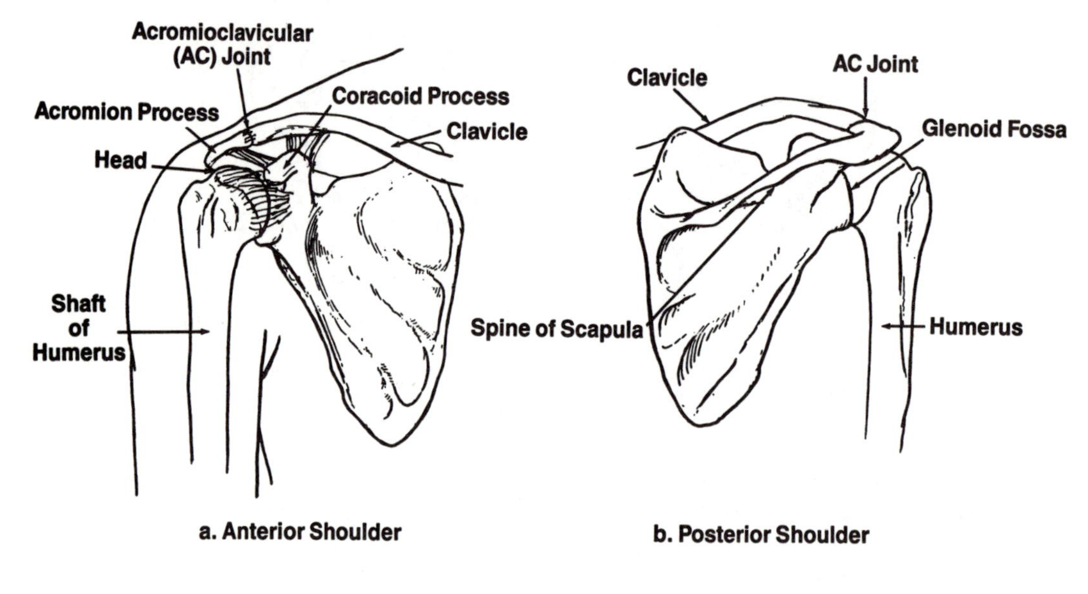

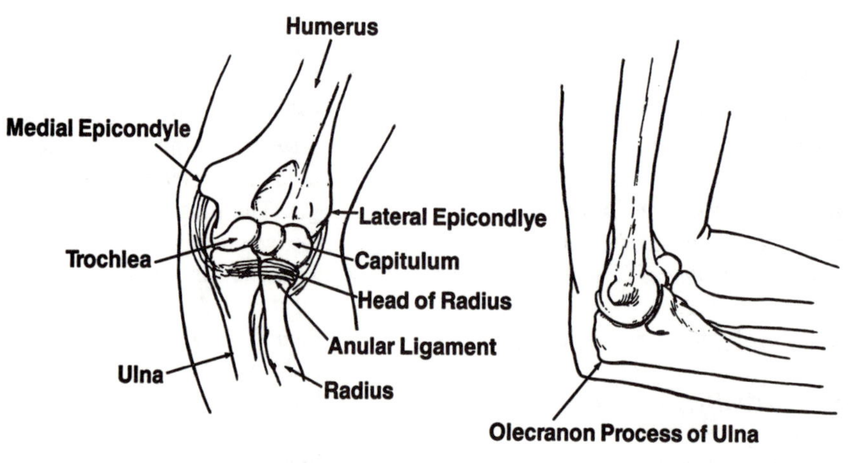

Fig. 4-12 Bones and Joints of Upper Extremity.

glenoid (glenoid = socket) fossa.

2. *Importance*:
   a. The clavicle and scapula constitute the "shoulder girdle".
   b. The small size of the glenoid fossa accounts for the common anterior shoulder dislocation.

C. *Humerus* (Fig. 4-12):
   1. *Parts*: head, shaft, medial and lateral epicondyles, trochlea (trochlea = pulley) capitulum (capitulum = little head).
   2. *Importance*:
      a. The "funnybone" is the area of the medial epicondyle where the ulnar nerve lies superficially.
      b. The lateral epicondyle is involved in "tennis elbow" (see end of chapter).

D. *Radius* (Figs. 4-12, 4-13):
   1. *Parts*: head (proximal), shaft, styloid process (distal and lateral).
   2. *Importance*: fractures of the distal radius are common in children from falling on an outstretched hand (Colle's fracture).

E. *Ulna* (Figs. 4-12, 4-13):
   1. *Parts*: olecranon (olecranon = elbow) process, shaft, head (distal), styloid process (dorsomedial).
   2. *Importance*: olecranon bursitis is common. A fracture of the styloid process frequently accompanies a Colle's fracture (above).

F. *Carpal bones* (carpals) 8 small wrist bones (Fig. 4-13):
   1. *Proximal Row* (radius to ulna): navicular (scaphoid), lunate, triquetrum, pisiform.
   2. *Distal Row* (radius to ulna): trapezium, trapezoid, capitate, hamate.
   3. *Importance*: A sprained wrist may involve a navicular (scaphoid) fracture. Ignoring such an injury can result in severe hand problems.

G. *Metacarpal bones* (metacarpals) (Fig. 4-13):
   1. *Parts*: 5 bones distal to carpals (see numbering system below).
   2. *Importance*: a common injury is a displaced fracture of the neck of the

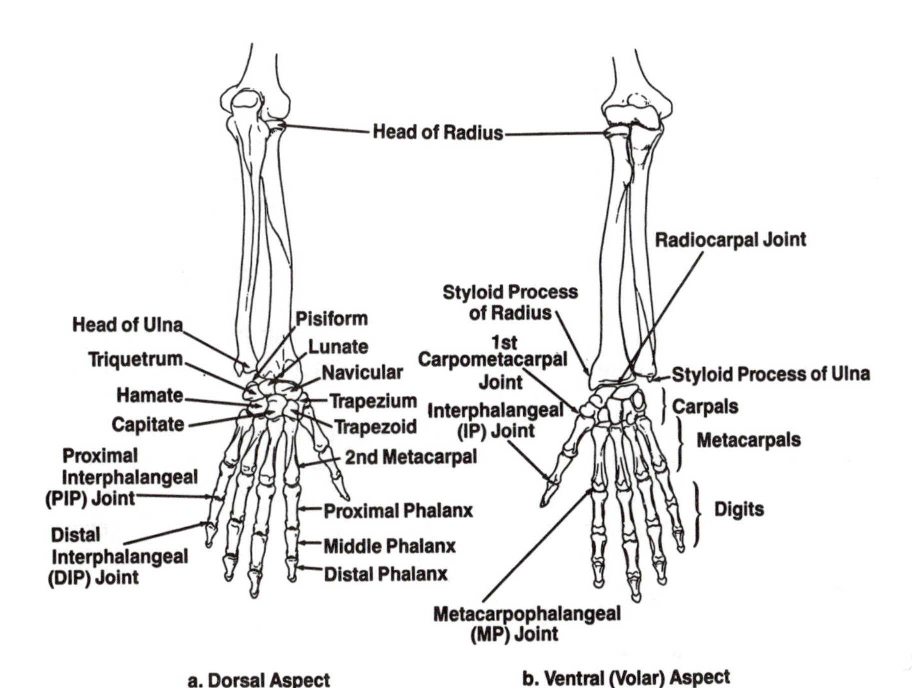

Fig. 4-13 Bones and Joints of Upper Extremity (cont'd).

5th metacarpal (boxer's fracture).

H. *Digits*: 5 digits, with 2-3 small bones per digit (Fig. 4-13):
   1. Each digit has a proximal, middle and distal phalanx (except the thumb, which has no middle phalanx).
   2. Numbering system: thumb is #1. Thumb: pollex, pollicis (L).

V. *Joints of the upper extremity*:

A. *Sterno-clavicular* (Fig. 4-10a):
   1. Bones: *sternum* and *clavicle*
   2. Type: gliding

B. *Acromio-clavicular* (AC) (Fig. 4-12a):
   1. Bones: *clavicle* and acromion process of the *scapula*
   2. Type: gliding
   3. *Importance*: AC separation is common from a fall on the outstretched hand or against the clavicle.

C. *Shoulder (gleno-humeral)* (Fig. 4-12a):
   1. Bones: head of the *humerus* and glenoid fossa of the *scapula*
   2. Type: ball-and-socket
   3. *Importance*: A commonly dislocated joint

D. *Elbow*: a composite joint, consisting of three separate articulations (Fig. 4-12c):
   1. *Humero-ulnar*:
      a. Bones: trochlea of *humerus* and trochlear notch of *ulna*
      b. Type: hinge
   2. *Radio-humeral*:
      a. Bones: capitulum of *humerus* and head of *radius*
      b. Type: combination pivot and hinge
   3. *Proximal radio-ulnar*:
      a. Bones: head of *radius* and radial notch of *ulna*
      b. Type: pivot (pronation-supination)
   4. *Importance*: Yanking the arm of a small child may cause displacement of the head of the radius from the annular ligament (*radial head subluxation*). The child will not move his arm. Reduction is accomplished by steady repeated supination to re-seat the radial head.

E. *Distal radio-ulnar* (Fig. 4-13):
   1. Bones: head of *ulna* and ulnar notch of *radius*
   2. Type: pivot (pronation-supination)

F. *Radiocarpal (wrist)* (Fig. 4-13):
   1. Bones: distal *radius* and proximal *carpals*
   2. Type: condyloid

G. *1st carpo-metacarpal* joint is a modified condyloid joint (saddle-joint) and is the only movable articulation between carpal and metacarpal (Fig. 4-13):

H. *Metacarpo-phalangeal (MP)* (Fig. 4-13):
   1. *Type*: condyloid
   2. *Importance*: involved in rheumatoid arthritis (see end of chapter)

I. *Interphalangeal* (Fig. 4-13):
   1. *Type*: hinge
   2. *Proximal interphalangeal (PIP)*:
      a. *Bones*: proximal and middle phalanges
      b. *Importance*: rheumatoid arthritis, as above
   3. *Distal interphalangeal (DIP)*:
      a. *Bones*: middle and distal phalanges
      b. *Importance*: involved in osteoarthritis (see end of chapter)
   4. *The thumb* has only an *interphalangeal (IP)* joint, between proximal and distal phalanges

VI. *Bones of the lower extremity*:

A. *The hip bone (os coxae)* consists of three fused bones — *ilium*, *ischium* and *pubis*. The term *pelvis, or pelvic girdle*, refers to the two hip bones plus the sacrum. *Parts* (Fig. 4-14):
   1. *Ilium* (superior): iliac crest, anterior and posterior spines.
   2. *Ischium* (inferior-posterior): ischial tuberosity, obturator (obturator = closure) foramen (between ischium and pubis).
   3. *Pubis* (inferior-anterior): the pubic symphysis (symphysis = fusion) is

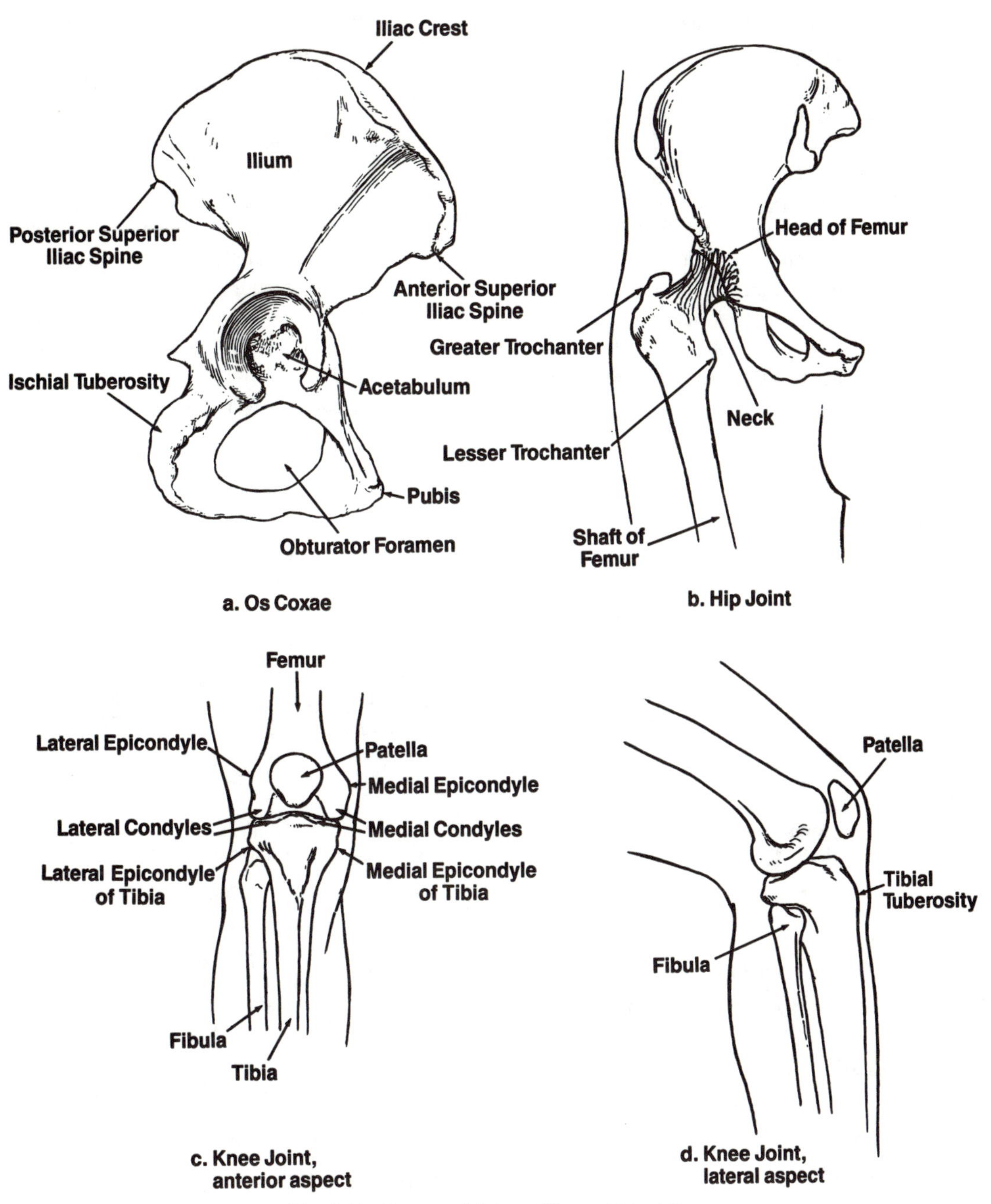

Fig. 4-14 Bones and Joints of Lower Extremity.

an amphiarthrosis which becomes more movable during childbirth.

4. *Importance*:
   a. The obturator nerve travels through the obturator foramen to the adductors of the thigh.
   b. The ilium, ischium and pubis unite at the hipbone socket, the *acetabulum*.
   c. A *bone marrow tap* is frequently performed at the iliac crest.

B. *Femur (thigh bone)* (Fig. 4-14):
   1. *Parts*: head, neck, greater and lesser trochanters (trochanter =

round ball) shaft, lateral and medial condyles and epicondyles, patellar surface.

2. *Importance*: A neck fracture is common in the elderly.

C. *Patella (knee bone)* (Fig. 4-14):

1. The *largest sesamoid bone* (a sesamoid bone is one imbedded in a tendon or joint capsule, usually near a digit). The patella is imbedded in the quadriceps tendon.

2. *Importance*: A dislocation is usually easily reduced by extending the leg.

D. *Tibia (shinbone)* (Figs. 4-14, 4-15):

1. *Parts*: medial and lateral condyles, tuberosity, shaft, medial malleolus (malleolus = little hammer).

2. *Importance*:

a) Most commonly fractured long bone (at ankle).

b) Painful tibial tuberosity in boys is usually benign (Osgood-Schlatter's disease).

E. *Fibula* lies lateral to the tibia: shaft, lateral malleolus (distal ankle portion) (Figs. 4-14, 4-15):

F. *Tarsal bones* (tarsals) (7): 1st, 2nd and

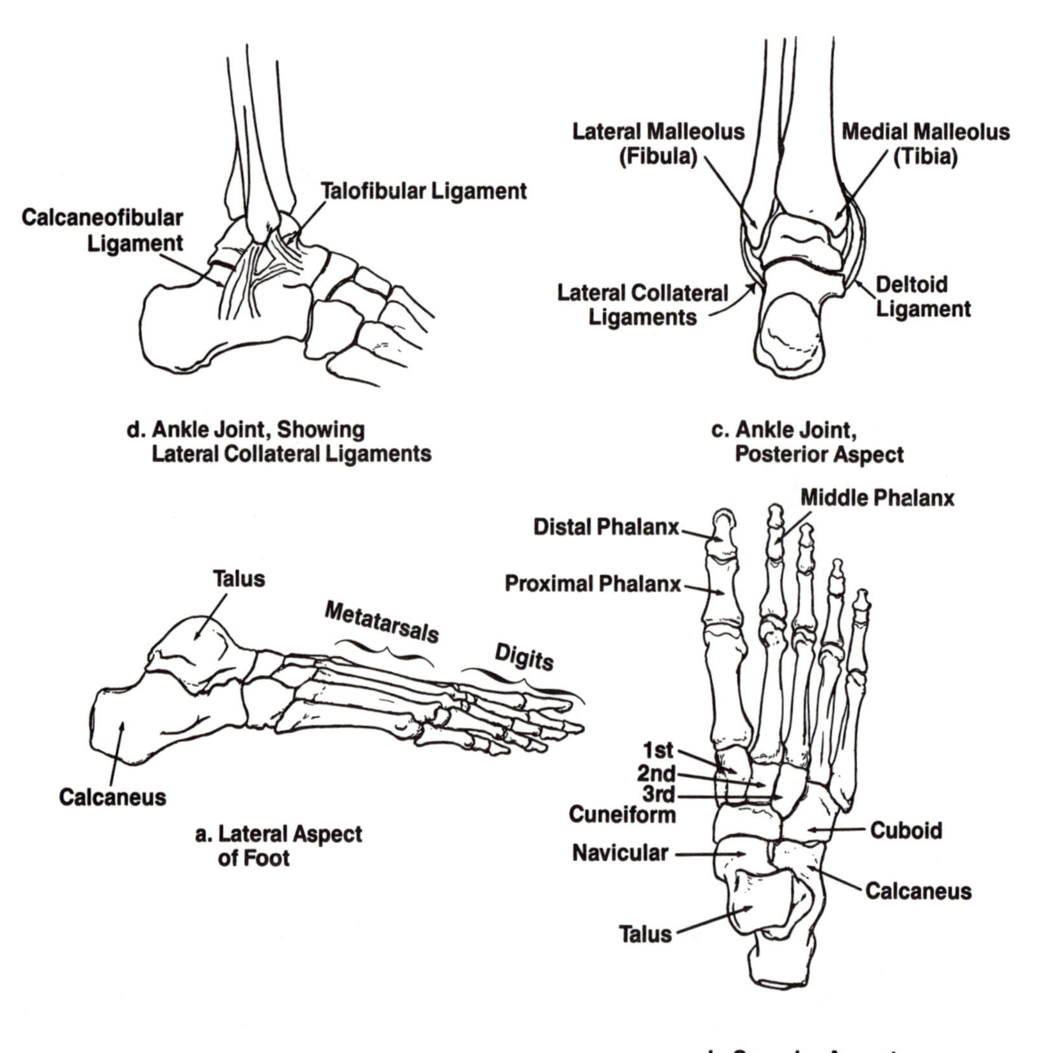

Fig. 4-15 Bones and Joints of Lower Extremity (cont'd).

3rd cuneiform, cuboid, navicular, talus, calcaneus (heel bone) (Fig. 4-15):

G. *Metatarsals and digits* (Fig. 4-15):
   1. Numbering system: Big toe is #1 (hallux, hallucis).
   2. *Importance*: fracture of distal phalanx (distal tuft) sometimes seen with a "stubbed toe"

VII. *Joints of the lower extremity*:

A. *Sacroiliac* (Fig. 4-16):
   1. Bones: *sacrum* and *ilium*
   2. Type: amphiarthrosis
   3. *Importance*: this slightly movable union is usually not involved, as was thought earlier, in low back pain.

B. *Hip* (Fig. 4-14):
   1. Bones: head of *femur* and acetabulum of *os coxae*
   2. Type: ball-and-socket
   3. *Importance*: After a fracture of the neck of the femur in the elderly, the leg is shorter and laterally rotated.

C. *Knee* (Fig. 4-17):
   1. Bones: femur, tibia, patella
   2. Type: hinge
   3. *Ligaments*:
      a. The *patellar* ligament encloses the patella and attaches to the tibial tuberosity.
      b. The *menisci* (cartilage cushions) (*lateral* and *medial*) lie between femoral and tibial condyles.
      c. *Cruciate* (cross) ligaments prevent forward and backwards slippage.
      d. *Collateral ligaments* (*lateral* and *medial*) prevent slippage from side to side.
   4. *Importance*:
      a. Ligamentous injuries are common in contact sports
      b. Twisting injuries, or a blow from the lateral side, may disrupt the medial meniscus, medial collateral, and sometimes the anterior cruciate ligament.
      c. Jamming a meniscus between the condyles of the femur and tibia may cause the knee to "lock".

D. *Ankle joint* (Fig. 4-15):
   1. Bones: medial malleolus of *tibia*, lateral malleolus of *fibula*, and *talus*.
   2. Type: hinge
   3. *Ligaments*:
      a. *Lateral collateral* (anterior and posterior talofibular, calcaneofibular)
      b. *Deltoid* (medial)
   4. *Importance*: a sprained ankle (a sprain is a stretching and tearing of ligamental fibers) is the most common injury affecting the lower extremity (see end of chapter).

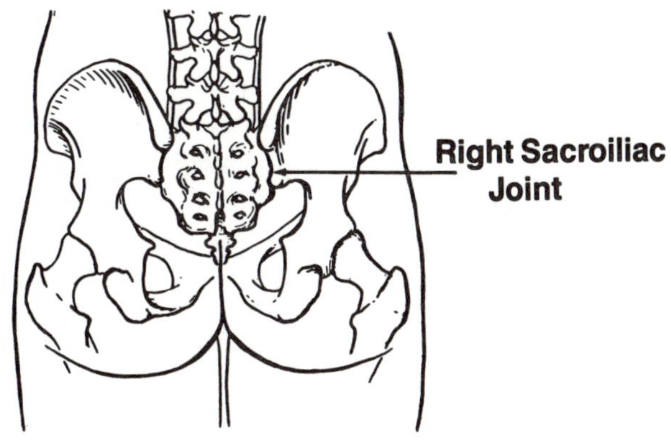

Fig. 4-16 Sacroiliac Joint.

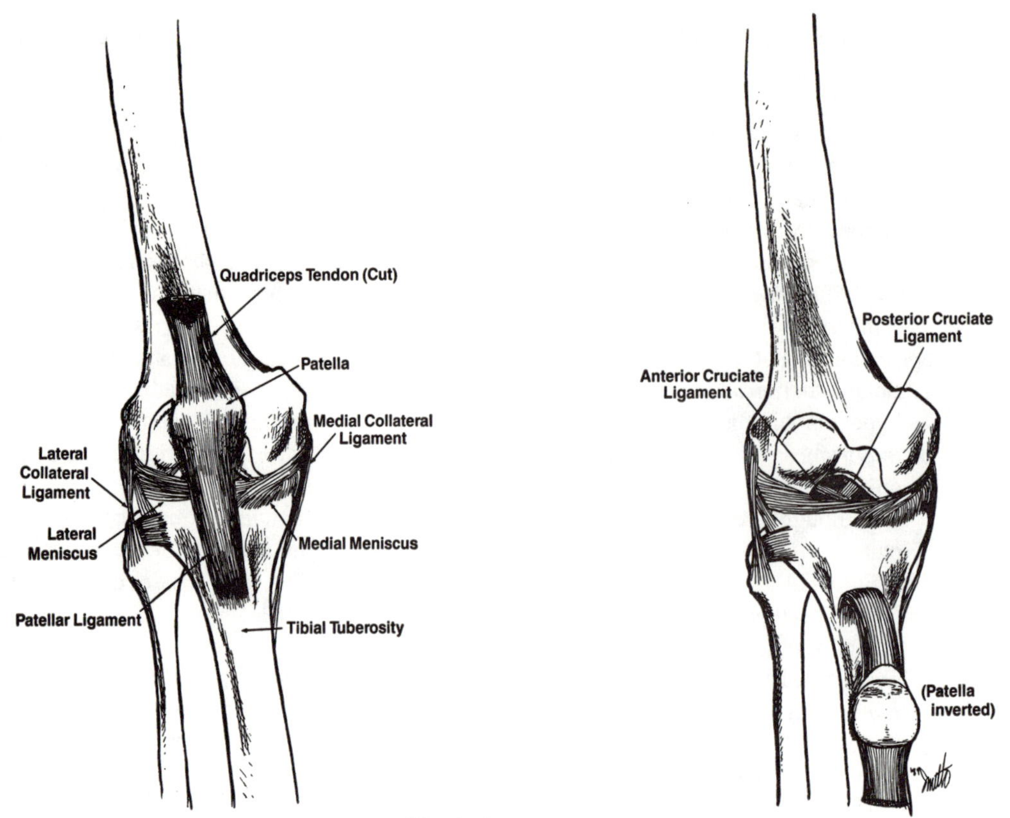

Fig. 4-17 Knee Joint.

VIII. *Common bone and joint disorders*:

A. *Arthritis*: The two most common forms of arthritis are *rheumatoid* and *osteoarthritis (degenerative joint disease)*.

1. *Rheumatoid arthritis* is a chronic, probably autoimmune, disease primarily affecting females (Fig. 4-18). Common manifestations include morning stiffness, joint pain and paresthesias. *The synovial membranes of the small hand joints such as the proximal interphalangeal (PIP) and metacarpophalangeal (MP) are thickened and damaged.* Primary treatment: an anti-inflammatory agent, such as aspirin.

2. *Osteoarthritis* is the most *common joint disease*, affecting principally the elderly, and secondarily seen in a previously injured joint (Fig. 4-19). It is "wear and tear" arthritis. The *joint cartilage degenerates.* The most commonly affected joints are the *distal interphalangeal (DIP)*, hip, intervertebral and sacroiliac. Primary treatment: an anti-inflammatory medicine.

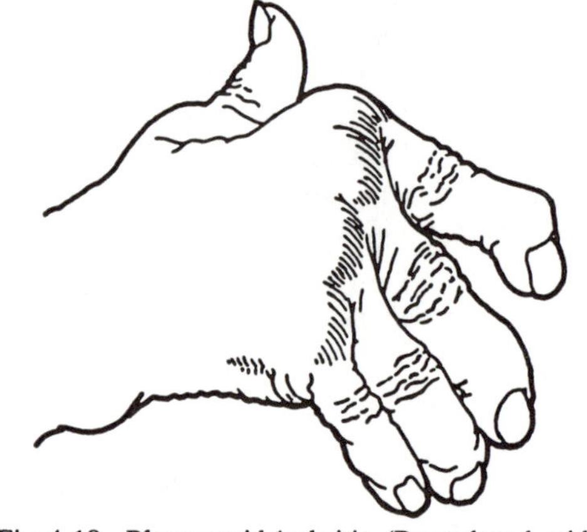

Fig. 4-18 Rheumatoid Arthritis. (Reproduced, with permission, from The Hand, American Society for Surgery of the Hand, 1978).

B. *Bursitis*, an inflammation of a bursa (from injury or sustained pressure),

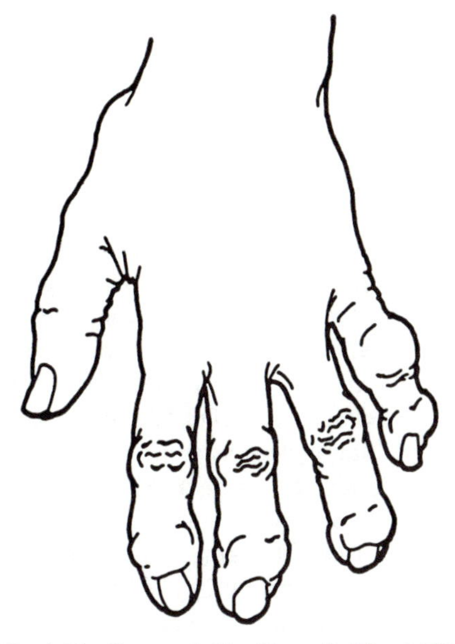

Fig. 4-19 Osteoarthritis. (See ref., Fig. 4-18).

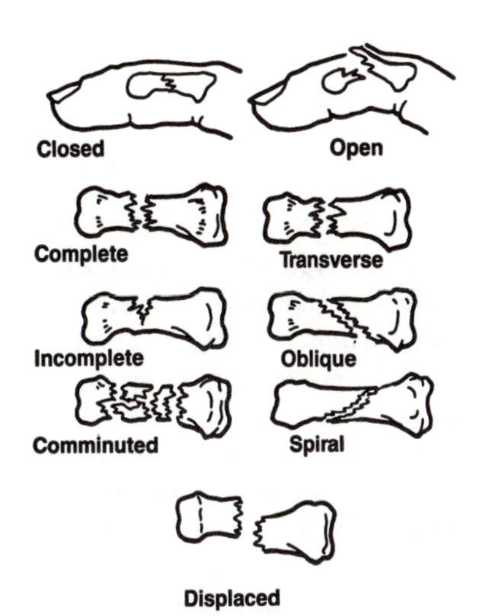

Fig. 4-20 Types of Fractures. (See ref., Fig. 4-18).

occurs most commonly in the knee and elbow. The area is painful, red and swollen. Primary treatment consists of the local aspiration of fluid, rest, elevation of the joint, and the use of an anti-inflammatory agent.

C. A *fracture is a broken bone*. Healing beings with hemorrhage and clot formation. *Granulation tissue* enters the clot, and *collagen* is deposited. *Fibroblasts* from the *periosteum* grow into the area. At the end of the first week, islands of cartilage bridge the fracture site, a *procallus*. Calcium is deposited in the cartilage matrix. Eventually the procallus becomes traversed by bone, and a rigid *bony callus* is formed (2-4 weeks). Healing is usually complete in *6 weeks*. Some types of fractures are shown in *Fig. 4-20*.

D. *Gout* is a metabolic disease. *Uric acid in the bloodstream is elevated and crystals are deposited in the tissues, particularly the joints*. A commonly affected joint is the *metatarsophalangeal joint of the big toe*. The feet are often involved because they are the most dependent areas. The joint is red, swollen and painful. A genetic predisposition exists. Most sufferers are middle-aged males. A frequent precipitating event is alcohol excess. Beer and wine contain high concentrations of purines, which are metabolized to uric acid. Primary treatment consists of an anti-inflammatory agent, rest and elevation of the affected part.

E. *A sprained ankle*. The typical case involves twisting inward (*inversion sprain*). The *lateral collateral ligaments are stretched and sometimes torn* (see Fig. 4-15). The tendons of the *peroneal muscles* may also be stretched or torn. The ankle is swollen and painful. Sometimes a fragment of bone is pulled off with the ligament or tendon (avulsion fracture). Primary treatment: an elastic bandage, pain medicine and rest. With severe sprains, a cast may be required.

F. *Tennis elbow* (lateral epicondylitis) is an *inflammation of the extensor/supinator forearm tendon fibers at their origin at the lateral epicondyle of the humerus*. It is seen in those who extend the wrist and grasp or rotate frequently (pronation-supination): tennis players, baseball pitchers, carpenters, pipe-fitters, plumbers, dog handlers. Primary treatment: a xylocaine injection, rest, heat, and an anti-inflammatory medicine.

-FIVE-

MUSCLES

I. *Characteristics of muscles*:
The three types of muscles were described in chapter 1: cardiac, skeletal and smooth. *Skeletal muscles* are stimulated to contract by *motor nerves*. If a nerve is severed, the muscles innervated by it die (atrophy). The area of contact between the motor nerve and the muscle is the *motor end-plate*, or myo-neural (neuromuscular) junction. *Heart and smooth muscle* are innervated by the *autonomic nervous system* — in this case nerves *modify* but do not initiate the impulse. *Heart muscle* and the *conducting system* are described in chapter 9. *Smooth muscle* and the *autonomic nervous system* are discussed in chapter 7.

A. *The motor end-plate* is a modified synapse consisting of a *terminal bud* of a nerve cell axon and a *muscle fiber*. When the nerve is stimulated (depolarized), the terminal bud releases *acetylcholine*. Acetycholine causes *depolarization of the membrane receptor* of the muscle. *Contraction follows*. The enzyme *cholinesterase* breaks down acetylcholine into choline and acetic acid, thus preventing sustained muscle contraction (*tetanus*) (Fig. 5-1). Drugs that block cholinesterase cause spasm; those that block acetylcholine cause paralysis. Some toxins and drugs that act on the motor end-plate are:

1. *Stimulators*:
   a. *Neostigmine*, a cholinesterase inhibitor, is used in the treatment of myasthenia gravis (see end of chapter).

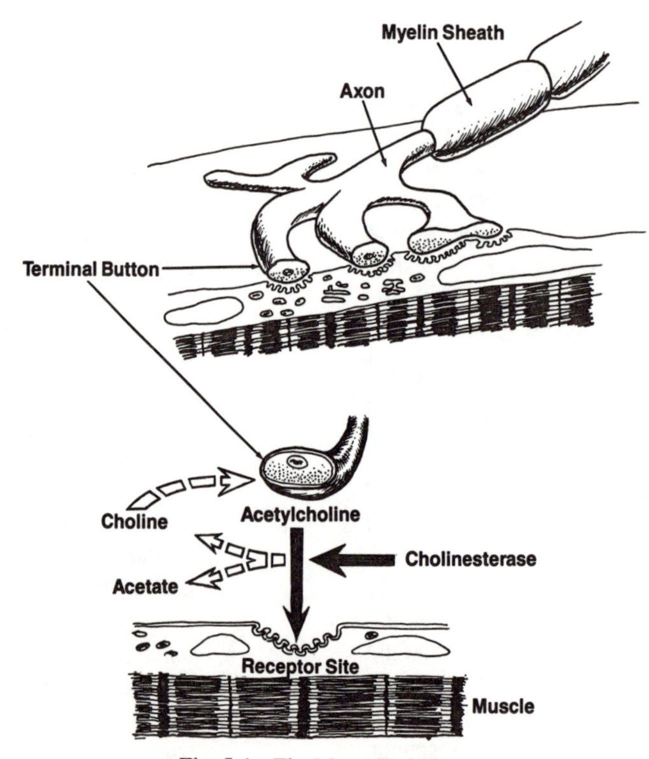

Fig. 5-1 The Motor End Plate.

b. *Tetanus toxin*, produced by the bacterium Clostridium tetani, causes an accumulation of acetylcholine, and spasm, by blocking cholinesterase.

2. *Paralytic agents (neuromuscular blocking agents)*:

   a. *Curare*, found in "bushrope" vines in the Amazon is rubbed into the grooves of arrow tips. It blocks acetycholine at the muscle receptor. A derivative, tubocurarine (the active substance), is used today as a muscle relaxant in surgery.

   b. *Botulism toxin*, from infection with the bacterium Clostridium botulinum, as well as *coral snake venom*, block acetylcholine.

   c. *Succinylcholine*, also used as a muscle relaxant in surgery, depolarizes the muscle receptor and prevents repolarization.

B. *Mechanical events of muscle contraction*.

As shown in Fig. 5-2, the *myofibril* is made up of units called *sarcomeres*. Each sarcomere consists of bands, lines and filaments formed by proteins. The *thick filaments* are *myosin*, and the *thin filaments* are actin (and lesser amounts of tropomyosin and troponin). Depolarization of muscle causes the following sequence:

1. *Calcium ions*, stored in the sarcoplasmic reticulum (the endoplasmic reticulum of the muscle cell), are released and move between filaments, interacting with proteins that prevent the formation of cross-bridges.
2. Cross-bridges are formed, *actin and myosin filaments slide over one another*, and shortening (contraction) of the myofibril takes place.

C. *An overview of glucose metabolism* in the body is shown in Fig. 5-3. Ingested

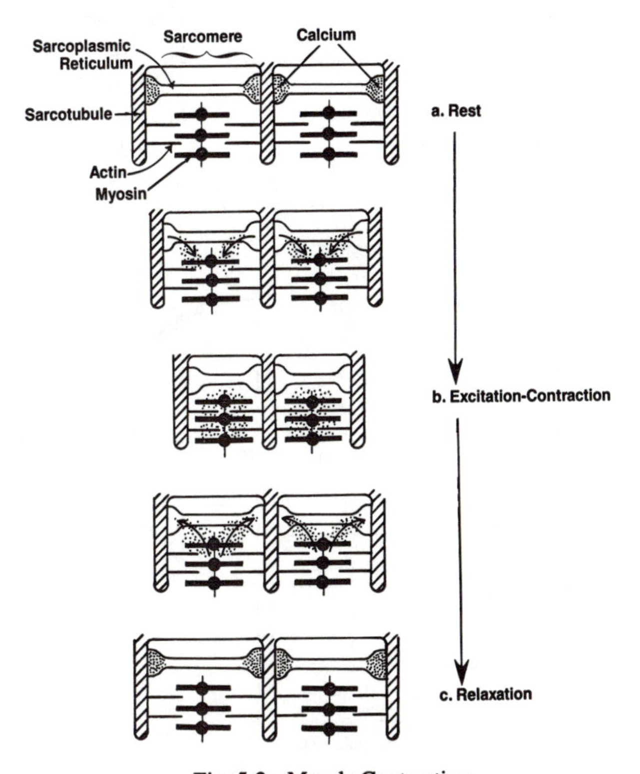

Fig. 5-2 Muscle Contraction.

glucose is carried to the tissues or is stored in the liver as *liver glycogen*, a long chain of glucose molecules. In tissues, including muscle, glucose is either used or stored as *muscle glycogen*. In order for glucose to enter the muscle cell, *potassium* and *insulin* are required. *Lactic acid* is the end product of muscle work, and is transported by the bloodstream to the liver, to be converted back to glucose, using oxygen (see also chapter 13).

D. *The metabolism of muscle contraction.*
   1. In the biological system, *phosphorous compounds* such as ATP, ADP and creatine phosphate (CP, phosphocreatine) are high-energy sources. *Work done by the body uses ATP. Glucose is the fuel for the manufacture of ATP.*
   2. *Glucose metabolism in the muscle cell.* As shown in Fig. 5-4, glucose in the sarcoplasm (cytoplasm of the muscle cell) is metabolized to pyruvic acid and enters the *mitochondria* if oxygen is present (*aerobic metabolism*). In the mitochondria, pyruvic acid undergoes a series of cyclical chemical changes resulting in the production of 36 molecules of ATP. This efficient system is the *citric acid, or Krebs cycle*. When oxygen is lacking, pyruvic acid does not enter the mitochondria, but rather is converted to lactic acid in the cytoplasm, producing *2 ATP's*. At first glance, the latter appears to be an inefficient system. It is, however, very effective when oxygen demand exceeds availability, as in sports races like sprinting or swimming dashes. This conversion of *glucose* to *lactic acid* is *anaerobic* (without oxygen) *glycolysis* (glucose breakdown).
   3. *Events involved in contraction and exercise.* Muscle contraction requires energy, supplied by ATP. All subsequent reactions are designed to replenish ATP. Skeletal muscle contains limited stores of ATP and CP. CP is quickly depleted, and restored mainly at rest. Chief fuels for replenishing ATP are *glucose* and *free fatty acids (FFA)* (Fig. 5-5).
      a. *After depolarization,* ATP is hydrolysed to ADP, releasing

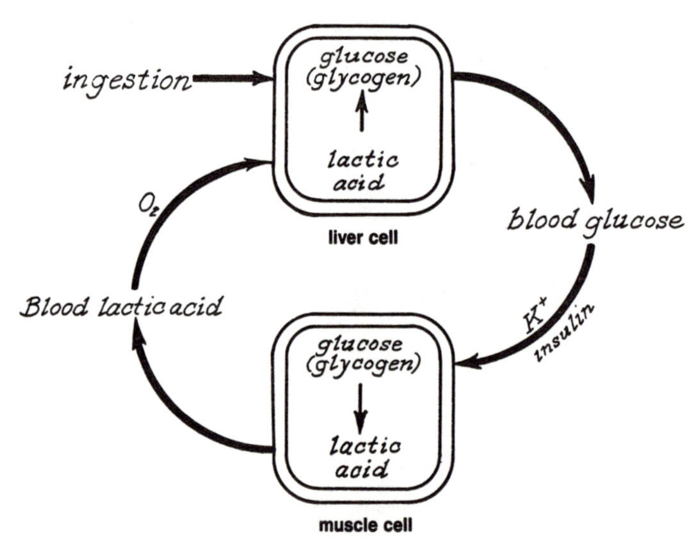

Fig. 5-3 Fate of Glucose in the Body.

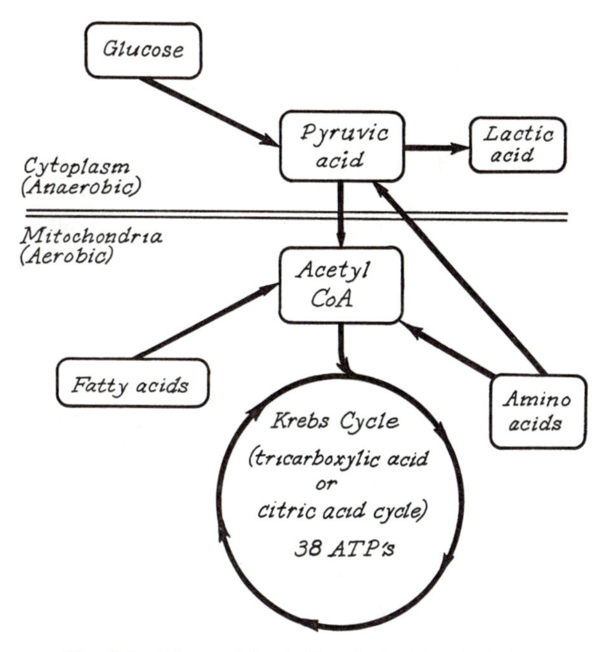

Fig. 5-4 Glucose Metabolism in the Muscle Cell.

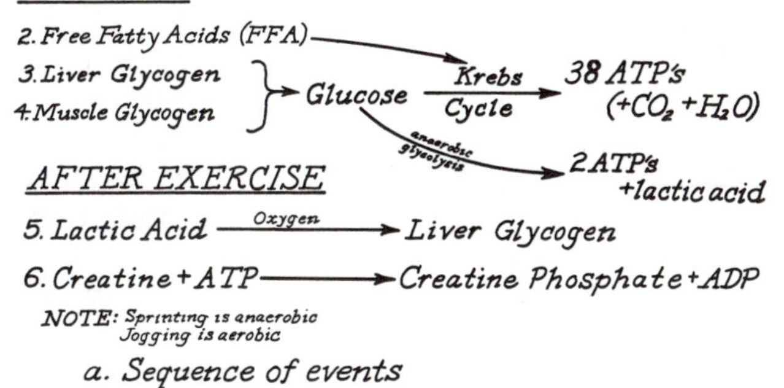

a. Sequence of events

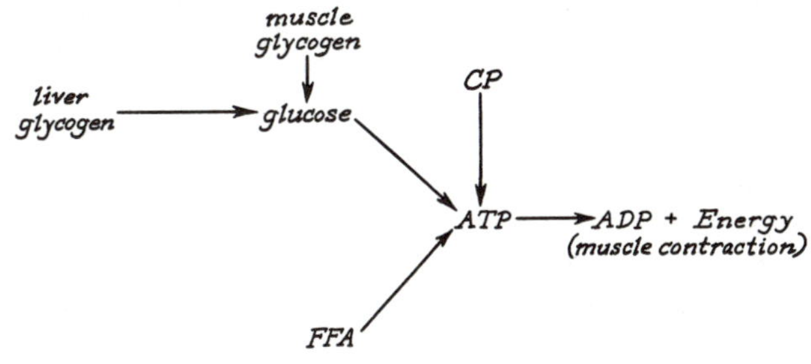

b. Summary of Sources of ATP in Muscle

Fig. 5-5 Muscle Contraction and Exercise.

7.3 kilo-calories of *energy* for contraction. *Myosin* molecules under the influence of *calcium* form cross-linkages with *actin,* producing shortening (Fig. 5-2).

b. *As exercise begins,* catecholamines stimulate liver and muscle glycogen to release *glucose* (glycogenolysis) and fats to release *FFA* (lipolysis). Glucose and FFA form acetyl CoA which enters the Krebs cycle in the mitochondria, producing *38 ATP's (aerobic metabolism). Carbon dioxide is blown off.*

c. *As exercise continues,* the contribution of FFA gradually diminishes, and ATP production begins to rely on *muscle glycogen.* As long as oxygen is available, 38 ATP's are manufactured via the Krebs cycle. When oxygen is depleted, a shift to *anerobic metabolism* occurs: glucose is converted to pyruvate, forming *lactic acid and 2 ATP's.* Eventually, not enough lactic acid is converted to glycogen in the liver, it begins to accumulate in muscle, and contraction ceases.

d. When the body switches to the manufacture of ATP anaerobically, it incurs an *oxygen debt.* As the breathing rate is increased after exercise, oxygen is restored to the body. The restoration of oxygen does two things: it allows conversion of lactic acid to glycogen in the liver, and restores the functioning of the Krebs cycle.

4. *Types of contraction:*

a. *Isotonic* is contraction with movement — "tension with movement".

b. *Isometric* is contraction with little or no movement — "tension *without* movement".

c. *Sports feature*: both types are used in the development of muscles. A muscle is best developed using isotonics with some isometrics. Bodybuilders use isometrics for strength and enlarging the muscle ("bulk"), then isotonics ("repetitions") to accentuate each muscle group ("definition").

E. *Hypertrophy and atrophy*:
Muscles do not regenerate to any appreciable degree. They may degenerate if not used (see "*disuse atrophy*" at the end of the chapter). Muscles enlarge (*hypertrophy*) if frequently used (although muscle fibers do not have the capacity for mitosis, they do have the ability to develop).

F. *Fiber-types.*

1. *Type I (red fibers)* have a *high myoglobin content* (a pigment similar to hemoglobin that transports oxygen from the bloodstream to the mitochondria in muscle cells, where the partial pressure of oxygen is low), *many mitochondria* and a *slow contraction time.*

2. *Type II (white fibers)* have *less mitochondria* and a *faster contraction time.*

3. *Each person has both types.* Which predominates is genetically determined. It is tempting to predict athletic potential on the basis of fiber-type percentage (some long-distance runners have a high percentage of red fibers; some sprinters have a high percentage of white fibers), but current thinking (reinforced by muscle biopsy sampling errors) suggests that this may not be justified.

G. *Accessory structures*:

1. A *tendon* is firm connective tissue *attaching muscle to bone.* A *ligament* is composed of similar tissue and *attaches bone to bone* (e.g., a joint capsule, Fig. 4-2). An *aponeurosis* is a flattened tendon.

2. *Fascia* are sheets or bands of connective tissue that surround muscle groups and other internal structures.
3. *Tendon-sheaths* are *tubular bursae* that surround tendons to prevent friction (e.g., in the hand) (Fig. 5-6).

II. *Aerobics,* a term coined by K.H. Cooper, an air-force doctor involved in physical fitness, refers to exercises that stimulate heart and lung activity over a certain time period. It means *cardiovascular fitness* and health, or increasing an oxygen load to the tissues. Generally, an exercise done for 1/2 hour without rest stresses the heart and lungs sufficiently to produce an aerobic effect. Aerobics increases the efficiency of muscle because *oxygen is better utilized* (through the Krebs cycle). The heart, a muscle pump, becomes more effective. It *contracts slower*, but pumps more blood per contraction. Fats are cleared from the bloodstream. This usually results in less arteriosclerosis, and thus a decreased susceptibility to a heart attack or stroke (many exercises advertised today as "aerobic", unfortunately, have only minimal fitness value). Those yielding the greatest cardiovascular benefit are jogging, running, cycling, swimming (skating, handball and cross-country skiing are on a par with swimming). Important concepts:

A. Energy from FFA is much higher than from glucose (1 gm of fat yields 9 k-cal; 1 gm of glucose yields 4 k-cal). Aerobics improves the body's ability to utilize fats, thus conserving glycogen.
B. It takes about 35 miles of fast walking or jogging to burn off about one pound of fat.
C. Aerobics, or general fitness, is often able to revascularize blocked coronary arteries, and is an important therapeutic modality after a heart attack (see chapter 10).

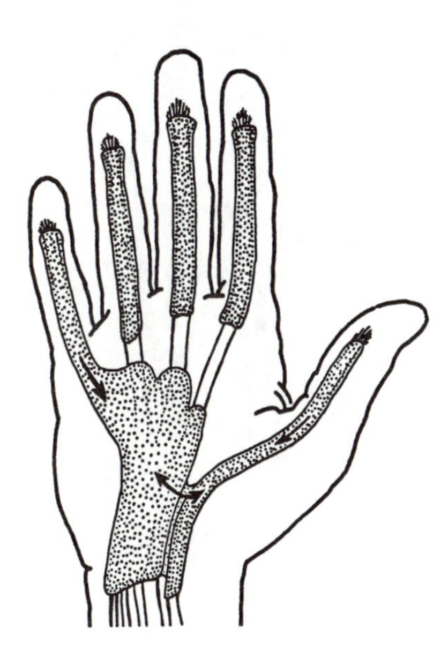

Fig. 5-6 Tendon Sheaths of Flexor Tendons. (Reproduced, with permission, from The Hand, American Society for Surgery of the Hand, 1978).

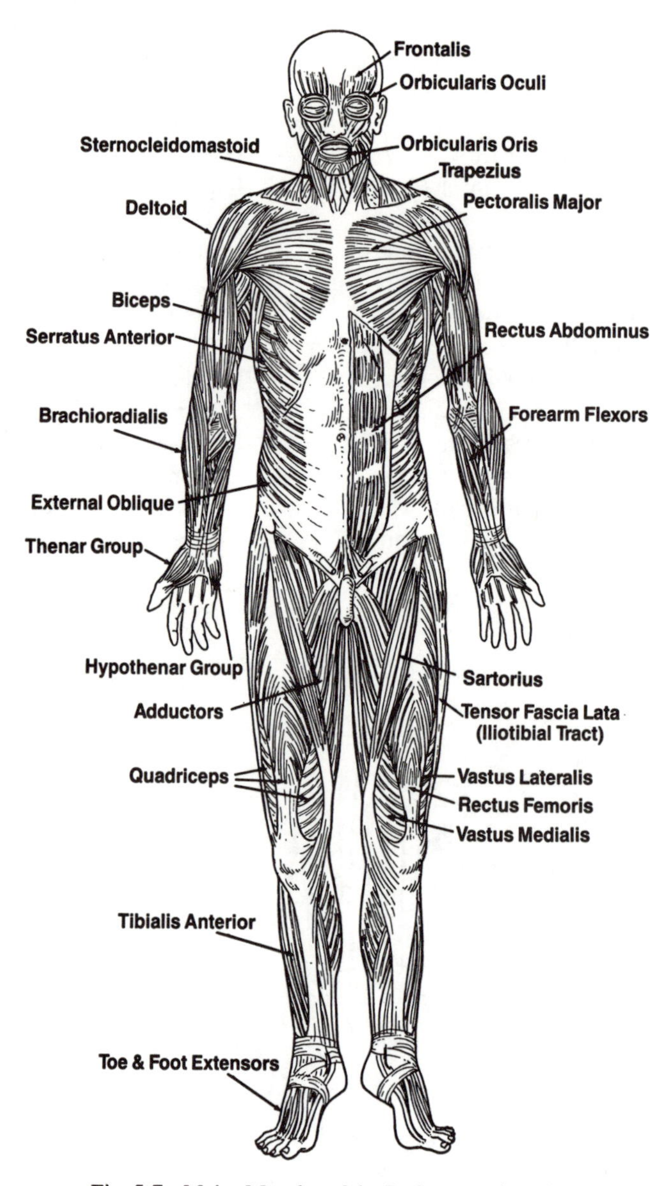

Fig. 5-7 Major Muscles of the Body (Anterior View).

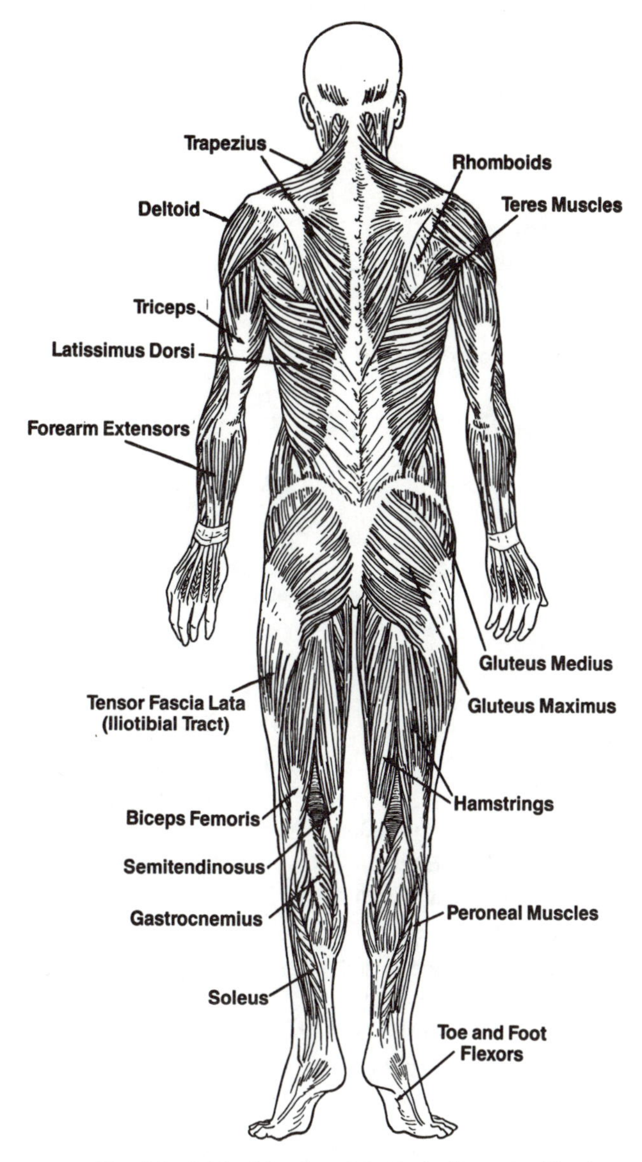

Fig. 5-8 Major Muscles of the Body (Posterior View).

Table 5-1: Major Muscles of the Upper Extremity

MUSCLE	BONES	MAIN FUNCTION	NERVE	IMPORTANCE
CHEST MUSCLES: (Fig. 5-7)				
Pectoralis major	Clavicle/sternum humerus	Flexes and adducts arm (Figs. 5-10, 5-11)	Anterior thoracic (nerve roots C5-T1)	Removed in radical mastectomy
Pectoralis minor	Ribs 3-5/scapula	Moves scapula forward	Anterior thoracic	Sometimes removed in radical mastectomy
BACK MUSCLES: (Fig. 5-8)				
Trapezius	(see neck muscles)			
Latissimus dorsi	Thoracic, lumbar, sacral vertebrae/ humerus	Extends and adducts arm (Figs. 5-10, 5-11)	Thoracodorsal	When developed gives "V-shape" to back
Rhomboids	(see muscles of thorax)			
SHOULDER MUSCLES: (Figs. 5-7, 5-8)				
Deltoid	Clavicle/scapula/ humerus	Abducts arm (Fig. 5-11)	Axillary	1. Main "shoulder muscle" 2. Intramuscular injection site
ARM MUSCLES: (Figs. 5-7, 5-8)				
Biceps	Scapula/Humerus/ radius	Flexes and supinates forearm (Fig. 1-3 a, d)	Musculocutaneous	Biceps tendon essential to palpate for location of brachial artery pulse for blood pressure (medial to tendon)
Brachialis	Humerus/ulna	Flexes forearm (Fig. 1-3a)	Musculocutaneous	
Triceps	Scapula/humerus/ ulna	Extension of forearm (Fig. 1-3a)	Radial	Intramuscular injection site
FOREARM MUSCLES: (Figs. 5-7, 5-8, 5-9)				
Anterior group Flexor carpi — Flexor digitorum superficialis — profundus —	Medial epicondyle of humerus/carpals/ metacarpals/ phalanges	Flexes and pronates hand, or flexes fingers (Fig. 1-3d, 5-12)	Primarily median	
Posterior group Extensor carpi — Extensor digitorum —	Lateral epicondyle of humerus/carpals/ metacarpals/ phalanges	Extends and supinates hand, or extends fingers	Radial	Extensor tendons are sites of "ganglia"
HAND MUSCLES: (Fig. 5-9)				
Thenar group	Carpals/1st metacarpal 1st proximal phalanx	Abducts/adducts/ flexes thumb (Fig. 5-13)	Median and ulnar nerves	Primate "opposition" of thumb

Table 5-1: Cont.

MUSCLE	BONES	MAIN FUNCTION	NERVE	IMPORTANCE
Hypothenar	Carpals/5th metacarpal/5th proximal phalanx	Abducts/flexes/ opposes little finger (Fig. 5-12, 5-13)	Ulnar	"Karate-chop" musculature
Lumbricales/ interossei	Run between fingers	Fine-movement of fingers/flexes 1st phalanx/extends 2nd and 3rd	Primarily ulnar	Flexion MP joints, extension IP joints make the "upstroke" in writing

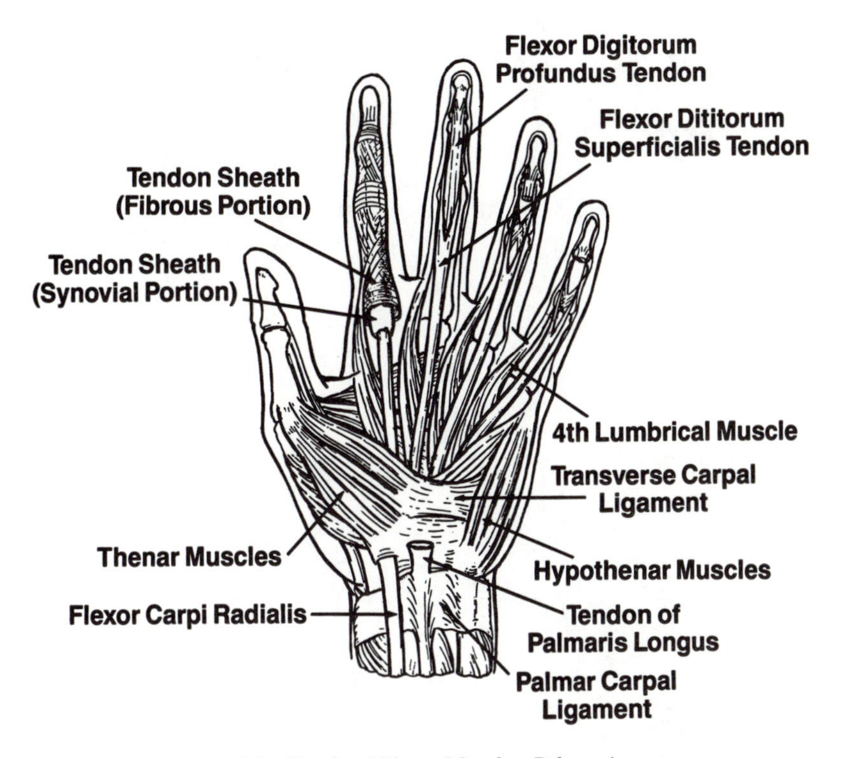

Fig. 5-9 Hand and Finger Muscles, Palmar Aspect.

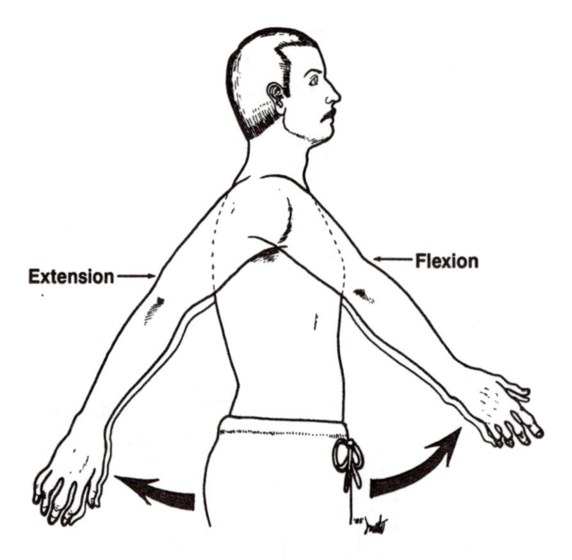

Fig. 5-10 Arm Flexion/Extension.

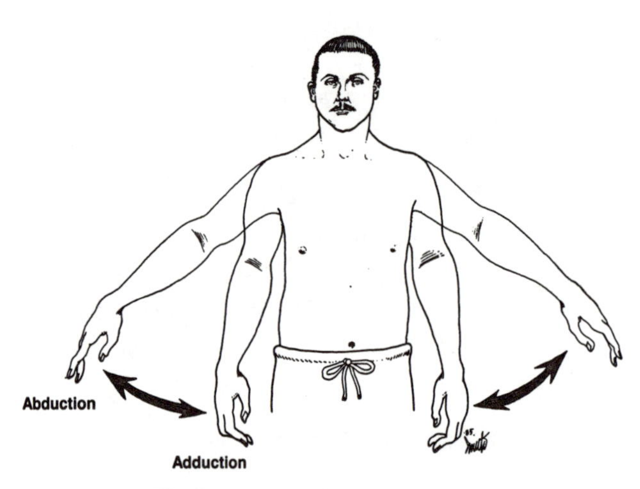

Fig. 5-11 Arm Abduction/Adduction.

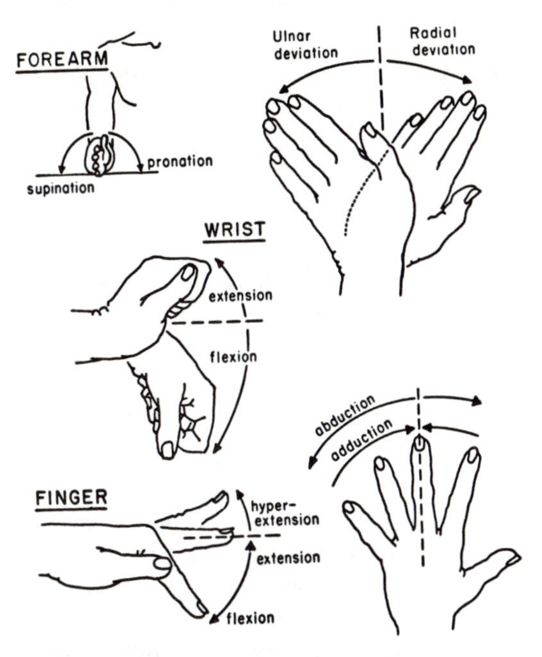

Fig. 5-12 Forearm, Wrist and Finger Movements. (See ref., Fig. 5-6).

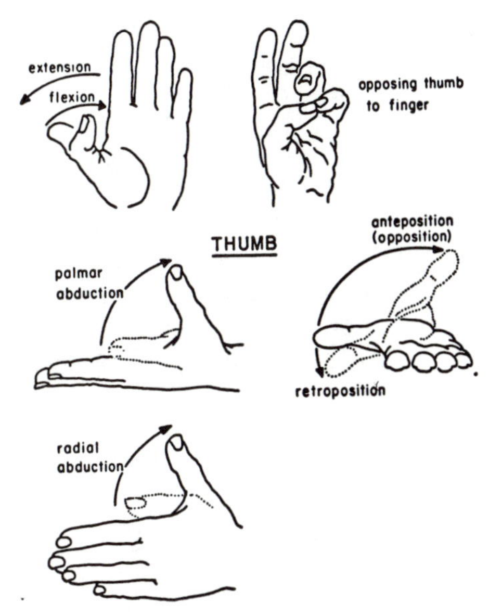

Fig. 5-13 Thumb Movements. (See ref., Fig. 5-6).

Table 5-2: Major Muscles of the Lower Extremity

MUSCLE	BONES	MAIN FUNCTION	NERVE	IMPORTANCE
ANTERIOR THIGH MUSCLES: (Fig. 5-7, 5-14)				
Quadriceps (vastus medialis, vastus lateralis, vastus intermedius, rectus femoris)	Ilium/femur/patella	Flexes thigh/extends leg (Fig. 5-15)	Femoral	1. Subject to "charleyhorse" 2. IM injection site
Sartorius	Ilium/tibia	Flexes thigh and leg	Femoral	"Tailor's muscle"
Iliopsoas	Ilium/lumbar vertebrae/femur	Flexes thigh	Femoral	"Filet mignon" cut of beef
MEDIAL THIGH MUSCLES: (Fig. 5-7)				
Adductor group (5 muscles)	Pubic bone/femur	Adduction of thigh (Fig. 1-3d)	Obturator	"Groin pull" common in football
GLUTEAL REGION: (Fig. 5-8)				
Gluteus maximus	Ilium/femur	Extends thigh (Fig. 5-15a)	Inferior gluteal	IM injection site
Gluteus medius	Ilium/femur	Abducts thigh (Fig. 1-3d)	Superior gluteal	IM injection site
Tensor fasciae latae	Ilium/iliotibial band	Abducts and flexes thigh	Superior gluteal	Iliotibial band attaches to tibia
Lateral rotators (6 muscles)	Ischium/femur	Laterally rotates thigh	Obturator	Muscle strain not uncommon
POSTERIOR THIGH MUSCLES: (Fig. 5-8)				
Hamstrings (3 muscles) Biceps Femoris (lateral) Semimembranosus, Semitendionsus (medial)	Ischium/tibia	Extends thigh/ flexes leg (Fig. 5-15)	Tibial	"Charleyhorse" common
ANTERIOR LEG MUSCLES: (Fig. 5-7)				
Tibialis anterior	Tarsals/metatarpals/ tibia	Extends foot (Fig. 5-16)	Peroneal	"Shin splints" in runners
Extensor digitorum and hallucis	Tibia/fibula/ phalanges	Extends foot and toes (Fig. 5-16)	Peroneal	
LATERAL LEG MUSCLES: (Fig. 5-8)				
Peroneal group	Tibia/fibula/tarsals/ metatarsals	Flexes and everts foot (Fig. 5-16, 1-3e)	Peroneal	Tendon sometimes injured in "sprained ankle"
POSTERIOR LEG MUSCLES: (Fig. 5-8)				
Gastrocnemius	Femur/calcaneus	Flexes foot (Fig. 5-16)	Tibial	1. Muscle cramp common 2. Achilles tendon subject to injury

Table 5-2: Cont.

MUSCLE	BONES	MAIN FUNCTION	NERVE	IMPORTANCE
Soleus	Fibula/tibia/ calcaneus	Flexes foot	Tibial	Muscle also inserts on Achilles tendon
Flexor digitorum and hallucis	Tibia/fibula/ phalanges	Flexes foot and toes	Tibial	

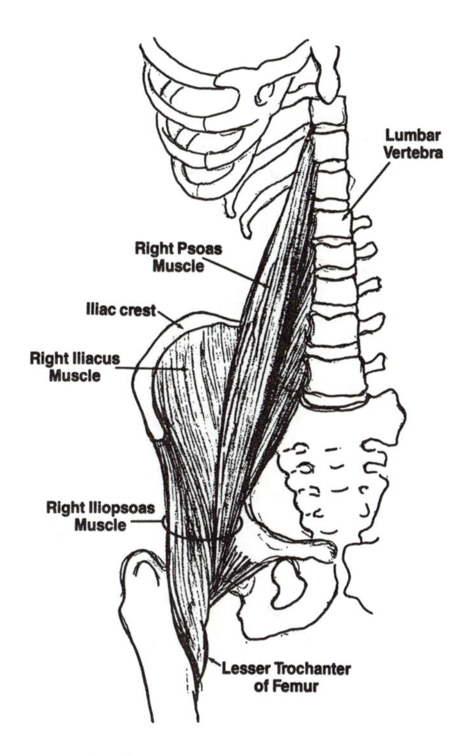

Fig. 5-14 Right Iliopsoas Muscle.

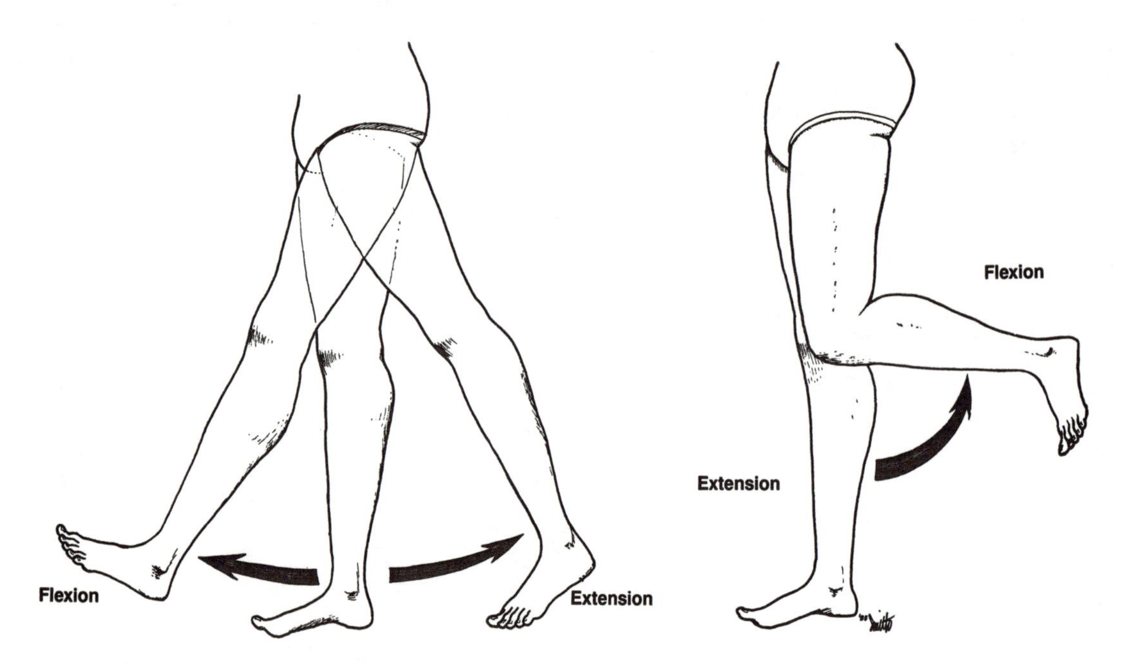

Fig. 5-15a Thigh Flexion/Extension.

Fig. 5-15b Leg Flexion/Extension.

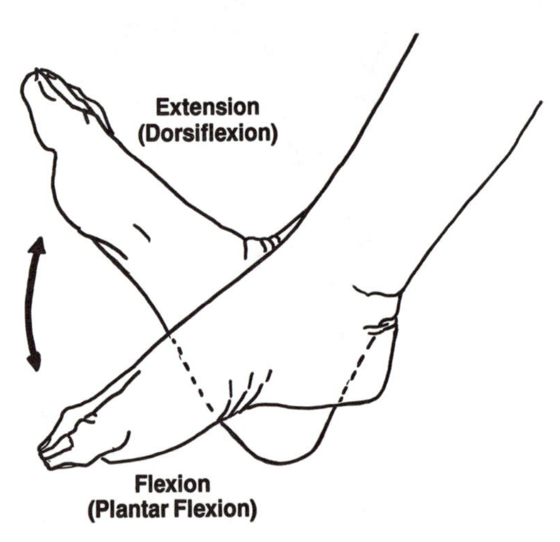

Fig. 5-16 Foot and Toe Flexion and Extension.

Table 5-3: Major Muscles of the Head and Neck

MUSCLE	BONES	MAIN FUNCTION	NERVE (CRANIAL or OTHER)	IMPORTANCE
FACIAL MUSCLES (Fig. 5-7)	Facial bones	Facial expression	VII	Stroke and Bell's palsy may paralyze one side of face
JAW MUSCLES (Fig. 5-17)		Muscles of mastication		
Masseter	Zygomatic/ mandible	Closes jaw	V	1. Goes into spasm in tetanus (lockjaw) and with antipsychotic drugs. 2. Strongest muscle in body
Temporalis	Temporal/mandible	Closes jaw	V	
Pterygoids	Sphenoid/maxilla/ mandible	Jaw forward and lateral for grinding	V	
NECK MUSCLES: (Figs. 5-7, 5-8, 16-1)				
Sternocleidomastoid (lateral)	Sternum/clavicle/ temporal	Flexes and rotates head/ elevates chin (Fig. 1-3b, 5-18a)	XI	Often involved in muscle strain in "whiplash" injury
Trapezius (posterior)	Occipital/scapula/ clavicle/vertebrae	Extension of head Movement of scapula up, down and medially (Fig. 5-18b)	XI	"Whiplash," as above
Suprahyoid (superior to hyoid)	Generally from mandible to hyoid bone	Elevates hyoid bone and tongue in swallowing. With fixed hyoid, opens mouth	V, VII XII, C-1	
Infrahyoid (inferior to hyoid)	Generally from hyoid to thyroid, sternum	Returns hyoid after swallowing/ depresses larynx	C1-3	Accessory muscles of speaking and respiration

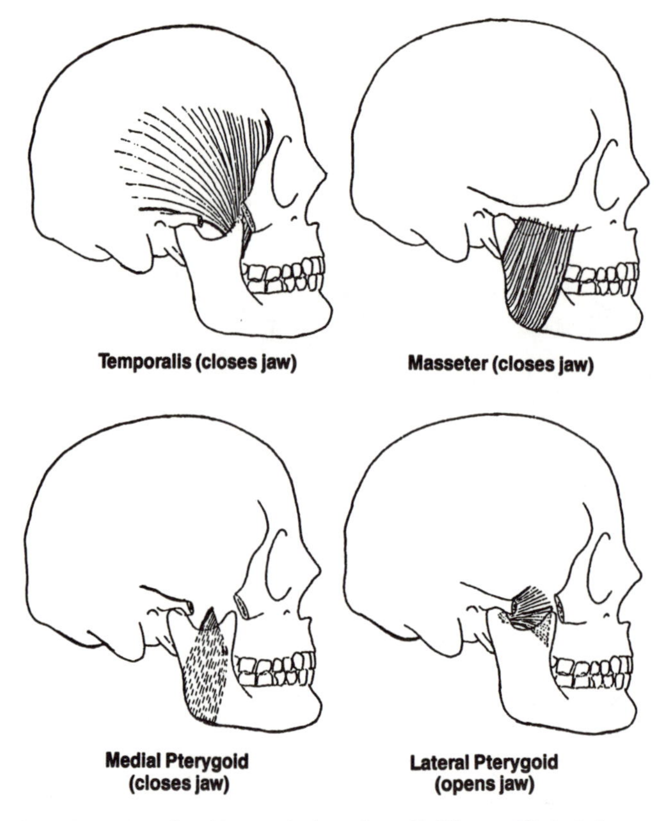

Fig. 5-17 Jaw Muscles. (Reproduced, with permission, from Goldberg, Clinical Anatomy Made Ridiculously Simple, MedMaster, 1984).

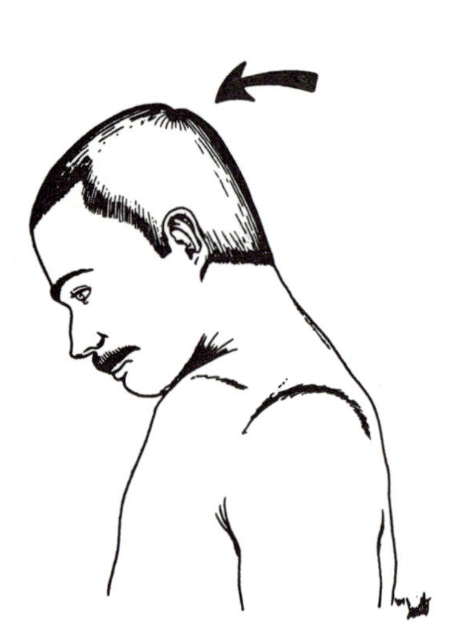

Fig. 5-18a Neck Flexion.

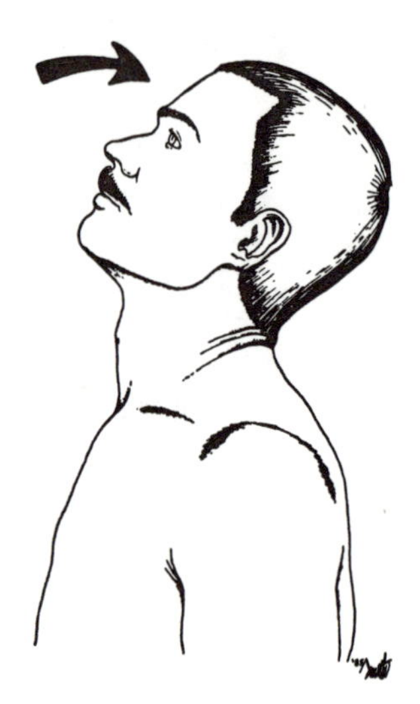

Fig. 5-18b Neck Extension.

Table 5-4: Major Muscles of the Thorax and Abdomen

MUSCLE	BONES	MAIN FUNCTION	NERVE	IMPORTANCE
ANTERIOR THORAX: (Fig. 5-7)				
Pectoralis major	(see upper extremity)			
Pectoralis minor	(see upper extremity)			
Serratus anterior	Ribs 1-8/scapula	Rotates scapula	Long thoracic	Injury to muscle or nerve can cause "winged scapula"
Intercostal muscles	(see Chapter 11)			
POSTERIOR THORAX: (Figs. 5-8, 5-19)				
Trapezius	(see Muscles of the Neck)			
Latissimus dorsi	(see Muscles of Upper Extremity)			
Rhomboids	Cervical, thoracic vertebrae/scapula	Moves scapula medially	Dorsal scapular	Muscle strain common from prolonged blackboard writing
Erector spinae (paravertebral muscles)	Deep muscles along vertebrae and ribs	Extension of torso (Fig. 5-20)	Spinal nerves	Strain of lumbar group is main cause of "low back pain"
ABDOMINAL MUSCLES: (Fig. 5-7)				
Lateral abdominal muscles *External oblique* (outer) *Internal oblique* (middle) *Transversus* (inner)	Lower ribs/iliac crest/pubic bone	Compresses abdomen/flexes torso (Fig. 5-20)	Intercostal nerves	1. Compression in defecation (valsalva maneuver) 2. Assist in breathing in COPD 3. Weakness above inguinal lig. can cause direct inguinal hernia
Rectus abdominis	Xiphoid/middle ribs/pubic bone	Compresses abdomen/flexes torso	Intercostal nerves	1. Umbilical hernia 2. Lateral separation of left and right muscle (diastasis recti)
Diaphragm (Fig. 5-21)	Xiphoid/lower ribs/ lumbar vertebrae	Inspiration	Phrenic nerves (C 3-5)	1. Openings for aorta, inf. vena cava and esophagus 2. Flattened in emphysema 3. Hiatal hernia: portion of stomach up through esophageal hiatus

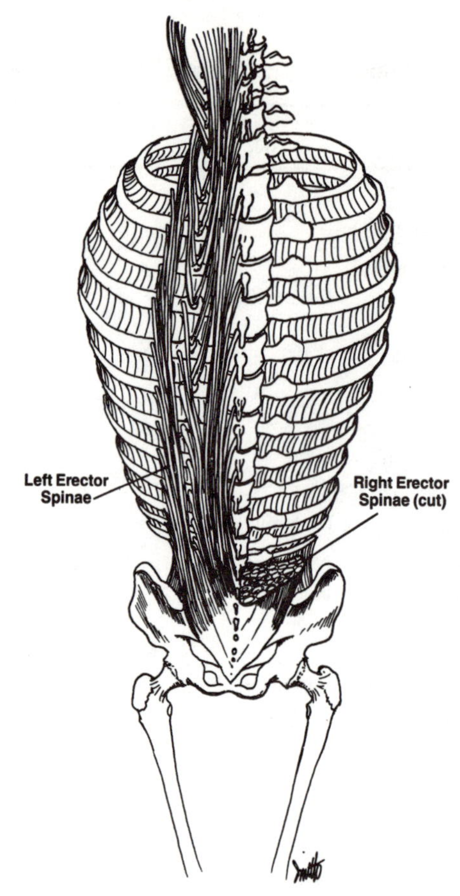

Fig. 5-19 Deep Back Muscles = The Erector Spinae (Paravertebral) Musculature.

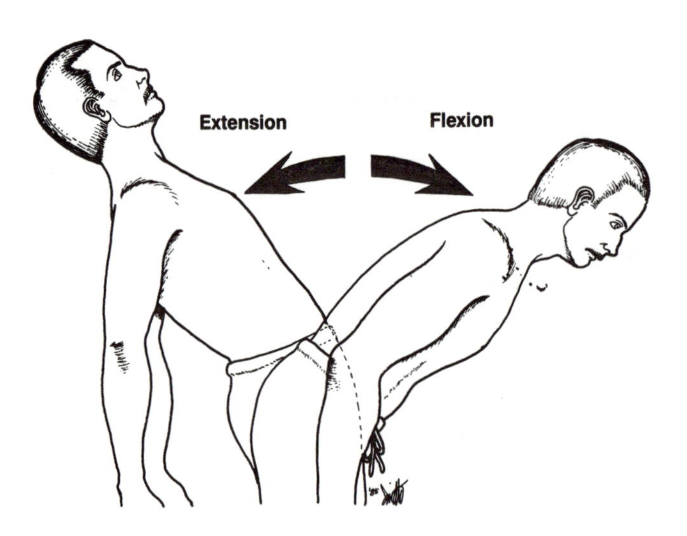

Fig. 5-20 Torso Flexion/Extension.

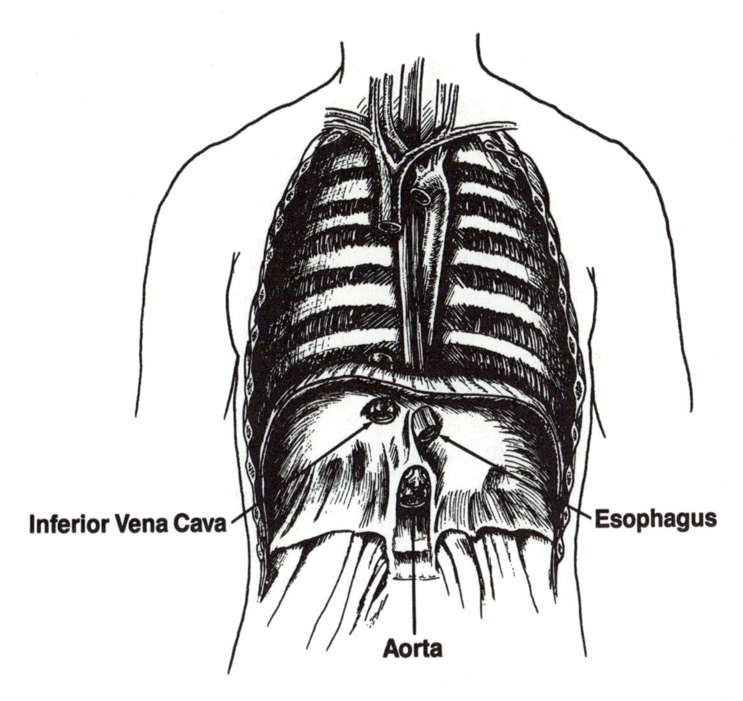

Fig. 5-21 The Diaphragm.

III. *Weak muscle groups* are those seldom used in daily activity. When there is contraction beyond the ability of the muscle to respond, or when a muscle is overstretched, the muscle is *strained* ("pulled muscle"). Common examples are a pulled hamstring in a sprinter, or a strained back when a person lifts at an odd angle. When weak muscles overstretch they respond by going into *spasm*, or sustained contraction, causing pain. Common weak muscle groups are:

A. *Adductors* of the thigh (Fig. 5-7). A *groin pull is an adductor strain*, usually from a twisting injury. Football players commonly encounter this when playing on artificial turf. Primary treatment consists of ice for 24 hours, then heat, an analgesic and an anti-inflammatory agent.

B. *Paravertebral back muscles* (erector spinae) (Fig. 5-19). The most common cause of *low back pain* is strain and spasm of the weak deep muscles of the back. The paravertebral group is a series of long muscles running along the vertebral column from the occipital bone to the sacrum. They attach to the vertebrae and ribs at various points, and have individual names (longissimus, iliocostalis, etc.). When stretched, particularly at odd angles, they go into spasm, causing severe pain. A common statement is: "I was reaching over a filing cabinet and then couldn't move". The spasm ranges from mild, to so severe as to completely immobilize the person. The spasm usually involves the lumbar group on one side, although the thoracic and cervical groups are sometimes affected. Occasionally, the iliopsoas is involved. Rarely does back pain involve an intervertebral disc. Primary treatment consists of an analgesic, a muscle relaxant and bed rest. Prevention involves strengthening the erector spinae by "good-morning" exercises (bending forwards and backwards with a barbell resting on the shoulders).

C. *Rhomboids* (Fig. 5-8).

D. *Iliopsoas* (Fig. 5-14).

E. *Abdominal muscles* (Fig. 5-7). The abdominal muscles are strengthened by sit-ups and leg raises.

IV. *Questions*:

A. Pushups develop which muscle?

B. Swimmers develop which muscle?

C. Which two exercises will tone and strengthen the breastline?

D. What is the best exercise for developing the quadriceps?

E. The hamstrings will develop with which exercise?

F. Deep knee bends with weights (squats) develop which muscle(s)?

G. Which exercise would strengthen the calf muscles?

Answers:

A. pectoralis major, B. latissimus dorsi, C. benchpresses or pushups, and dumbbell crossovers, D. deep knee bends with weights (squats), E. running or cycling, F. quadriceps and iliopsoas, G. running or cycling.

V. *Common muscle disorders*:

A. *A charleyhorse is a muscle bruise, accompanied by ruptured fibers and blood vessels, resulting in a large clot (hematoma)*. Commonly the result of a blow, prevalent in contact sports, it may occur in sprinters because of increased strain on the *quadriceps*. It may also involve the *hamstrings*. Primary treatment: ice, a compression dressing and no weight-bearing.

B. A *cramp* is a *severe muscle spasm*. Often the result of a blow to the muscle, or a strain, it sometimes occurs without any apparent reason. The *gastrocnemius* (calf muscle) is a common site. Other causes of cramping are salt loss from perspiring, temperature changes, muscle fatigue, and an impaired circulation, particularly in the elderly. Primary treatment: steady hand pressure on the muscle in the neutral position, and gentle massage.

C. *Disuse atrophy. Muscles, when not used, lose tone and atrophy*. This condition is seen in debilitated people, eventually leading to permanent crippling and incapacitation. Primary treatment: massage and range-of-motion exercises.

D. A *mallet finger* occurs when a baseball strikes the fingertip. The *extensor tendon is severed*. Pull of the flexor tendon results in *flexion of the DIP joint*. Primary treatment: splinting the joint in extension for several weeks (Fig. 5-22).

E. *Muscular dystrophy* is a genetic disease in which *connective tissue infiltrates muscle*, with a loss of normal muscular activity. Progressive deformity accompanied by muscle atrophy occurs. Patients become weaker and are ultimately confined to chairs or beds. Some may reach middle age with supportive care. Primary treatment consists of physical therapy and orthopedic devices.

F. *Myasthenia gravis* is a disorder of the *myoneural junction*. It is thought to be an *auto-immune* disease in which *antibodies are formed that destroy motor end plate receptors*. Acetycholine thus becomes unavailable to depolarize the muscle. The result is *muscle weakness*. In many cases the *thymus* is enlarged, and thymus-derived lymphocytes are found at the endplate.

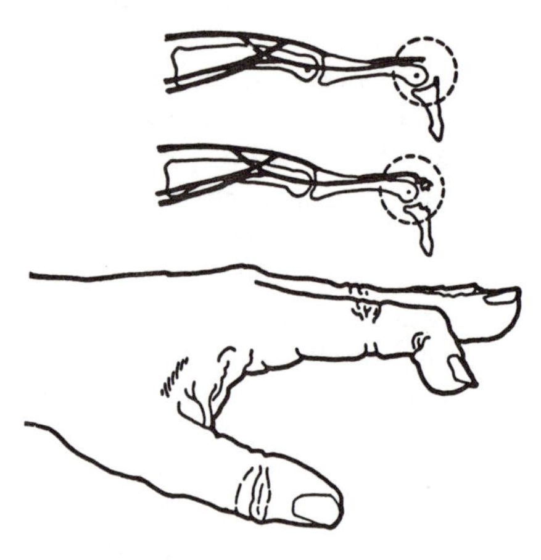

Fig. 5-22 Mallet Finger. (See ref., Fig. 5-6).

Signs and symptoms include easy fatigability, eyelid lag and double vision. Primary treatment consists of an anticholinesterase agent such as neostigmine (Prostigmine). The drug inhibits the breakdown of acetycholine, thus increasing its concentration and duration of action at the endplate. Thymectomy (removal of the thymus) is also useful.

G. A *rotator cuff injury* (the rotator cuff is tendons of the teres minor, infraspinatus, supraspinatus and subscapularis muscles, forming a cuff around the neck of the humerus. They stabilize the head of the humerus in the glenoid fossa, and abduct and rotate the arm) may occur in young athletes from throwing, tackling or wrestling, and in older people from degenerative tendon changes (Fig. 5-23). Arm abduction against resistance causes shoulder pain. Primary treatment includes resting the arm, an antiinflammatory agent and the local injection of an analgesic. A rotator cuff tear may require surgical repair.

H. *Shin splints* is pain in front of the leg and ankle, seen in new joggers (or occasional sprinters). The cause is commonly *inflammation of the periosteal attachment (periostitis) of the tibialis anterior muscle to the tibia.* Primary treatment: local heat,

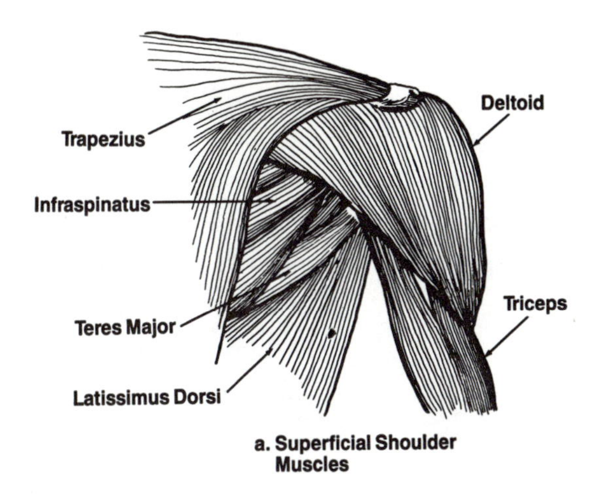

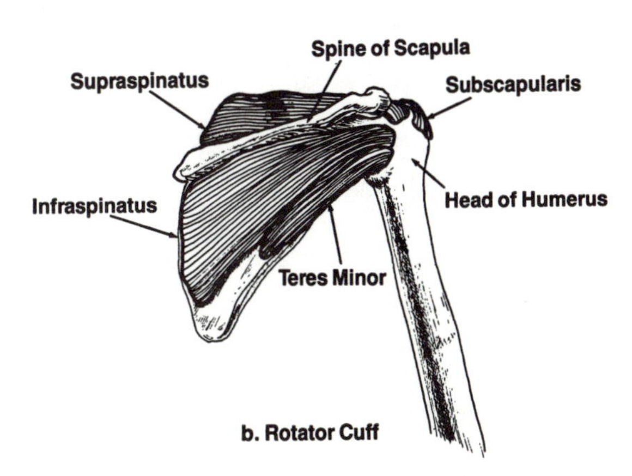

Fig. 5-23 Rotator Cuff (posterior view).

rest, and, perhaps, strapping.

I. *Tenosynovitis* is a *bacterial infection of a tendon sheath*, usually involving a finger. The digit is painful, red and swollen. If either the thumb or little finger is involved, the infection may spread to the wrist, since the sheaths of the thumb and little finger communicate with those of the wrist (see Fig. 5-6). Hospitalization with intravenous antibiotics is usually necessary.

J. *Whiplash* may be an acute cervical sprain, involving ligamental stretching or tearing, or the more common *neck muscle strain*. In most cases, the *sternocleidomastoids* and/or *trapezii* are involved. The head hyperextends (e.g., in a rear-end auto accident). Gently squeezing the sternocleidomastoids and/or trapezii elicits pain. Primary treatment is a soft cervical collar (to take pressure off the muscles), an analgesic and a muscle relaxant (see also p. 37).

-SIX-

THE BRAIN AND SPINAL CORD

The *central nervous system* (CNS) is the *brain* and *spinal cord*.

I. *Definitions*.
   A. *Sulcus*: groove
   B. *Gyrus*: ridge
   C. *Fissure*: large groove
   D. *White matter*: myelinated nerve fibers
   E. *Tracts*: collections of nerve fibers in the CNS having a common function.
   F. *Grey matter*: nerve cell bodies in the CNS. There are about 14 billion in the brain.
   G. *Nuclei*: collection of nerve cell bodies inside the CNS (has nothing to do with the nucleus of a cell).
   H. *Ganglia*: collection of nerve cell bodies outside of the CNS (the basal ganglia of the brain is the exception).
   I. *Neuroglia* (glia): glia means glue. These are the connective-tissue type cells of the CNS. There are about a trillion in the brain (Fig. 6-1).

II. *The brain*.
   A. The *cerebral hemispheres (cerebrum)* constitute the majority of brain substance and consist of a large area of *white matter* (nerve fibers) surrounded by the *cortex*, an outer thin mantel of nerve cell bodies (Fig. 6-2).

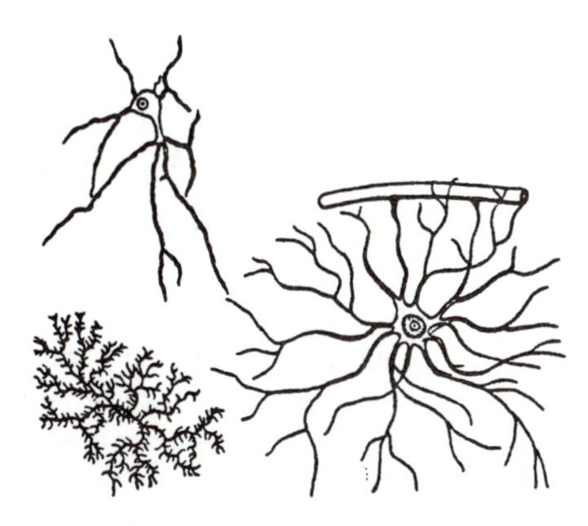

Fig. 6-1 Neuroglia (Glia). (Reproduced, with permission, from "Nervous Tissues, I, Cells", General Biological, Inc., 1957).

1. *Central sulcus*: separates motor and sensory areas
2. *Longitudinal fissure*: divides cerebrum into right and left halves
3. *Transverse (lateral) fissure*: separates temporal lobe from the rest of the brain
4. *Corpus callosum*: white matter connecting the two hemispheres
5. *Basal ganglia*: collection of nerve cell bodies lying within the white matter of cerebrum (Fig. 6-3).
   a. *Function*: motor coordination; forms an important component of the extrapyramidal system.
   b. *Importance*: Parkinson's disease is a disorder of the basal ganglia (see end of chapter).
6. *Right and left hemispheres*.
   Nearly all right-handed people have a dominant left hemisphere. Most left-handed people also have a dominant left hemisphere (70%). The term *dominant* means language. The right hemisphere (*non-dominant*) is concerned with *spacial organization, imagination, creative and intuitive abilities* (e.g., music).
7. *Functional centers of the cerebral cortex* (Fig. 6-2a).
   a. The *frontal lobe* contains the *precentral gyrus*, the *voluntary motor area* of the brain. Near the precentral gyrus in the dominant hemisphere is *Broca's area*, which controls the *motor aspects of speech*. Broca's area processes information from Wernicke's area and relays it to the pre-central gyrus (see below).
   b. The *parietal lobe* contains the *postcentral gyrus*, the *sensory area* of the brain.

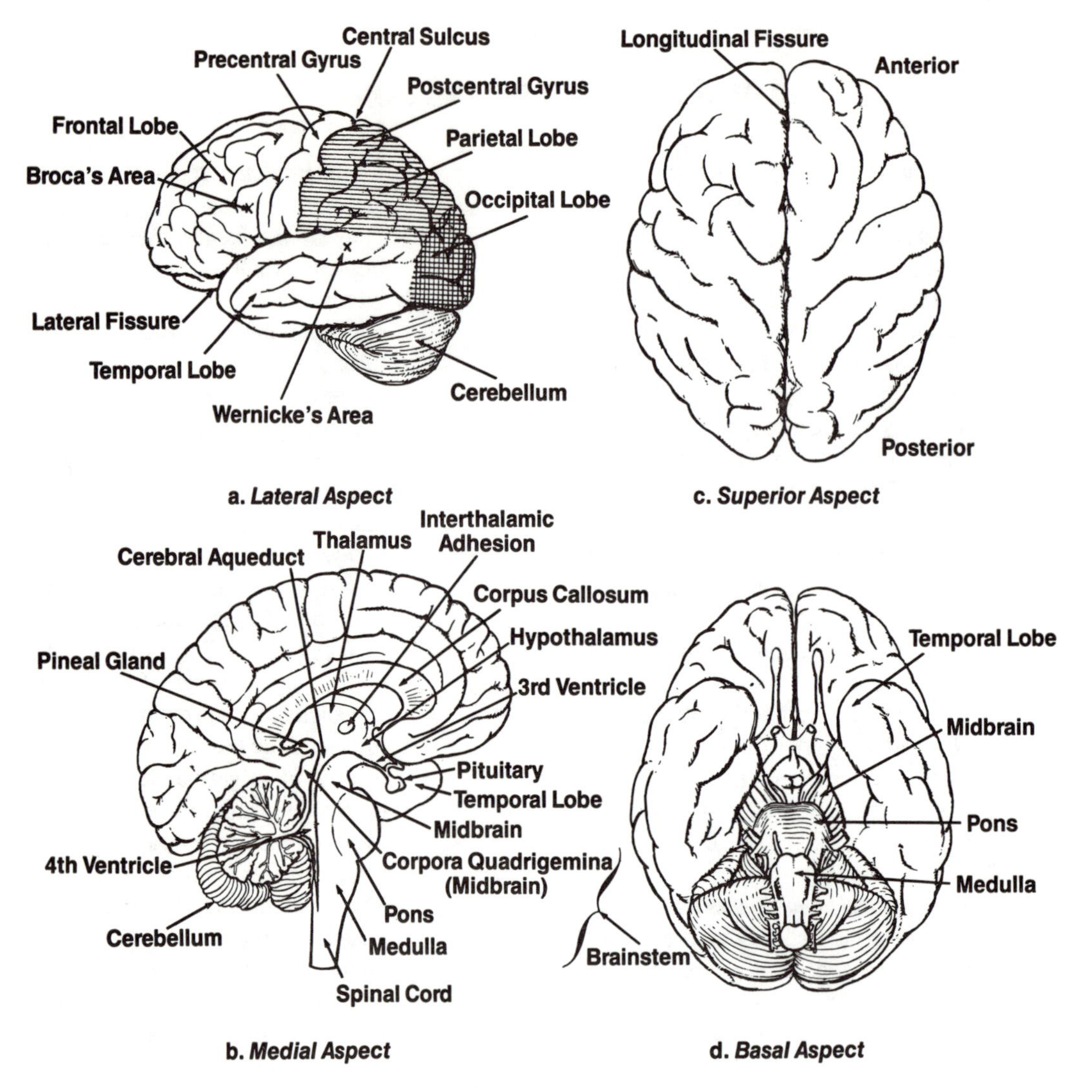

Fig. 6-2 Brain.

   c. The *temporal lobe* contains the *auditory area*. In addition, the superior temporal gyrus of the dominant hemisphere contains *Wernicke's area*, involved with the *comprehension of language*.
   d. The *occipital lobe* contains the *visual area*.

8. *Association areas* constitute the *majority of the cerebral cortex* and are involved with *higher learning, or intelligence*. There are 3 major areas: *frontal*, *temporal* and *parieto-occipital*:
   a. *Learning* in a basic and primitive sense can be thought of as the best and simplest way to solve a problem. Anatomically and physiologically, learning is the utilization of multiple synaptic pathways to solve problems. Advanced learning takes place in the association areas of the cerebral cortex (some learning also takes place in the brain stem). Learning requires memory.
   b. *Memory* is the storing of infor-

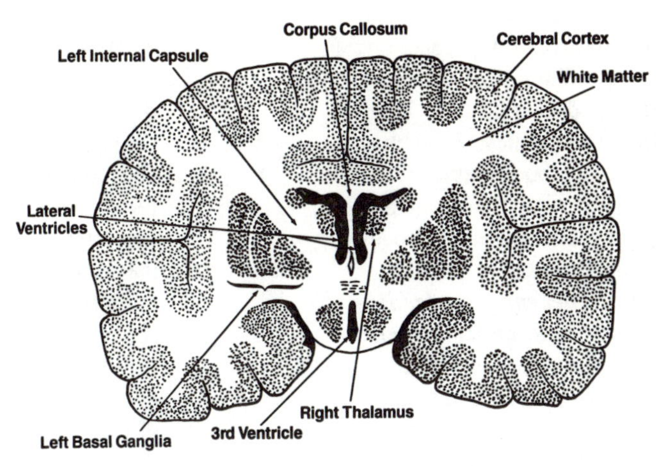

Fig. 6-3 Frontal Section of Brain.

mation in the brain. There are 2 types, *recent* and *long-term*. Recent memory is fragile and unstable, and disappears without reinforcement. When long-term memory takes place, neuronal synthesis of *RNA* increases. RNA, remember, is the template for protein synthesis. Long-term memory involves protein synthesis and is very stable. The *temporal lobes* are involved with long-term memory.

9. *Drugs affecting the cerebral cortex.*
   a. *The stimulants* are caffeine, the amphetamines and cocaine. Caffeine (in coffee and tea) stimulates learning by increasing synaptic transmission in the cortex. The amphetamines and cocaine release catecholamines (primarily norepinephrine and dopamine) from sympathetic neurons (sympathomimetic effect, see next chapter). The effect on the CNS ranges from a feeling of well-being (euphoria) to paranoid psychosis.
   b. *The depressants* (alcohol, narcotics, minor tranquilizers and barbiturates) act by blocking norephinephine and dopamine. The paradoxical effect of alcohol as a "stimulant" may be appreciated if one views the cerebral cortex as a personality structure (ego) in which the censoring force (superego) is depressed, releasing primitive biological impulses (id) from inhibitory control. Depressants are also *anesthetics*. The cortex is first depressed, then the more primitive centers (brainstem) as the dosage increases. The prototype is alcohol intoxication.
   c. The *hallucinogens* are lysergic acid diethylamide (LSD), phencyclidine (PCP), mescaline (peyote) and tetrahydrocannabinol (marijuana). LSD blocks the neurotransmitter serotonin. PCP block acetylcholine. These drugs seem to alter brain function by randomly stimulating and blocking neurotransmitters. Smell, for instance, may be "seen", color is "heard", etc. PCP, the most dangerous of the hallucinogens, uncouples sensory pathways in the brain, producing a sensory deprivation syndrome and creating an astonishing increase in body

strength, accompanied by an acute schizophrenic reaction. Because of high lipid-solubility and the production of long-acting metabolites, PCP may remain in body tissues for months or even years, causing lingering intermittent episodes of violence and psychosis.

d. *Physical dependency (addiction)*, seen with the *depressants*, means that when the drug is withdrawn there is severe autonomic excitability. The person thus requires the drug to be "normal".

e. *Tolerance*, seen with both *stimulants* and *depressants*, means that larger and larger doses of the drug are required for the same effect. This makes amphetamines quite dangerous, since physical dependency does not exist (thus no warning system), but tolerance does. Consequently, the person may approach the lethal dose without being aware of it.

B. *The brainstem* is the primitive brain, and contains *centers for vital functions* as well as centers for cranial nerves, discussed in the next chapter (Figs. 6-2b, d).

1. The *thalamus* is a collection of nerve cell bodies deep to the white matter of the cortex. It is the main *relay station for sensory fibers* (except olfactory) coming to the cerebral cortex, and is concerned with *crude sensation and pain* and *muscular coordination* (Fig. 6-3).

2. The *hypothalamus* lies between the thalamus (above) and pituitary (below) (Fig. 6-26). Functions:
   a. *Controls pituitary* by producing "releasing-hormones" (see chapter 12).
   b. *Temperature center*
   c. *Sexual center*
   d. *Thirst and hunger center*
   e. *Rage and fear center*

3. The *pineal body (gland)* lies on the dorsal surface of the brainstem (Fig. 6-2b). *Function*: The pineal appears to act as a *biological clock*, regulating length-of-day information (*circadian rhythm*). It secretes the hormone *melatonin*. Daylight inhibits its synthesis, and darkness seems to stimulate it. Melatonin may be involved with the *onset of puberty*, by triggering the release of luteinizing hormone (LH) from the pituitary. Nighttime levels of melatonin decrease during sexual maturity, and LH levels rise.

4. The *midbrain* is located between the thalamus and the pons (Fig. 6-2b).
   a. *Functions*:
      1) contains centers for visual and auditory reflexes
      2) contains cranial nerve nuclei
   b. *Importance*: a part of the reticular activating system (defined in D)

5. The *pons* lies between the midbrain and the medulla (Figs. 6-2b, d).
   a. *Functions*:
      1) regulates the rhythmic discharge of the respiratory center of the medulla
      2) contains cranial nerve nuclei
   b. *Importance*: center for "REM" sleep (see below)

6. *The medulla* connects the pons and spinal cord (Fig. 6-2b, d).
   a. *Functions*:
      1) Cardiac center: regulates heartbeat
      2) Vasomotor center: regulates blood pressure
      3) Respiratory center: regulates breathing
      4) Contains cranial nerve nuceli
   b. *Importance*: an injury to the medulla is often fatal, since it controls vital functions.

C. *The cerebellum* lies in the posterior cranial fossa of of the skull and is composed of an outer cortex and inner white matter, similar to the cerebrum (Figs. 6-2b, d).
   1. *Functions*: contains centers for balance, equilibrium and muscular coordination
   2. *Importance*: lesions of the cerebellum are rare, except for *cerebellar degeneration from chronic alcoholism*, resulting in balance and coordination difficulties.

D. *The reticular formation and the reticular activating system (RAS).*
   The *reticular formation*, the primitive inner core of the spinal cord and brainstem, is involved in the regulation of respiration, blood pressure, heart rate, endocrine secretion, conditioned reflexes, learning and consciousness. Incoming stimuli are integrated by the reticular formation. A portion of it, the *reticular activating system (RAS)* of the midbrain, is responsible for the *arousal reaction*. Epinephrine and amphetamines stimulate RAS conduction. Anesthesia and barbiturates depress conduction in the RAS. Patients with large lesions of the RAS are *comatose*.

E. *Sleep has stages numbering one through four*, the latter being deepest. Normally a person sleeps progressively lower, going from stage one through four. At the deeper levels, RAS activity is depressed in the pons and medulla. Then, at intervals of 90 minutes or so, the eyes begin to move quickly. The EEG shows a pattern similar to wakefulness, called "*rapid-eye-movement" (REM) sleep*. REM sleep appears to be associated with dreaming. Repeatedly waking a person at the beginning of REM sleep produces anxiety and irritability. If the person is then allowed to sleep, there is more than the usual REM sleep and dreaming for a few nights to "catch up". This is "*REM rebound*". Many drugs, particularly *sleeping pills and tranquilizers*, suppress REM and *Stage 4* sleep. Stopping the drug may result in "REM rebound", often with nightmares. Therefore, it is important to use medicines with minimal REM rebound.

F. *The meninges are three connective-tissue membranes surrounding the brain and spinal cord*. From *outside to inside*, the following spaces and structures are encountered (Fig. 6-4):
   1. *Epidural space*: between skull and dura mater.
   2. *Dura mater*: thick, tough, outer covering of the CNS.
   3. *Subdural space*: space between dura and arachnoid.

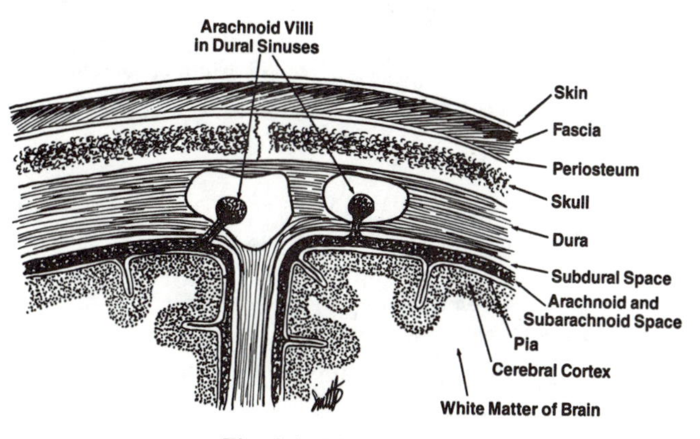

Fig. 6-4 The Meninges.

4. *Arachnoid* (spider): middle, web-like membrane of the CNS, containing many vessels.
5. *Subarachnoid space*: between arachnoid and pia, contains cerebrospinal fluid; ends at vertebral level S-2.
6. *Pia mater*: thin inner layer directly adherent to the brain and cord.

G. *Cerebrospinal fluid (CSF)* is produced by the filtration of blood through a capillary network (choroid plexus) lying in cavities (ventricles) in the brain. From the ventricles, CSF travels via several small openings into the *subarachnoid space*. It flows around the brain and cord and filters back into the *venous system* at the dural sinuses (Fig. 6-5).

Importance:

1. Acts as a *fluid cushion* and may have a nutritive function.
2. Brain tumors and hypertension may cause an *increased CSF pressure in back of the eye (papilledema)*. This may be detected with an ophthalmoscope.
3. If CSF drainage is blocked, enlargement of the ventricles and brain may occur (*hydrocephalus*).
4. *Examination of the CSF* after a lumbar puncture may show infection, such as bacterial or viral *meningitis*. A lumbar puncture is done below L-2 to avoid the spinal cord, which ends between L-1 and L-2 (Fig. 6-6). Examination of the CSF may reveal *bleeding*, such as

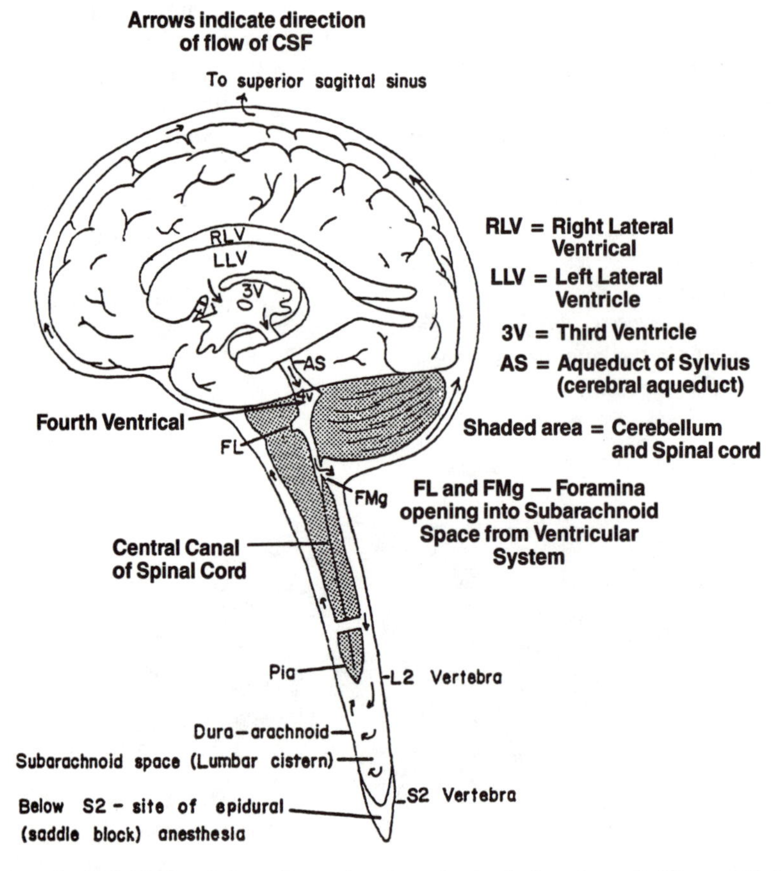

Fig. 6-5 Cerebrospinal Fluid Circulation. (Reproduced, with permission, from Goldberg, Clinical Anatomy Made Ridiculously Simple, MedMaster, 1984).

occurs in a subarachnoid hemorrhage (CSF shows a yellow color), or *increased pressure* occuring in a brain tumor.

H. *Vessels of the brain* (Figs. 6-7, 6-8):

1. *The arterial supply* to the brain is from the *internal carotid* and *vertebral arteries*. The two internal carotid arteries become the two *middle cerebral arteries* in the brain. The two vertebral arteries fuse at the pons to form the *basilar artery*. These three arteries connect (anastomose) at the midbrain as the *circle of Willis*. The circle insures vascularization of the brain in spite of occlusion of either the carotid or basilar arteries. The blood supply of the carotids is the *carotid system*; that of the vertebrals and basilar the *vertebro-basilar system*.
2. *Venous drainage* from the brain is by several veins, as well as the *dural sinuses*, spaces in the dura that drain to two deep veins, the *internal jugular veins*.

III. *The spinal cord* extends from the foramen magnum to slightly above the 2nd lumbar vertebra (L-2). *Grey matter* is on the *inside* and *white matter lies on the outside* (the opposite of the brain). *Peripheral nerves* feed in and out of the cord at various levels (Fig. 6-9, 1-10).

A. *The grey matter* forms an "H" configuration inside the cord. The posterior, or dorsal, portion of the H (*dorsal horns*) consists primarily of *incoming sensory fibers*, as well as the cell bodies of interneurons (interneurons form connecting links to various parts of the cord). The anterior, or ventral, portion of the H (*ventral horns*) consists of *motor nerve cell bodies* whose axons travel to *skeletal muscles* (Fig. 6-10).

B. *The white matter* is made up of *myelinated nerve fibers (tracts)* going to (ascending) and coming from (descending) the brain (Fig. 6-10).

1. *Sensory (ascending) tracts* (Fig. 6-11).
   a. *The pain and temperature pathway*. Receptors originate in the dermis of the skin, and other organs. Nerve fibers enter the spinal cord at various levels, cross over to the other side, and ascend to the *thalamus* as the *lateral spino-thalamic tract*.

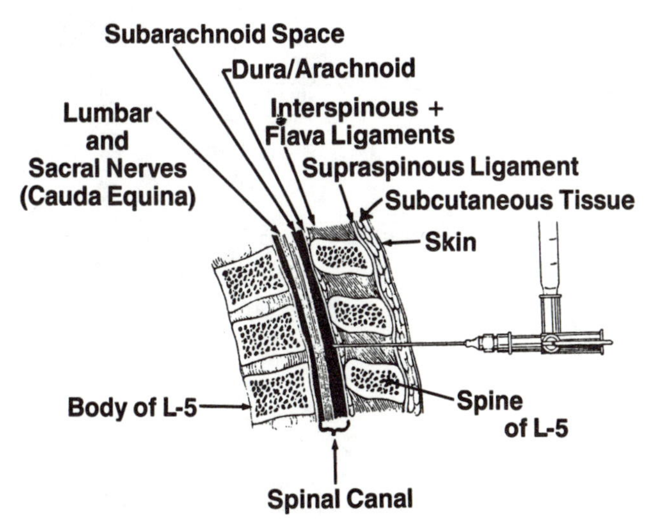

Fig. 6-6 Tissues in Lumbar Puncture. (Reproduced, with permission, from Suratt, Manual of Medical Procedures, Mosby, 1982).

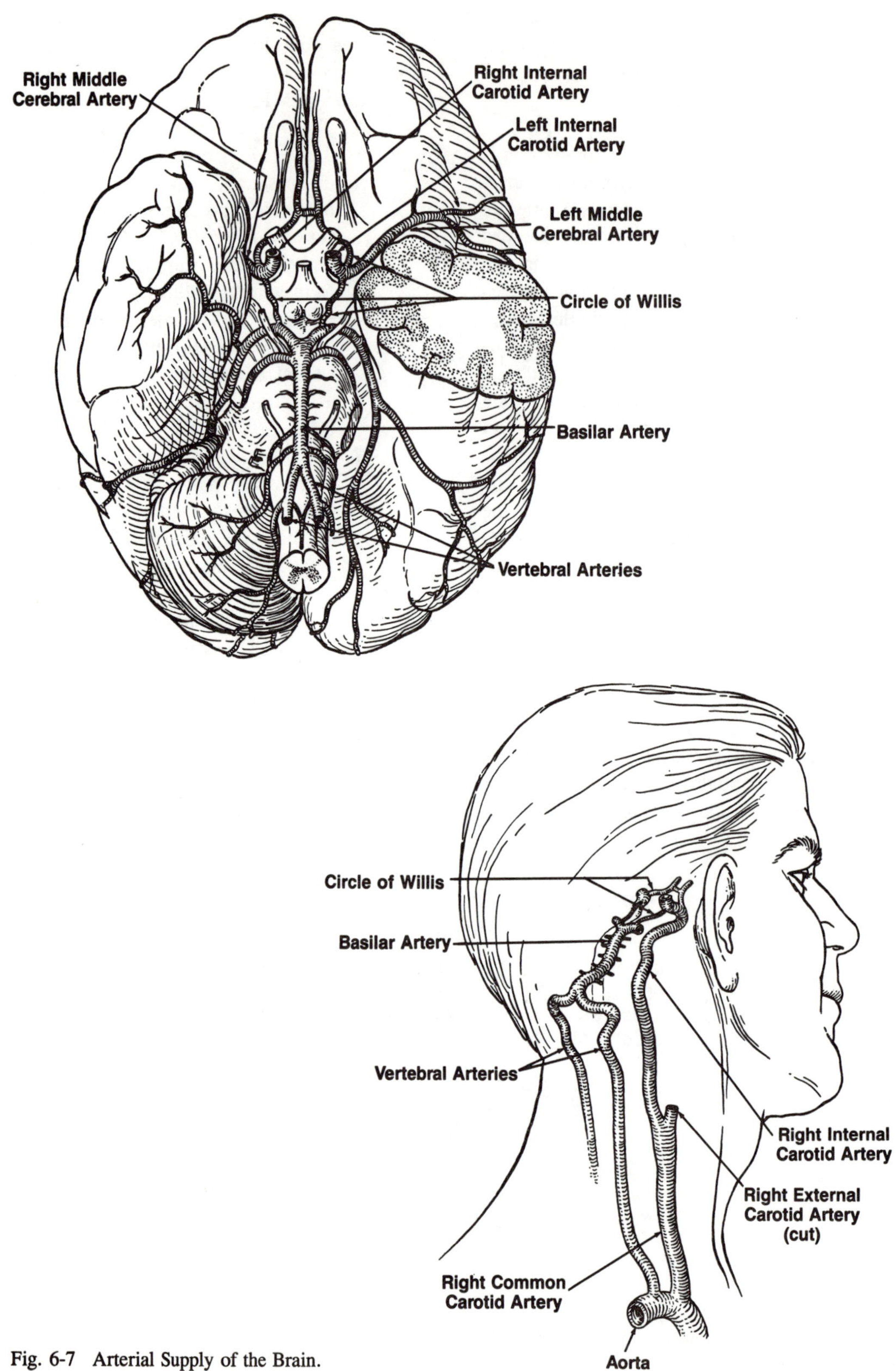

Fig. 6-7 Arterial Supply of the Brain.

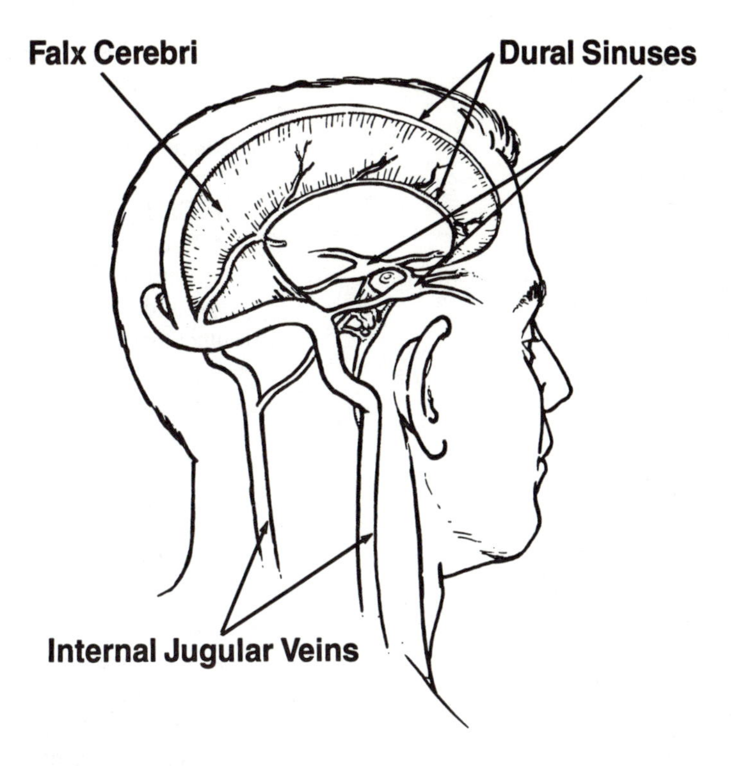

Fig. 6-8 Venous Drainage of the Brain.

From here, fibers run to the *postcentral gyrus* of *the parietal lobe (the sensory center).*

b. *The proprioceptive pathway. Proprioception* means the awareness of the positions of the arms and legs. Proprioceptive fibers accompany those for *deep touch* (pressure), *two-point discrimination* and *vibratory sense*. Receptors are located in muscles, tendons, ligaments and joints. The neurons run, as with the pain and temperature fibers, to the spinal cord.
   1) Some fibers synapse directly with motor neurons in the ventral horns, forming a *deep tendon reflex (DTR)* arc (e.g., knee-jerk, ankle-jerk).
   2) Other fibers run up the dorsal part of the spinal cord as the *dorsal columns*. Crossing over occurs in the medulla, and fibers ascend to the thalamus. From the thalamus, neurons travel to the postcentral gyrus.
   3) Proprioceptive loss may be diagnosed in several ways:
      a) Loss of knowledge of finger or toe position. The patient closes his eyes, the examiner moves the patient's finger or toe up or down, and asks him to state its position. He will be unable to tell whether the toe or finger is up or down.
      b) Loss of two-point discrimination. The patient cannot distinguish whether there are two points touching a portion of the body.
      c) Loss of vibratory sense. The patient cannot feel the vibration of a tuning fork held against a bone.

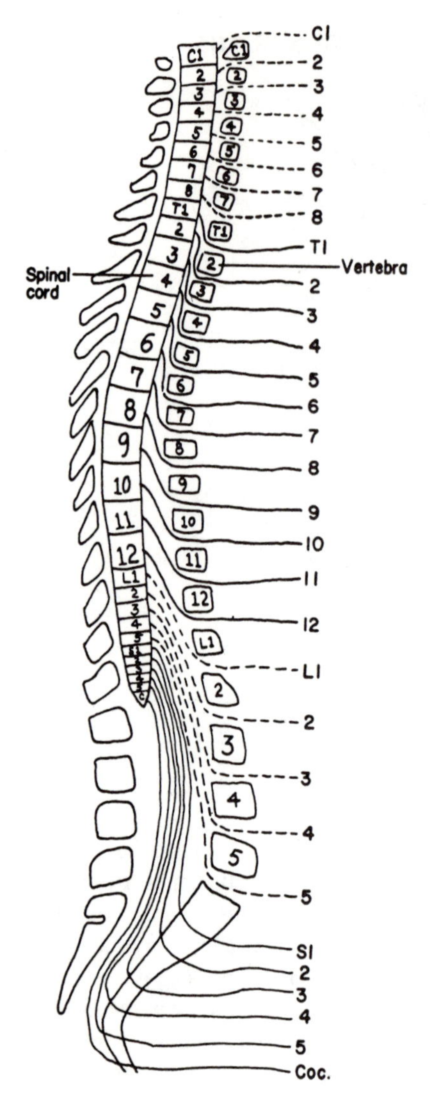

Fig. 6-9 Vertebrae, Spinal Cord and Spinal Nerves. (See ref., Fig. 6-5).

d) Loss of the ability to recognize common objects by touch (astereognosis).

2. *Motor (descending) tracts* (Fig. 6-11).

The main motor tract is the *pyramidal (corticospinal) tract* and is concerned with *fine, skilled movements* (descending motor pathways other than the pyramidal tract, designated as the *extrapyramidal system*, are concerned with *gross movements and posture*).

a. *The pyramidal tract* is the important part of the *voluntary motor system*. Fibers originate in the *pre-central gyrus* of the *frontal lobe* of the cortex and descend to the medulla. Here they cross over to the opposite side and then descend down the cord as the *lateral corticospinal tract*. The axons synapse with motor neurons of the ventral horns at various levels. *Axons of ventral horn cells* run to *skeletal muscles*. Beginning in the brain, the first series of cells are "*upper motor neurons*". The cells of the spinal cord ventral horns are "*lower motor neurons*".

1) Upper motor neuron lesions result in paralysis or paresis

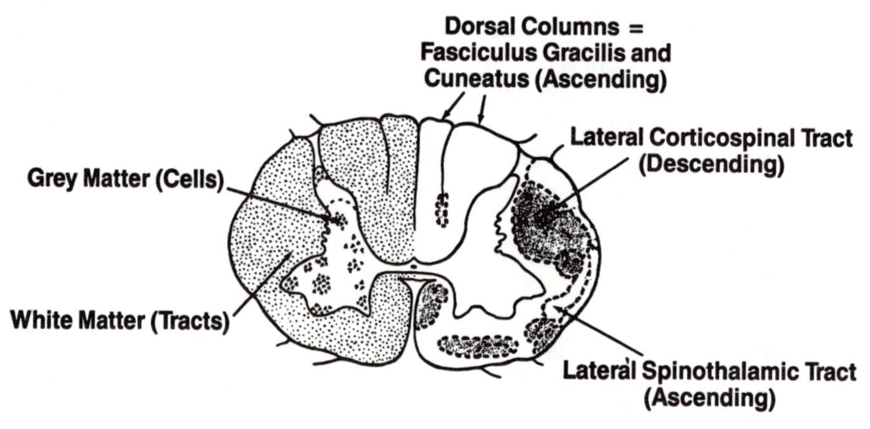

Fig. 6-10 Cross Section of Spinal Cord.

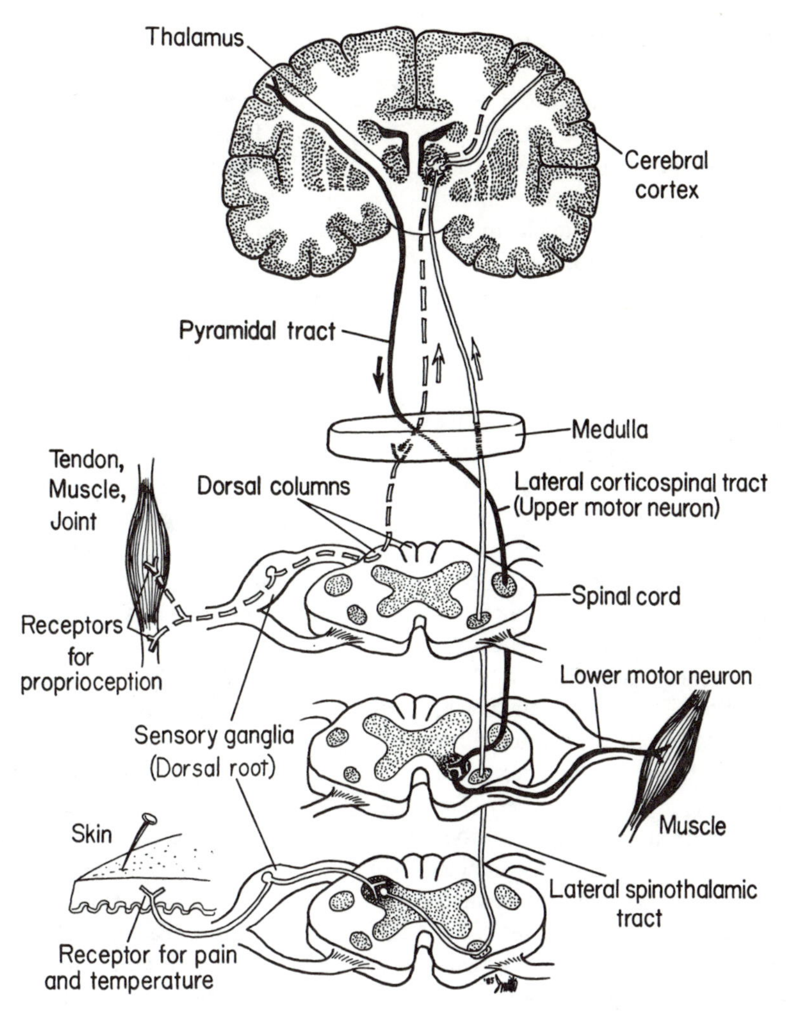

Fig. 6-11 Three Major Sensory and Motor Pathways.

of the spastic type (*spastic paralysis*).

2) Lower motor neuron lesions result in paralysis or paresis of the flaccid type (*flaccid paralyis*). Reflexes are underreactive, and there are muscle tremors (fasciculations).

b. *A stroke* involving one side of the *brain* causes sensory and motor loss in the opposite side of the body. Why? Because *tracts cross over* to the opposite side in the spinal cord or medulla.

IV. *Common disorders of the central nervous system*:

A. *Incidence of most common neurological diseases in the USA* (in order of frequency):

1. Cerebrovascular disease
2. Epilepsy
3. Parkinson's disease
4. Multiple sclerosis
5. Brain tumors

B. *Definitions*:

1. *Aphasia*: language difficulties — defects of speech, comprehension of speech, or the written word.

2. *Ataxia*: loss of muscular coordination
3. *Coma*: state of depressed brain function in which the patient cannot be aroused. Causes include disease and trauma. Both cerebral hemispheres or the reticular formation must be affected.
4. *Syncope* (fainting): brief period of unconsciousness due to compromised blood supply to the brain, often from emotional shock.

C. *Alzheimer's disease* is a degenerative mental disorder, commonly seen after the age of 50. Other terms used to describe Alzheimer's disease are presenile dementia (ages 50 to 65) and senile dementia (after age 65). *Memory difficulties, language problems and errors of judgment predominate.* There is general progressive intellectual decline. The cause is unknown. *Atrophy of neurons* is present, with a reduction of several neurotransmitter substances, particularly *acetylcholine*. Long-axon cholinergic neurons leading from the base of the brain to the cerebral cortex and hippocampus (an infolding of the inferior surface of the temporal lobe, essential for memory) are particularly vulnerable (Fig. 6-12). Treatment trials with a cholinesterase inhibitor and a choline-containing substance (lecithin) have been encouraging.

D. *Brain chemistry and behavior.*
1. *General. Most of the cerebral cortex involves behavior*, which utilizes multiple synaptic pathways in the association areas. A change in neurotransmitter concentrations at various synapses will cause a change in behavior. Mental illness of the *schizophrenic type* (the most common) is thought to involve an *increase in dopamine*. An increase in dopamine causes hallucinations. Certain types of *depressions* are thought to result from *blocking norepinephrine and dopamine* in the brain.
2. *Primary treatment*. The *anti-psychotic tranquilizers* such as chlorpromazine (Thorazine) block dopamine, and are the preferred treatment agents for most psychoses. Certain depressions respond to *anti-depressant drugs*, such as imipramine (Tofranil) or amitriptyline (Elavil). Antidepressants *increase norepinephrine and serotonin* in certain synapses by block-

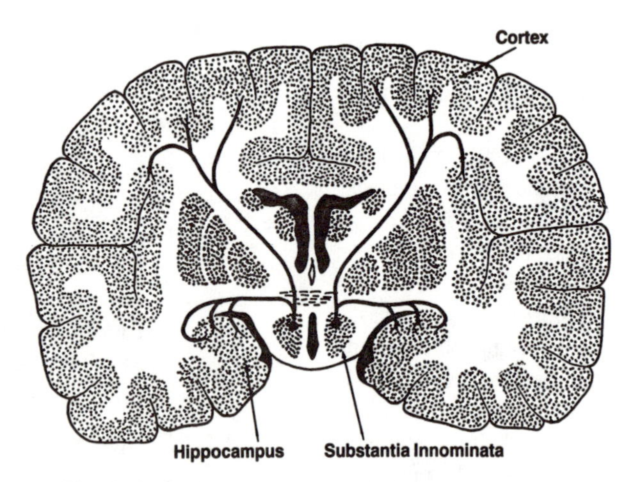

Fig. 6-12 Pathways involved in Alzheimer's disease.

ing cholinergic neurons (anti-cholinergic response), similar to the effect of atropine.

3. *Toxicity of the anti-psychotic and anti-depressant drugs.*
   a. *Extrapyramidal reactions*, caused by blocking dopamine with an anti-psychotic tranquilizer, is a frequent problem. Common "*extrapyramidal side effects" (EPS)* include neck muscle spasm (torticollis), an inability to speak (dysarthria), facial grimacing and other involuntary movements. These symptoms are treated with an *anti-parkinson* drug such as diphenhydramine (Benadryl) or benztropine (Cogentin), a parasympathetic blocking agent.
   b. *Overdosage from antidepressant agents causes atropine-like symptoms,* such as tachycardia, pupillary dilation, and even life-threatening arrhythmias and convulsions. Treatment is with a cholinergic drug such as physostigmine.

E. *Brain tumors.* Several interesting and important facts about brain tumors are:
   1. Most brain tumors are not primary; that is, they do not originate in the brain. They are *metastatic* from malignant tumors elsewhere in the body. The most common primary sites for the source of metastatic disease are the *lung* in males, and the *breast* in females.
   2. Sometimes, neurological signs and symptoms *precede* the discovery of breast or lung cancer.
   3. Primary brain tumors are derived from *glia cells*, not from neurons.
   4. The malignancy potential of most primary brain tumors is low. However, a tumor has no space in which to expand; consequently, brain substance is compressed, often with fatal results.
   5. *Some signs and symptoms of compression*:
      a. Motor or sensory loss, commonly on one side of the body
      b. Awkwardness of movement (ataxia)
      c. Personality changes
      d. Later, headache may appear because of pressure on the dura.
   6. Primary treatment is surgical removal, when possible

F. *Cerebral palsy* is a general term for *brain damage before, during, or shortly after birth.* The damage may involve the entire brain, but is often limited to the pyramidal and extrapyramidal tracts, with consequent *spasticity of muscles of the extremities.* Speech difficulties and convulsions may occur. Mental retardation is present in about 30 percent of cases. Primary treatment is symptomatic. An individual with spasticity and speech difficulties may *not* be mentally retarded; in fact, good intellectual function is often present. Primary treatment is symptomatic.

G. *Cerebrovascular disease* is a gradual buildup of *arteriosclerotic lesions in arteries of the neck or brain.* Arteries commonly affected are the common carotid, internal carotid and middle cerebral. Complications of cerebrovascular disease are blood clots and hemorrhage.
   1. *A transient ischemic attack (TIA)* is a "*pre-stroke*" *condition*, usually lasting less than 24 hours. Platelet emboli break off from ulcerated atherosclerotic plaques at the carotid artery bifurcation. A TIA may also be caused by thrombi forming on plaques in the cerebral arteries. Sometimes emboli arise from the heart.

a. *Carotid TIA* (most common): transient blindness in one eye, aphasia, numbness or weakness of the hand or foot ("clumsy-hand").
b. *Vertebro-basilar TIA*: slurred speech, dizziness, ataxia, syncope, numbness around lips.
c. *Primary treatment*: arteriography is used to assess the condition. Either medical or surgical treatment is instituted. Current medical therapy consists of an *anti-platelet medicine* such as aspirin. Surgical treatment involves excision of the thickened, atheromatous tunica intima. The plaque is pulled free and the vessel wall is cleared of debris (*endarterectomy*). The tunica intima later regenerates.

2. *A cerebrovascular accident (CVA) (stroke) is an abrupt loss of brain function caused by a vascular lesion that lasts longer than 24 hours, often permanently.* Common arteries involved are the *internal carotid and middle cerebral*, producing upper motor neuron and sensory losses on the opposite side of the body. A partial (*hemiparesis*) or total (*hemiplegia*) motor deficit may occur. The *limbs are spastic*, with the forearm in flexion and the leg in extension (because of the unequal innervation of extensors and flexors of the arm and leg).
   a. *Characteristics of various types of strokes*:
      1) Thrombosis (50%): (lowest mortality rate) usually occurs during sleep.
      2) Hemorrhage (40%): (highest mortality rate)
         a) Intracerebral is associated with *hypertension* and is usually accompanied by a headache.
         b) Subarachnoid hemorrhage takes place from a ruptured aneurysm (may be congenital) at the circle of Willis; it often occurs during physical exertion and is accompanied by a headache and stiff neck.
      3) Embolism (less than 10%) occurs during waking hours, often clearing in a few days. The source of emboli is usually the heart.
   b. A test for damage to the pyramidal system is the *Babinski sign*: stroking the plantar aspect of the foot produces extension and fanning of the toes (normally flexion occurs). It is thought that extension occurs because the purposeful fine movements of the pyramidal system have been disrupted, leaving only the extrapyramidal neurons (gross movement) intact.
   c. *Behavioral changes* (association areas of the cortex) are present. If the *left cerebral hemisphere is involved, language difficulties* occur (*aphasia*). A *right, or non-dominant, hemisphere stroke produces inattention and unconcern.* Confusion is present with both types.
   d. *Diagnosis* is confirmed by computerized assisted tomography (*CAT scan*—a computerized x-ray of the brain), or by *cerebral angiography* (injection of a dye into the carotid artery to visualize vessels of the brain). A new method of detecting abnormalities, including early stroke, is nuclear magnetic resonance (NMR) imaging (also known as magnetic resonance imaging, or MRI). The subject is placed in a

magnetic field and the magnetic resonances of hydrogen nuclei in body water and fat produce images of internal body structures. The process is then computerized, similar to the CT scan. The advantage is that no radiation is involved, so it is much safer than the CT.

e. Primary treatment is supportive; rehabilitation is important.

H. *Epilepsy* is characterized by an abrupt *alteration in brain function, ranging from a mild behavior change to a general convulsion*. Synapses in the cerebral cortex spontaneously fire in bizarre fashion. Causes include heredity, head trauma, stroke, brain tumor, and encephalitis.

1. A *grand mal* seizure begins with an "aura", or "funny" feeling, taste or smell. Consciousness is lost, and tonic-clonic contractions of skeletal muscle take place (*tonic*: continuous tension; *clonic*: series of spasms). After the episode, the patient is confused and sleep follows (post-ictal state) (*ictal*: seizure, or stroke).
2. A *petit mal* attack is a blank spell without convulsions, seen primarily in children.
3. *Status epilepticus*: sustained seizures
4. Primary treatment for most types of epilepsy includes phenytoin (Dilantin) and phenobarbital. In an emergency situation, intravenous diazepam (Valium) is used.

I. *Head injury*. A blow to the head may or may not involve a fracture. The fracture itself is usually not important. More significant is the possibility that intracranial bleeding may have occurred.

1. A *concussion* and *contusion* of the brain are indistinguishable clinically. A *concussion is brain injury with no significant damage to nerve tissue; a contusion is a bruise of the brain,* a more serious condition. In the typical case of *concussion,* the blow produces a loss of consciousness for a few seconds, with a return to normal in a minute or two. In the typical case of *contusion,* the loss of consciousness is often more profound. Amnesia (the temporary disruption of synaptic pathways in the brain) before or after the event is sometimes present, usually resolving in a day or two. Hospitalization is required. In both cases, a headache may be present, sometimes lasting for weeks.
2. *Subdural hematoma: collection of venous blood between the dura and arachnoid,* compressing the brain (Fig. 6-13a).
3. *Epidural hematoma: hemorrhage of the middle meningeal artery* on the undersurface of the temporal bone. Blood collects between the skull and dura, compressing the brain. This type of bleeding often accompanies a *linear skull fracture of the temporal bone* (Fig. 6-13b).
4. The *diagnosis* of a subdural and epidural hematoma, as well as intracerebral hemorrhage, involves the use of the *CAT scan* or arteriography. Sometimes a *widened pupil* is present from impingement of the clot on the 3rd cranial nerve. The *most important physical sign in both cases is deterioration of the mental status,* which reflects brain stem damage or bleeding and pressure on the association areas of the cortex.
5. *Primary treatment* of the hematomas is surgical evacuation of the clot. If the mental status deteriorates from brain swelling, the treatment is hyperventilation (to blow off CO2, a potent vasodilator), a diuretic and the upright position to decrease blood flow to the brain.

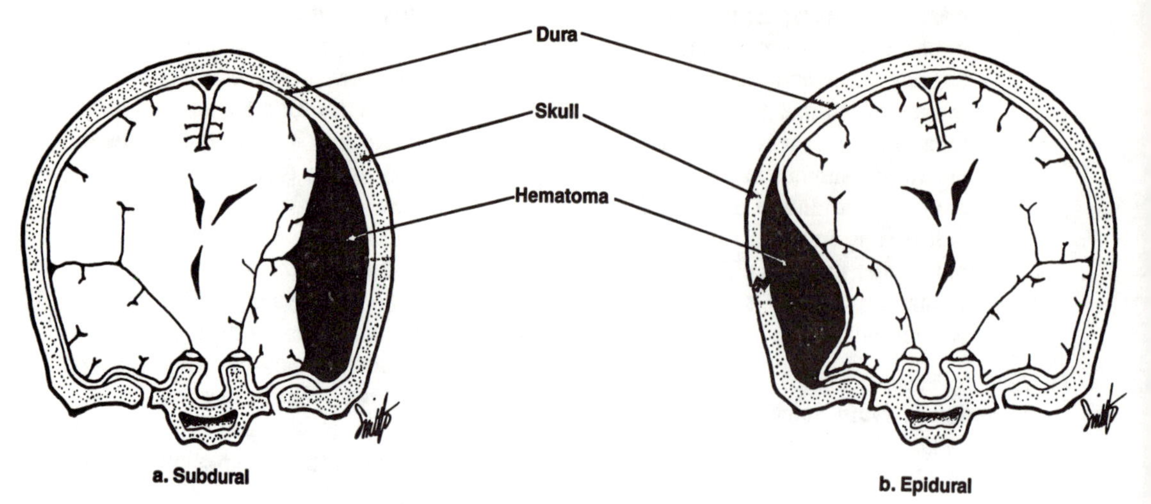

Fig. 6-13 Subdural and Epidural Hematoma.

J. *Headache* is a common symptom with a multitude of causes. *Pain is produced from pressure on the sensory nerves of the vessels, the meninges or the muscle-tendon-bone unit* (the brain has no sensory innervation; therefore headaches do not originate in the brain):
   1. *Tension* (the most common type) is believed to be caused by a *muscle-tendon strain* at the origin of the trapezius and deep neck muscles at the occipital bone, or the origin of the *frontalis* muscle on the frontal bone (occipital and frontal headaches).
   2. *Migraine* and *cluster* (probably a migraine variant). Pain is believed to be caused by *dilation of the cranial vessels*. The pain is knifelike, throbbing and unilateral. A visual prodrome (e.g., flashing lights) is believed to be caused by vasoconstriction preceding the vasodilation and pain. Cluster headaches come in "clusters" with remissions and exacerbations lasting weeks or months; they often occur at night and are associated with a red eye.
   3. *Primary treatment:* an analgesic. Migraine may be refractory to treatment once the headache is established. Prevention of migraine is sometimes possible with ergotamine (a vasoconstrictor).

K. *Infections of the central nervous system are usually bacterial or viral.* Common signs and symptoms are headache, fever and stiff neck. A lumbar puncture shows white cells in the CSF.
   1. *Encephalitis* is an *infection of the brain*. The most common cause is a viral infection. Primary treatment is symptomatic. An anti-viral drug is used if the organism is herpes virus.
   2. *Meningitis* is an *infection of the meninges*. Common causes are bacterial (pneumococcus) and viral. Primary treatment: an appropriate antibiotic, if bacterial. If viral, treatment as in 1.
   3. *Myelitis* is an *infection of the spinal cord and/or brain stem*. Before the vaccine, poliomyelitis virus was the main cause. *Polio* affects the *ventral horn cells* of the spinal cord and causes flaccid paralysis (lower motor neurons) of muscles supplied by the involved nerves. Primary treatment: supportive.

L. *Multiple sclerosis* is a disease of autoimmune and/or viral etiologies in which *myelin degenerates* (thus nerve impulse conduction is faulty) in random areas of the central nervous system.

1. Symptoms may include paresthesias, incontinence, vision problems, speech and gait difficulties.
2. Exacerbations and remissions are frequent, because the axon is preserved even though the myelin degenerates.
3. Primary treatment: supportive

M. *In Parkinson's disease*, the neurotransmitter *dopamine is deficient in the basal ganglia* (and in an area of grey matter in the midbrain called the substantia nigra) (Fig. 6-3). The result is an *exptrapyramidal motor loss* (similar effects can be produced by drugs which block dopamine transmission, such as the antipsychotic tranquilizers — see section C):

1. Major manifestations are a "pill-rolling" tremor at rest, rigidity of limb muscles (cogwheel rigidity), a shuffling gait and a masklike face.
2. Primary treatment. Three types of drugs have an effect (see Fig. 7-12):
   a. L-dopa (levodopa) is the *precursor of dopamine* and crosses the blood-brain barrier, restoring the dopamine pathway.
   b. Amantadine (Symmetrel) *releases dopamine* at the synapse.
   c. Benztropine (Cogentin) and trihexyphenidyl (Artane) are *anticholinergics*.

-SEVEN-

NERVES

The *peripheral nervous system* consists of the *cranial* and *spinal nerves*, as well as the *autonomic nervous system* which innervates the heart, smooth muscle and glands.

I. *Twelve cranial nerve pairs* enter (sensory) or leave (motor) the *brainstem* at various levels (except for I and II, which lie in front of the brainstem). Centers for cranial nerves III and IV are located in the midbrain; those for cranial nerves V, VI, VII and VIII are in the pons; those for IX, X, XI and XII are in the medulla (Fig. 7-1).
   A. Table 7-1 indicates the name and number of the cranial nerve, its type, function, common tests and abnormalities.
   B. *Importance*: cranial nerve disorders may occur following a stroke (hemorrhage, thrombosis), tumor, or trauma. A cranial nerve abnormality may help locate a lesion.

II. *Spinal nerves*, similar to the cranial nerves, enter and leave the *spinal cord* at various levels (see Fig. 6-9). They are all *mixed* (sensory and motor fibers running in the same nerve) and generally accompany arteries. The dorsal root is sensory, and the ventral root is motor. Sensory fibers arise from receptors (e.g., dermis of skin) and travel to the dorsal roots of the spinal cord. Motor fibers originate from ventral horn cells and innervate skeletal muscles (see Fig. 1-10). A *plexus is a network of several spinal nerve roots*.
   A. *Locations of spinal nerves* (Fig. 7-2):
      1. *Cervical* (neck): 8 spinal nerve pairs (C-1 to C-8)
      2. *Thoracic* (chest): 12 spinal nerve pairs (T-1 to T-12)
      3. *Lumbar:* 5 spinal nerve pairs (L-1 to L-5)

      } leave through *intervertebral foramina*

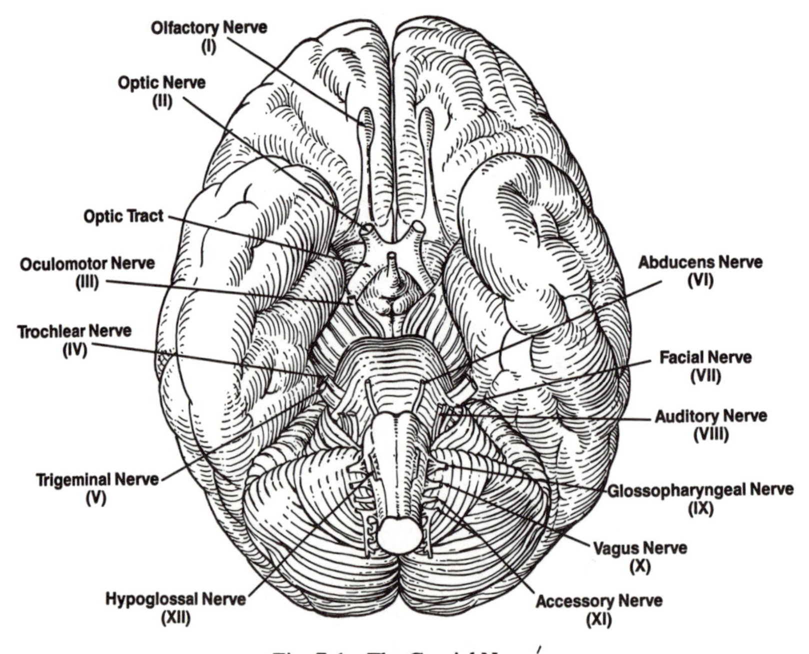

Fig. 7-1 The Cranial Nerves.

Table 7–1: Cranial Nerves

NAME	TYPE	FUNCTION	TESTS AND ABNORMALITIES
I. Olfactory	sensory	smell	coffee, tobacco
II. Optic	sensory	vision	a. visual acuity b. pupillary reaction (aff) c. visual fields
III. Oculomotor IV. Trochlear VI. Abducens	motor motor motor	eye movement: upward and inward eye movement: downward eye movement: lateral	a. pupillary reaction (eff); dilated fixed pupil b. eyeball: "down and out" eye: medial upward dev. eye: medial dev.
V. Trigeminal	sensory motor	from skin of face plus cornea to masseter & other chewing muscles	loss of sensation on one side of face inability to clench teeth on one side
VII. Facial	sensory motor	taste on anterior portion of tongue to muscles of facial expression	inability to grimace on one side of face
VIII. Auditory *(acoustic)* (vestibulocochlear)	sensory	hearing, equilibrium	*hearing:* a. watch-ticking b. Weber and Rinne tests c. startle-reflex: handclapping *equilibrium:* a. vertigo b. nystagmus
IX. Glossopharyngeal	sensory motor	taste on posterior portion of tongue to pharyngeal muscles	a. deviation of uvula towards the unaffected side b. loss of gag reflex
X. Vagus	sensory motor	from thoracic & abdominal organs to pharyngeal & laryngeal muscles & thoracic & abdominal viscera	 (as in IX) (plus hoarseness)
XI. Spinal Accessory	motor	to sternocleidomastoid and trapezius muscles	a. inability to shrug one shoulder or b. move chin to one side against pressure from examiner's hand
XII. Hypoglossal	motor	tongue movement	deviation of tongue to affected side

4. *Sacral:* 5 spinal nerve pairs (S-1 to S-5) leave through sacral foramina
5. *Coccygeal:* 1 spinal nerve pair.

B. *Sensory nerve roots* (cutaneous innervation) are shown in Fig. 7-3.

C. More than one *motor root* supplies each muscle, and it is usual to name the nerve to the muscle, rather than the nerve root. However, certain roots seem to be an index to the

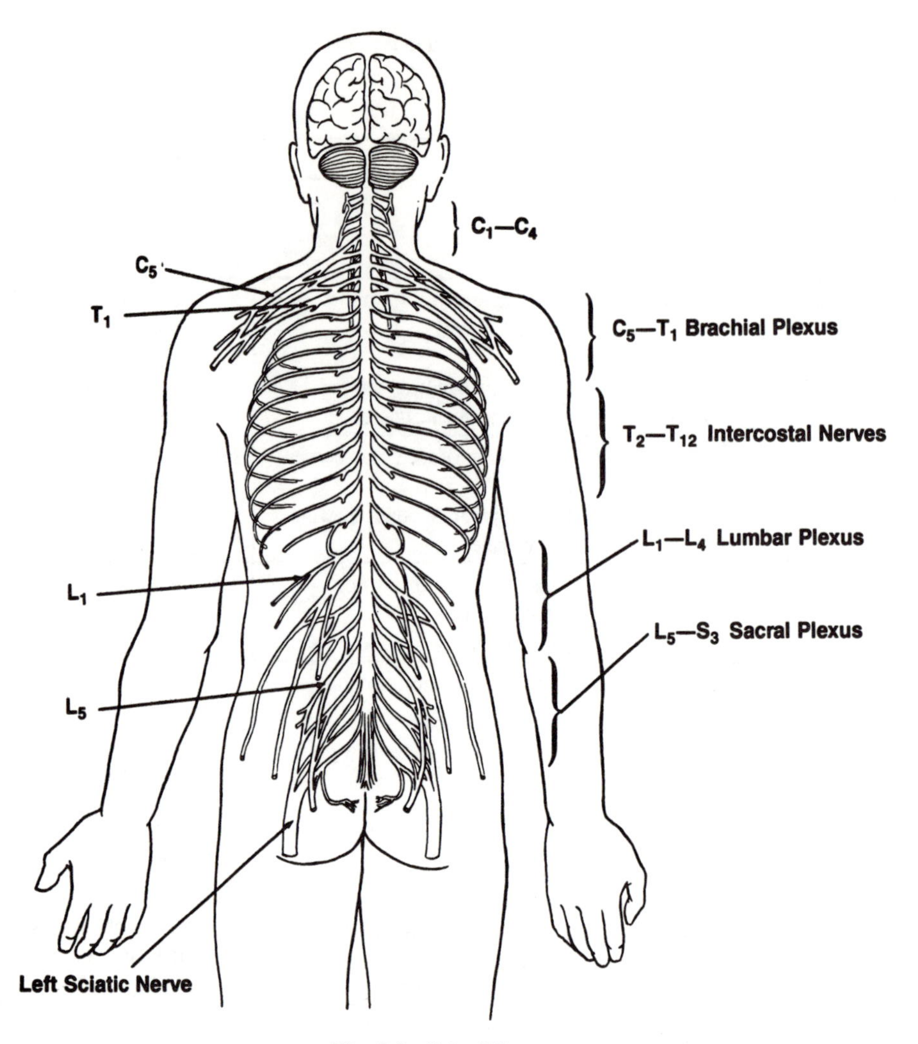

Fig. 7-2 Spinal Nerves.

functioning of a muscle group; if a particular root is impaired, the muscle is weak (see Reflexes, Section III).

D. *Important nerves of the upper extremity*:
   1. The *brachial plexus* is formed from nerve roots *C-5 to T-1* (Fig. 7-4).
   2. *Motor branches* (Fig. 7-5):
      a. *Axillary nerve* to deltoid muscle.
      b. *Musculo-cutaneous nerve* to biceps, brachialis and coracobrachialis muscles.
      c. *Radial nerve* to triceps and forearm extensors.
      d. *Median nerve* to forearm flexors, thenar muscles and first two lumbricales on the radial side.
      e. *Ulnar nerve* to interosseous muscles and hypothenar group.
   3. *Sensory* innervation of the hand (Fig. 7-6):
      a. *Radial nerve*: dorsal aspect of the lst, 2nd and 3rd digits.
      b. *Median nerve*: ventral aspect of the lst, 2nd and 3rd digits.

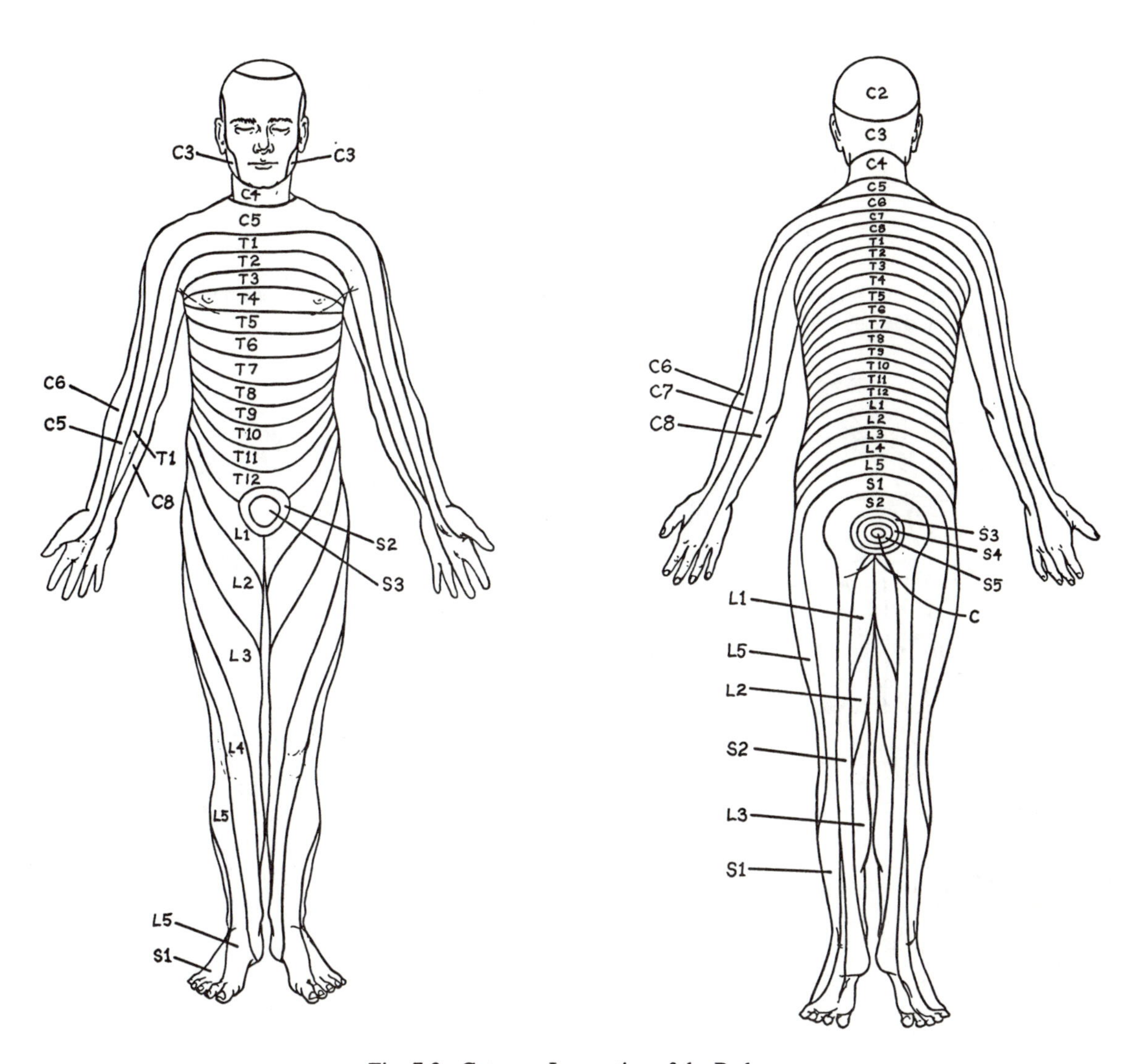

Fig. 7-3 Cutaneus Innervation of the Body.

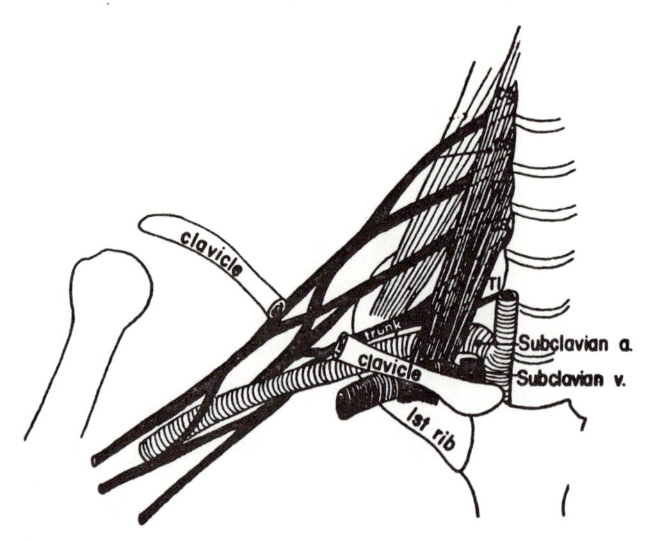

Fig. 7-4 The Brachial Plexus. (Reproduced, with permission, from Goldberg, Clinical Anatomy Made Ridiculously Simple, MedMaster, 1984).

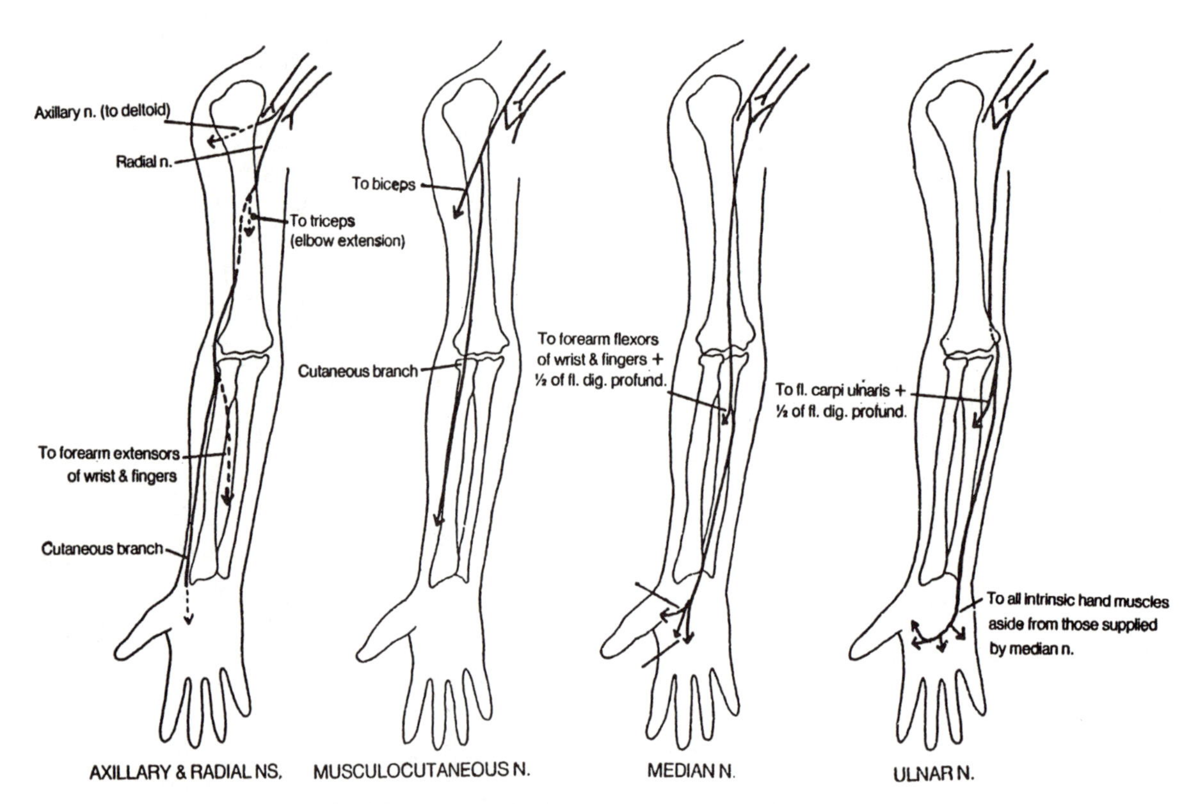

Fig. 7-5 Nerves of the Upper Extremity. (See ref., Fig. 7-4).

c. *Ulnar nerve*: dorsal and ventral aspects of the 4th and 5th digits. Striking the "*funny-bone*" (which is *not* funny) means striking the ulnar nerve at the medial epicondyle of the humerus (tingling in digits 4 & 5).

E. *Thoracic nerves (intercostal nerves)*

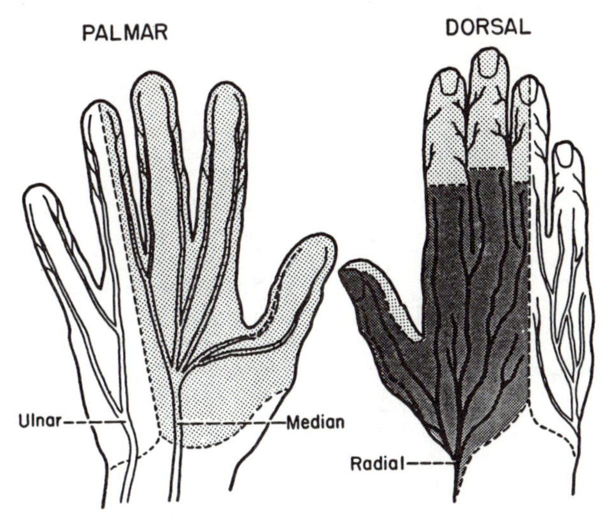

Fig. 7-6 Sensory Innervation of the Hand. (Reproduced, with permission, from The Hand, American Society for Surgery of the Hand, 1978).

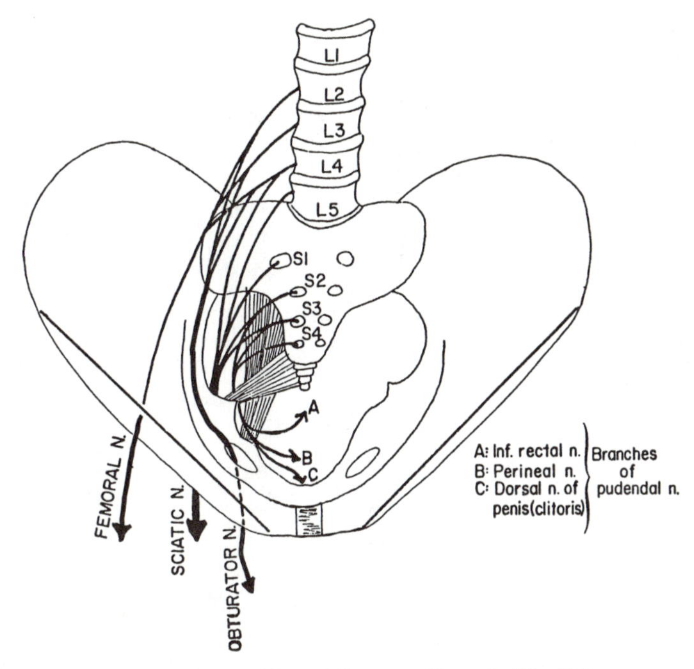

Fig. 7-7 Lumbar and Sacral Plexuses. (See ref., Fig. 7-4).

1. *Motor*. to intercostal muscles (breathing muscles)
2. *Sensory*: from skin of thorax

F. *Important motor nerves of the lower extremity* (Fig. 7-7, 7-8)

1. The *lumbar plexus (L-1 to L-4)* forms the *femoral nerve* which innervates the quadriceps, and the *obturator nerve* which innervates the adductor muscles.
2. The main branch of the *sacral plexus* (L-5to S-3) is the *sciatic nerve* (the largest nerve of the body); it becomes the tibial and peroneal nerves at the popliteal fossa.

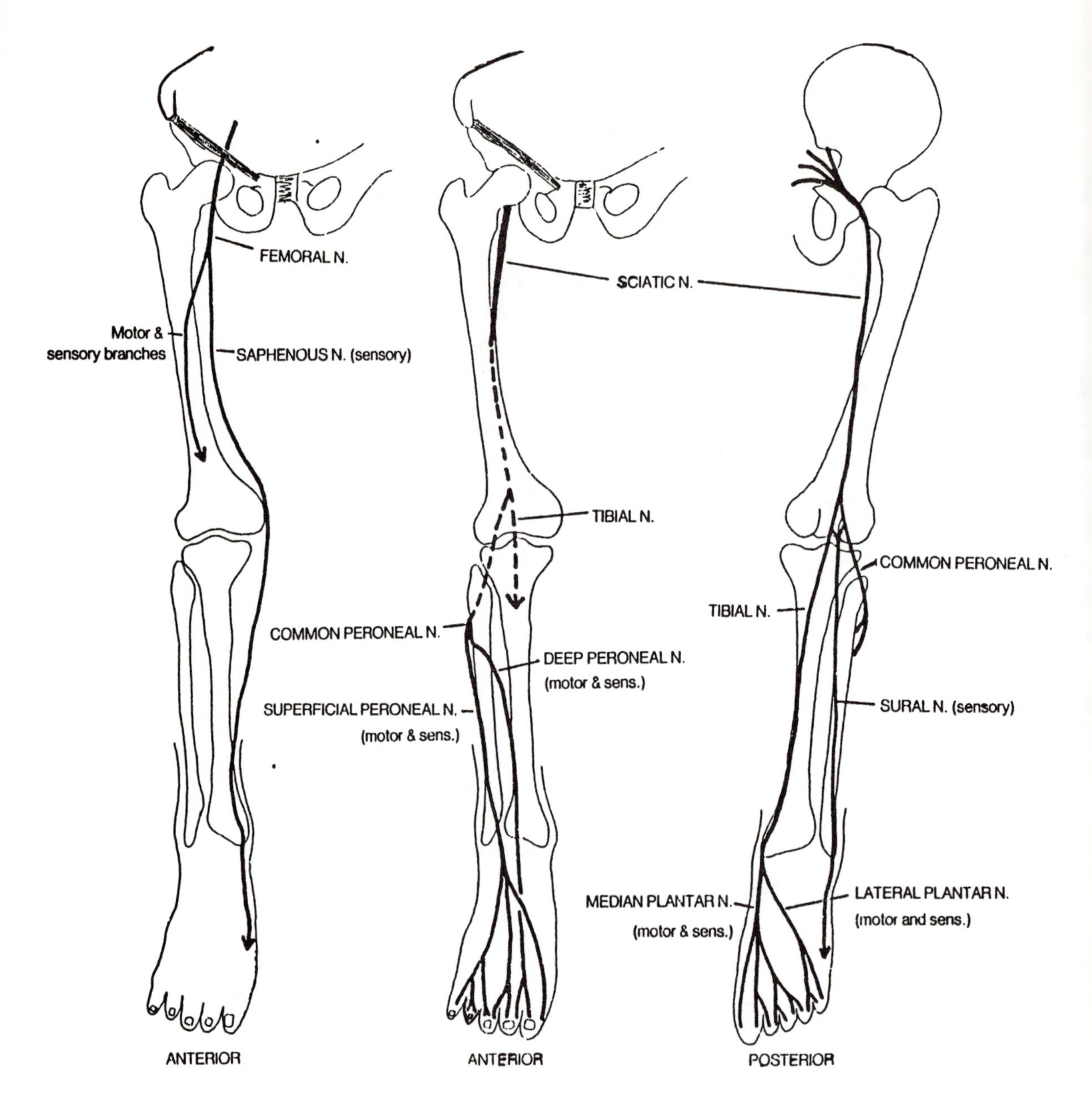

Fig. 7-8 Nerves of the Lower Extremity. (See ref., Fig. 7-4).

a. The *tibial nerve* innervates the hamstrings, calf muscles (foot flexion) and toe flexors.
b. The *peroneal nerve* innervates the tibialis anterior, peroneal muscles and the foot and toe extensors.

G. *Injury to a spinal nerve* results in sensory loss (*anesthesia*) in skin supplied by the sensory roots, and *flaccid paralysis* of the muscles supplied by the motor components. For example, damage to the radial nerve (C5-8) at the forearm would result in anesthesia of the dorsal portion of the first three fingers and paralysis of the hand and finger extensors ("wrist drop").

III. *Reflexes* involve the spinal cord, and sometimes the brain (conditioned reflexes). *Most are polysynaptic*, utilizing *sensory neurons*, *interneurons* and a *motor neuron response*. A simple thing such as burning one's finger utilizes many interneurons conducting up and down the spinal cord and brain. The only true monosynaptic reflexes are *deep tendon reflexes* (DTR's), involving just sensory and motor neurons. DTR's are important, because they evaluate the sensory nerve, the spinal cord, the motor nerve and the muscle group innervated by the nerve. The tendon is struck, the receptor in the tendon is stimulated, the impulse travels along the sensory nerve to the cord, and a synapse takes place with the cell bodies of lower motor neurons in the ventral horns. The impulse travels along the motor axon to the motor end-plate, and the muscle contracts. A defect at any point will interfere with the reflex (Fig. 7-9). Important deep tendon reflexes are:

A. *The biceps and triceps reflexes* evaluate spinal cord levels *C-5 and C-6*, respectively. They also help evaluate the brachial plexus and the biceps and triceps muscles.
B. The *patellar reflex (knee-jerk) evaluates spinal cord level L-4*. It also helps evaluate the lumbar plexus (specifically, the femoral nerve) and the quadriceps muscle.
C. The *achilles reflex (ankle-jerk) evaluates spinal cord level S-1*. It also helps evaluate the sacral plexus and the gastrocnemius and soleus muscles. If absent and accompanied by low back pain which radiates to the leg and foot, there may be a slipped disc at the level of L-5/S-1.

IV. *The autonomic nervous system*:
The autonomic system is an involuntary response system which affects internal organs. It may be modified by voluntary control (Fig. 7-10).
A. *Characteristics*:
   1. Innervates smooth muscle, heart and glands.
   2. Consists of two major divisions, *sympathetic* and *parasympathetic*, usually with opposite effects.
   3. Is a two neuron, motor system chain.
   4. First neurons (preganglionic) arise within the brain stem or cord.
   5. Second neurons (postganglionic) lie in ganglia outside the cord and send axons to the organs.
   6. Is regulated by the hypothalamus.
B. *Divisions*:
   1. *Sympathetic*:
      a. Originates in the *thoracic and upper lumbar regions* of the spinal cord.
      b. Most of the second neurons lie in a ganglionic chain (*sympa-*

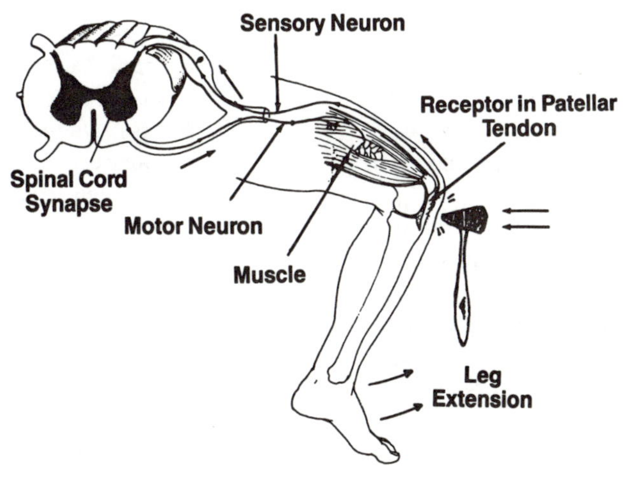

Fig. 7-9 Deep Tendon (Stretch) Reflex. (Modified and reproduced, with permission, from Hoppenfeld, Orthopedic Neurology, Lippincott, 1977).

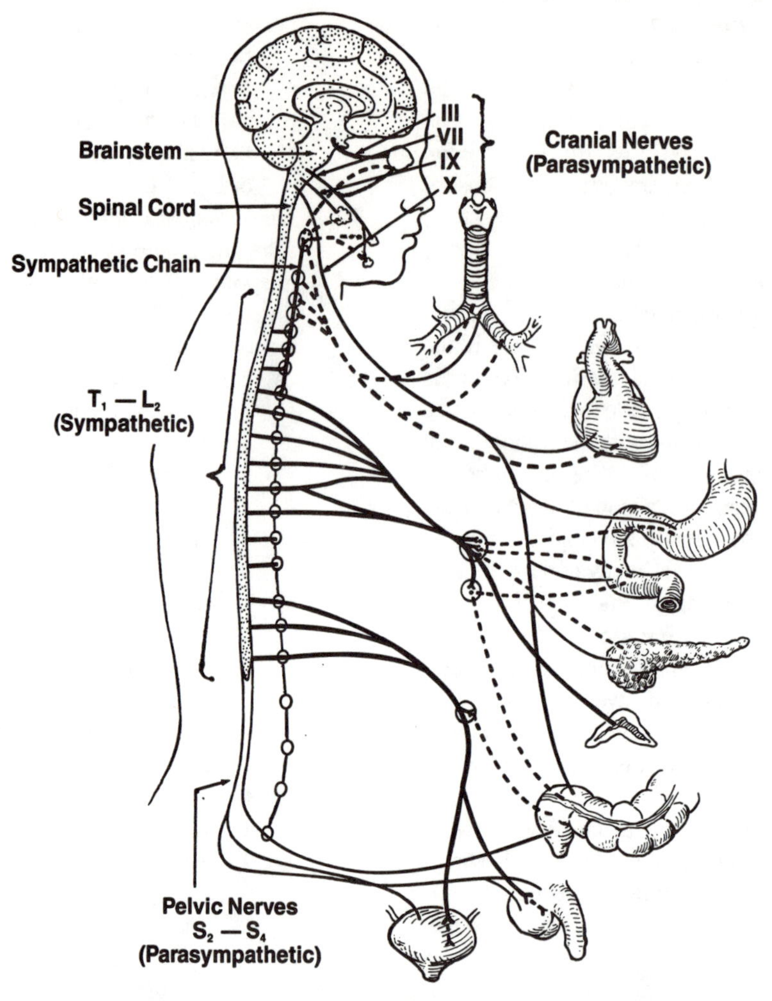

Fig. 7-10 The Autonomic Nervous System.

thetic chain) located along the vertebral column.

c. Is the *adrenergic* system.

d. Is the "*fight-or-flight*" division.

2. *Parasympathetic*:

a. Originates in the *brain stem* (fibers travel with cranial nerves III, VII, IX and X) and the *sacral region* of the spinal cord.

b. The second neurons lie in ganglia close to the organ innervated

c. Is the *cholinergic* system

d. Is the *vegetative* ("energy conserving") division.

C. *Synaptic transmission at autonomic junctions* (Fig. 7-11): Sympathetic and parasympathetic fibers innervate the same organ. The area on the organ receiving the post-ganglionic nerve is the *receptor site*.

1. *Synaptic transmitter substances*:

a. As seen in Fig. 7-11, preganglionic axons, and postganglionic parasympathetic axons, secrete acetylcholine (ACh) (see also Fig. 5-1).

b. Postganglionic sympathetic axons secrete norepinephrine (NE) (an exception is ACh

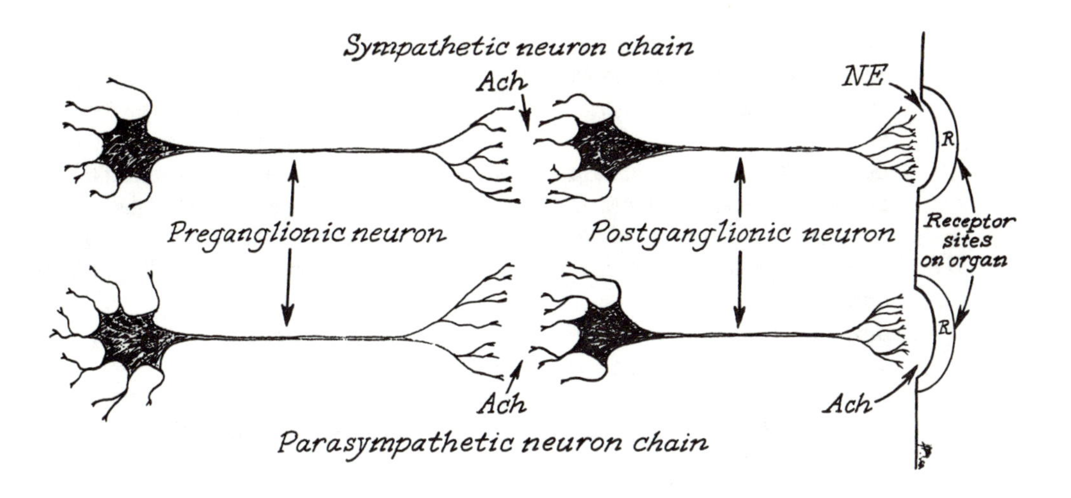

Fig. 7-11 Autonomic Synapses.

secreted by sweat gland neurons) (Fig. 7-12).

2. *Types of receptors*:
   Both *cholinergic* and *adrenergic receptors* are at the organ receptor site. Divisions of cholinergic receptors exist (muscarinic and nicotinic), but the adrenergic divisions are the more important clinically. *Receptors for sympathetic nerves* are divided into two main types, *alpha* and *beta*. Beta is subdivided into *beta-1* and *beta-2*. Both beta-1 and beta-2 receptors are found in many organs, but one or the other usually predominates.

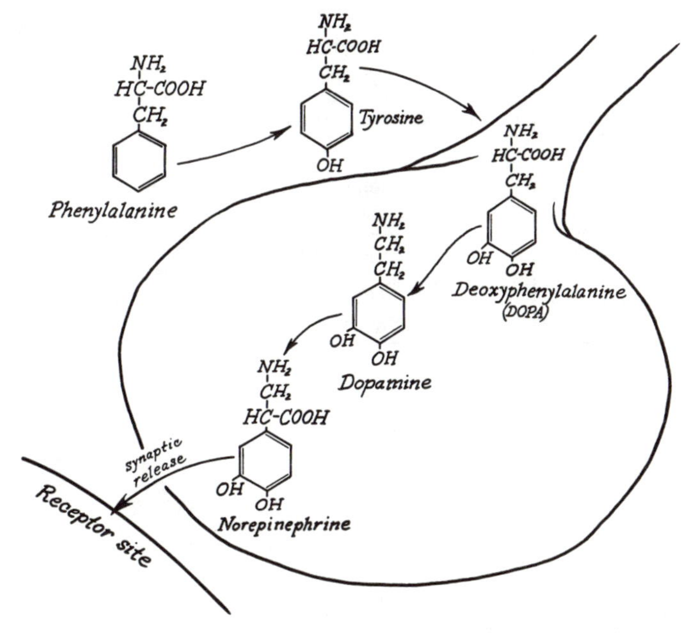

Fig. 7-12 Synthesis and release of norepinephrine from sympathetic neuron.

a. *Alpha receptor stimulation* produces contraction of the smooth muscle of vessels (*vasoconstriction*).
b. *Beta-1* receptors predominate in the *heart*. *Stimulation of beta-1 receptors causes an increase in the heart rate and force of contraction.*
c. *Beta-2* receptors predominate in *bronchial* smooth muscle. *Stimulation of beta-2 receptors causes opening of the bronchial airway (broncho-dilation).*
d. As an example, epinephrine stimulates beta-1 and beta-2 receptors. It is used to stimulate the heart in cardiac arrest, but also opens the airway in asthma.

3. *Blocking agents.*
Sympathetic blocking agents have effects similar to parasympathetic stimulation; parasympathetic blocking agents have effects similar to sympathetic stimulation. For example, the beta-1 blocking agent propranolol (Inderal) slows the heart rate.

D. *Main function of the autonomic system* (Table 7-2):
As seen in table 7-2, sympathetic effects are usually the opposite of parasympathetic effects. It is simpler to learn one system, since the other is the opposite. Memorizing the "fight-or-flight" response seems to be the more practical. These responses make fighting, or fleeing, possible: e.g., pupillary dilation to improve vision, a

Table 7-2: Main Functions of the Autonomic Nervous System

ORGAN	SYMPATHETIC RESPONSE (Adrenergic)	PARASYMPATHETIC RESPONSE (Cholinergic)
1. Eye (iris)	Pupil dilates	Pupil constricts
2. Lungs	Bronchi dilate	Bronchi constrict (stimulates secretion)
3. Heart	Rate increases	Rate decreases
4. GI Tract:	Motility and secretion decreases	Motility and secretion increases
Sphincters	Contract	Relax
5. Liver	Glycogenolysis	—
6. Urinary Tract:		
Bladder	Relaxes	Contracts
Sphincter	Contracts	Relaxes
7. Adrenal Medulla	Adrenalin secretion (cholinergic fibers)	—
8. Male Sex Organs	Ejaculation	Erection
9. Glands:		
Sweat	Stimulates secretion (cholinergic fibers)	—
Salivary	—	Stimulates secretion
Lacrimal	—	Stimulates secretion
10. Blood Vessels:		
Coronary and Skeletal muscles	Dilates	—
Skin and GI tract	Constricts	—

greater heart rate for increased cardiac output, bronchodilation to improve breathing, and slowing digestive responses so as to reduce interference with fight or flight.

E. *Drugs affecting the autonomic system*: Many drugs are used for a specific autonomic effect. As mentioned, *epinephrine* is both a *beta-1 and beta-2 agent (agonist)*. Several *selective beta-2 agonists* have been developed to alleviate the bronchospasm of asthma. Metaproterenol (Metaprel, Alupent), terbutaline (Brethine) and albuterol (Ventolin) have fewer cardiac (beta-1) side effects than epinephrine. *Beta blocking agents* such as propranolol (Inderal), metaprolol (Lopressor), and nifedipine (Procardia) are useful in the treatment of cardiac arrhythmias, angina and hypertension. Some beta-blockers are non-selective — in addition to blocking beta-1 receptors, they also partially block beta-2 receptors, and may cause bronchospasm (see chapter 9). Many autonomic drugs produce undesirable side effects. The use of epinephrine to dilate the bronchi in asthma causes tachycardia, and may precipitate an attack of angina or an arrhythmia in a susceptible individual. An *atropine-type drug* (parasympathetic blocking agent) that dilates the pupil may produce dry mouth, urinary retention, tachycardia and constipation. *Opiates* have a sympathetic effect (sympathomimetic action) on the gastrointestinal tract, and produce constipation. This is seen in addicts, but also in the multitude of people who use prescription drugs containing *codeine*. *Depressant drug withdrawal* (heroin, barbiturates, alcohol) produces a variety of both sympathetic and parasympathetic responses (parasympathetic: increased lacrimation, diarrhea; sympathetic: tachycardia, pupillary dilation).

V. *Pain* is a protective device for the body. A complicated array of fiber types, minor pathways and pain thresholds exist, but in simple terms, sensory fibers are stimulated and the impulse travels up the spinal cord in the lateral spinothalamic tract to the thalamus. It then travels to the parietal lobe of the brain. Pain fibers from internal organs run mostly in sympathetic nerves, although some travel in parasympathetic and cranial nerves. Generally, the *perception of pain occurs in the thalamus;* the quality, or *interpretation, of pain occurs in the cortex.* A painful stimulus one day may not be painful the next. Anesthetics (including alcohol, barbiturates, etc.) first depress cortical neurons, so that although the thalamus and other lower centers indicate pain, one "doesn't care."

A. A *pain-inhibiting system* exists in the body. *Receptors for opiates* (morphine, etc.) are present along the pain pathway. Internal, or *endogenous, opiates* (endorphins, enkephalins), produced by the body block pain impulses in various portions of the pathway, probably as a protective device. The neurotransmitter "substance P", secreted by pain fibers in the dorsal horns of the spinal cord, is blocked by an enkephalin. Stimulation of large sensory fibers sometimes inhibits impulses in remote areas of the body.

B. Referred pain is pain felt in a part of the body remote from the area causing it. Commonly, pain is referred from an internal organ to the area of skin which developed near the same embryonic segment. As an example, the pain of *angina pectoris* is often referred to the *ulnar nerve distribution of the left arm.* In this case, sensory fibers from both the heart and the ulnar nerve synapse with second neurons in the same developmental area of the spinal cord. The message is misinterpreted as coming from the ulnar distribution.

C. Acupuncture may be a variation of these mechanisms.

VI. *Common disorders of the peripheral nervous system*:

A. *Definitions*:

1. *Neuralgia*: severe nerve pain
2. *Neuritis*: inflammation of a nerve
3. *Neuropathy*: nerve disease

B. *Bell's palsy* (acute facial paralysis) is *paralysis of the muscles of one side of the face* from an *inflammation of the facial nerve* at or near its origin in the CNS. The cause is unknown, but herpes simplex is thought to play a role. The paralysis is usually partial, and lasts about 3 to 6 weeks. Primary treatment: eye drops or ointment and oral corticosteroids.

C. *Carpal tunnel syndrome* is caused by *compression of the median nerve* as it passes under the transverse carpal ligament at the palmar aspect of the wrist. The cause is unknown. Commonly seen in postmenopausal women, it is also seen in conditions where fluid retention causes swelling of the hand and wrist. Symptoms are palmar pain, numbness or paresthesias in the first three digits. Primary treatment: sometimes surgical opening of the transverse carpal ligament is necessary (Fig. 7-13).

D. *Guillain-Barre syndrome (infectious polyneuritis)* occurs a month or two after a viral infection. Lymphocytes and macrophages invade the myelin sheath, causing *partial demyelination*. The person develops tingling in his hands or feet. Motor weakness, decreased DTR's and mild sensory loss occur. Often an ascending paralysis begins in the legs and moves upward. Facial weakness is common. Sometimes respiratory support must be given. Complete recovery in a few weeks is the rule.

E. *Herpes zoster (shingles)* is a *viral disease consisting of multiple groups of vesicles or crusted lesions that follow a cutaneous nerve distribution*, usually on one side of the trunk, although lesions may occur on the face, neck and elsewhere. Like herpes simplex, it is usually self-limiting. This virus also causes chickenpox. Primary treatment consists of an analgesic and steroid to decrease the inflammation around the cutaneous nerve (Fig. 7-14).

F. A *slipped disc is a rupture of an intervertebral disc. Pain is caused by impingment of the prolapsed disc on a nerve root*. L-4 (between vertebrae L-4 and L-5) and L-5 (between L-5 and

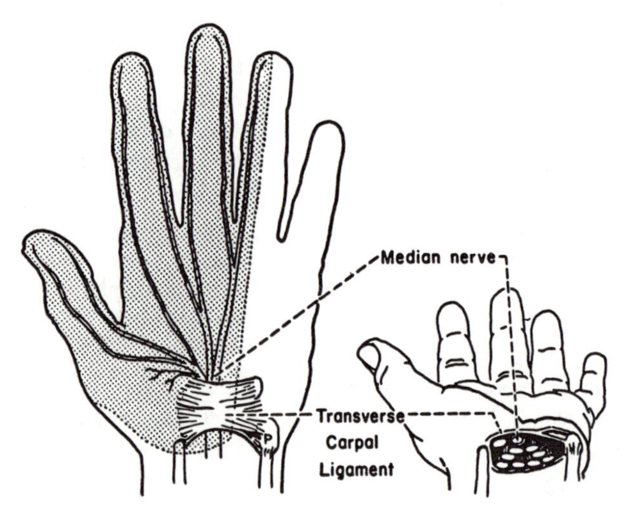

Fig. 7-13 Carpal Tunnel Syndrome. (See ref., Fig. 7-6).

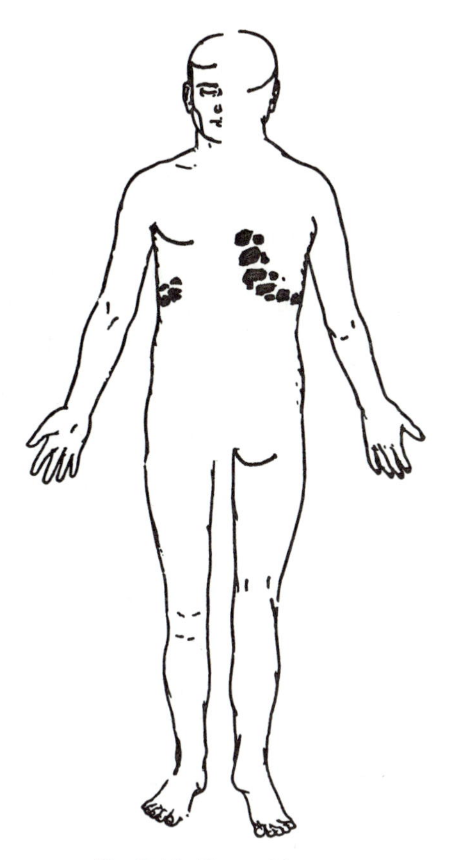

Fig. 7-14 Herpes Zoster.

sacrum) discs are most commonly involved, affecting nerve roots L-5 and S-1, respectively. Pain often radiates along the entire sciatic nerve distribution (*sciatica*), from the gluteal region down the lateral side or back of the thigh to the leg and foot. Often caused by back strain or injury, it is occasionally precipitated by coughing or sneezing. Primary treatment: bed rest, anti-inflammatory medicine and surgery if conservative measures fail. An injection of chymopapain, an enzyme derived from the papaya fruit,

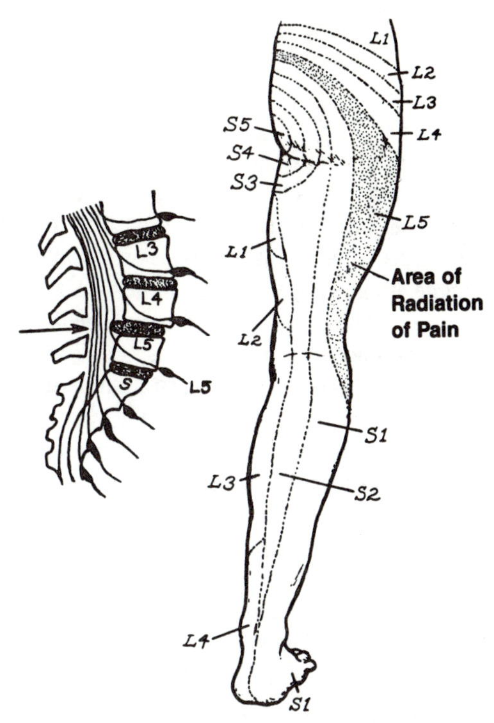

Fig. 7-15 Slipped Disc at L 4-5.

is sometimes used to dissolve the protein center of the disc as an alternative to surgery (Fig. 7-15).

G. *Trigeminal neuralgia* (tic douloureux). *Sudden severe pain in the jaw area on one side of the face is the hallmark of trigeminal neuralgia.* Often, the pain is triggered by chewing, or simply touching the face. *Pain fibers of the trigeminal nerve are stimulated.* The cause is unknown. One hypothesis is that it is related to a viral infection of the upper portion of the trigeminal nerve. Primary treatment is the analgesic carbamazepine (Tegretol). Sometimes surgical intervention is necessary.

-EIGHT-

THE EYE AND EAR

The *special senses* are taste, smell, hearing, equilibrium and vision. An absence of smell (anosmia) rarely causes problems. The loss of taste is seldom total, as taste buds on the anterior tongue are innervated by cranial nerve VII, and those on the posterior tongue by cranial nerve IX. The most important sense-organs in man are the eye (vision) and the ear (hearing and equilibrium). In lower animals, hearing, vision and smell are important aspects of learning and behavior. In man, the special senses are underplayed in favor of the functioning of the large association areas of the brain.

I. *The eye.*

A. *External anatomy*. The *orbit* is the bony framework for the *eyeball*, or *globe* (Fig. 8-1). Eyebrows are present along the superior margin of the orbit (shaving the eyebrows may have a lasting effect. In some people they will grow back; in others there is sparse or no growth). Upper and lower *lids* and *lashes* protect the eyeball. The opening between the lids (the *palpebral fissure*) sometimes closes with a black eye. The *white of the eye* is the *sclera*. The *colored portion* (blue, brown, green) is the *iris*. The *pupil* is the *hole* in the middle of the iris. The *transparent cornea* lies over the iris. The corner of the eye nearest the nose contains a pink fleshy area, the lacrimal caruncle. Below and above the caruncle on the lid margins, are the two openings for the drainage of tears into the lacrimal sac.

B. *The lacrimal apparatus* (Fig. 8-2). The *lacrimal gland* is located superior and lateral to the eyeball; it manufactures tears which drain down into the inner corner of the eye. The ducts drain tears to the *lacrimal sac*, which drains into the *nasal cavity* by way of the *naso-lacrimal duct*. A flow of tears, when crying, thus makes your nose run. *Function*:

1. *Lubrication*
2. *Prevents the growth of microorganisms*. Tears contain albumin, globulins and lysozyme. Much of the antimicrobial action is from the gamma-globulin (antibody) fraction.

C. *The extrinsic eye muscles* (Fig. 8-3).

1. Innervated by cranial nerves III, IV and VI.
2. *Function*: movements of the eyeball, or *extra-ocular movements (EOM's)*. *Importance*: a lesion of a nerve (rare), or a weak muscle, may produce loss of movement in one or more planes. A "wandering" or "lazy" eye, seen commonly in children, involves a weak or elongated eye muscle. It can usually be easily corrected, either by eye exercises or by surgical shortening of the muscle.

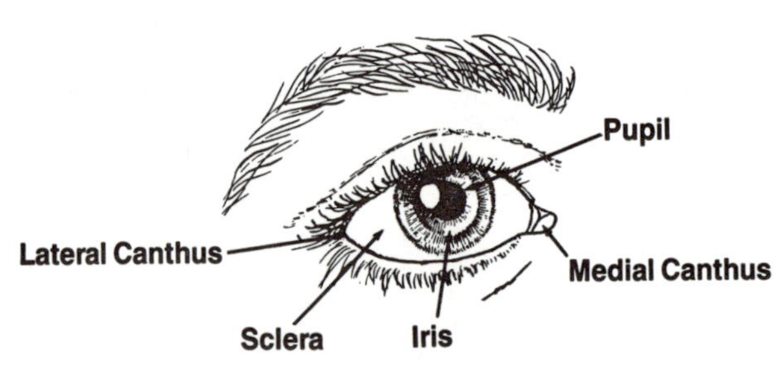

Fig. 8-1 Right Eye.

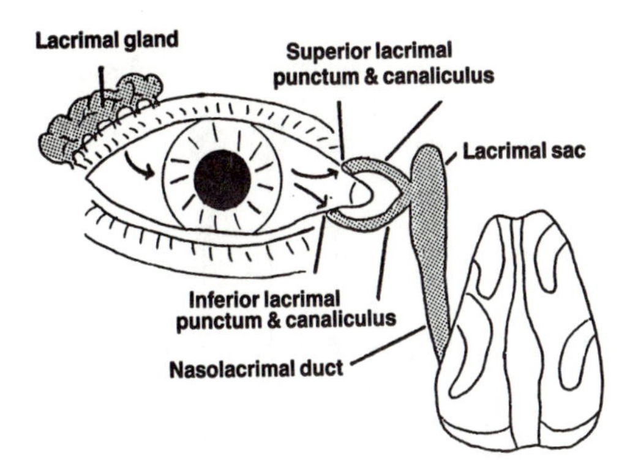

Fig. 8-2 Right Lacrimal Apparatus. (Reproduced, with permission, from Goldberg, Clinical Anatomy Made Ridiculously Simple, MedMaster, 1984).

D. The *conjunctiva* is the thin mucous membrane lining the inner eyelids and anterior aspect of the eyeball, except for the cornea. *Function*: Produces a mucous substance that assists tears in keeping the conjunctiva and cornea moist. *Importance*: Inflammation of the conjunctiva (pinkeye) is the most common eye disorder besides refractive errors.

E. *The eyeball* (bulb, globe) is one inch in diameter, and lies in the *bony orbit* imbedded in a protective cushion of *fat* (Fig. 8-3). Layers of the eyeball are:
   1. *Sclera and cornea (outer layer)*:
      a. The *sclera* is tough and hard.
      b. The *cornea* is the anterior continuation of the sclera over the iris and pupil.
      c. The *extrinsic eye muscles* attach to the sclera.
   2. *Choroid, iris and ciliary body (middle layer)*:
      a. The *choroid* is vascular and pigmented.
      b. The *iris* is a *colored muscular ring* that dilates or constricts, changing the *pupil size*.

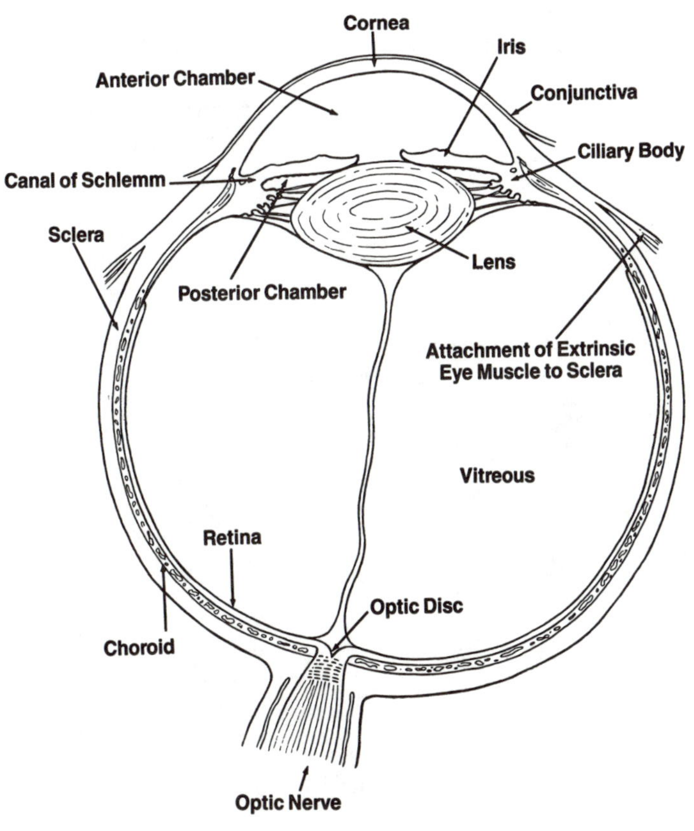

Fig. 8-3 Anatomy of the Eyeball.

c. The *ciliary body* is a muscular ring holding the *lens* in place.

3. *Retina (inner layer) (Fig. 8-4)*:

   a. The *retina* is tissue-paper thin and consists of *light receptor neurons*, the *rods* and *cones*, as well as two additional layers of neurons.

   b. Axons of the innermost layer of neurons (the *ganglion cells*) converge at the back of the eyeball to form each *optic nerve* (cranial nerve II). The area where the optic nerve leaves the eyeball contains no rods and cones, thus no vision, and is the *blind spot*.

F. *Optic nerve* (Fig. 8-5):

Both optic nerves run posteriorly from the eyeball and come together anterior to the pituitary gland as the *optic chiasm*. Some fibers cross to the opposite side, and some continue on the same side. After crossing, they become the *optic tracts*, which spread out as the *optic radiations* to the *occipital lobes* of the cerebral cortex.

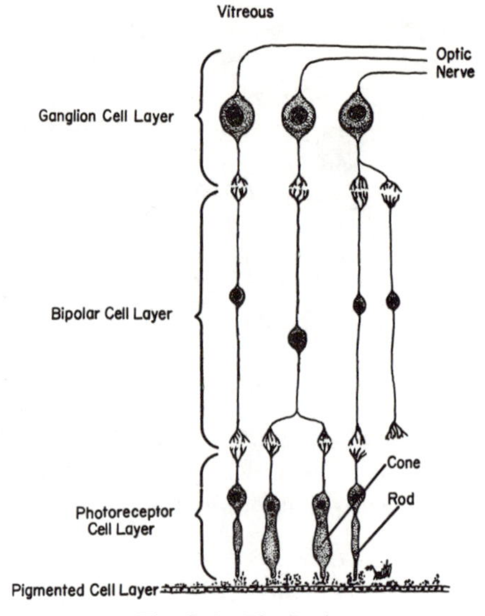

Fig. 8-4 The Retina.

G. *Refractive structures* (Fig. 8-3):

1. The *cornea* is the main refractive structure. About 50% of the pop-

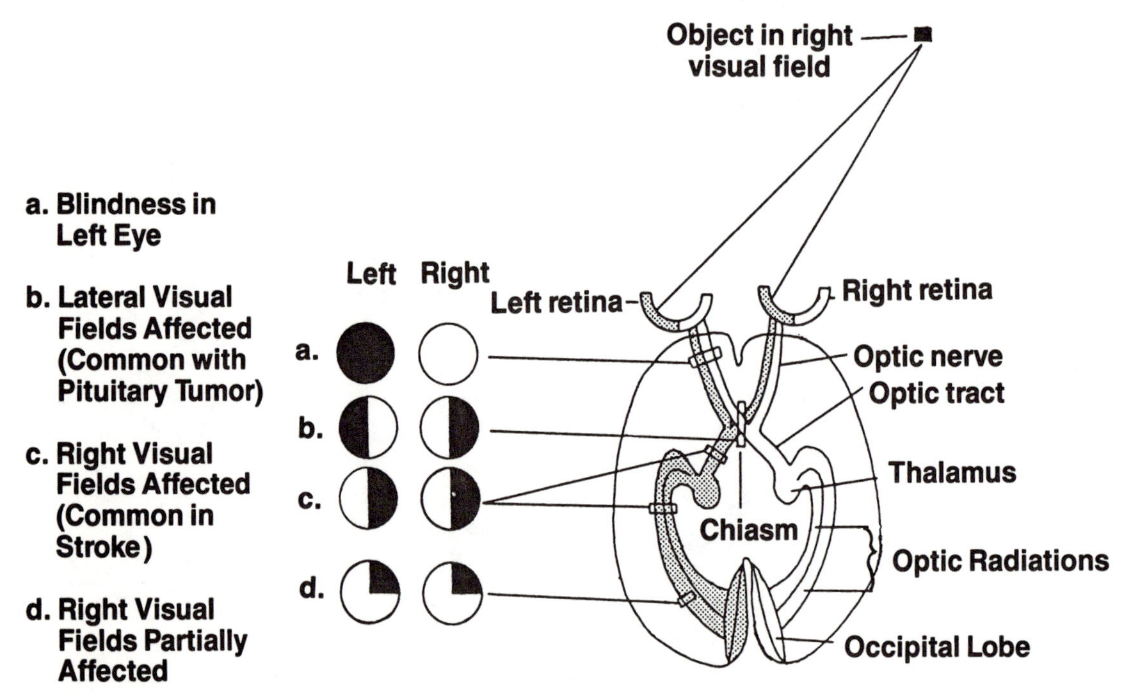

Fig. 8-5 The Optic Pathway. (Reproduced, with permission, from Goldberg, Clinical Neuroanatomy Made Ridiculously Simple, MedMaster, 1984).

ulation have a distorted cornea (not round). The result is blurred vision in one or more planes. This is *astigmatism*.

2. *Aqueous humor* is a clear fluid produced by ciliary body capillaries. It flows from the posterior chamber (behind the iris) through the pupil into the anterior chamber (in front of the iris). It is reabsorbed into the venous system at the junction of iris and cornea.
3. The *lens* is a clear structure that fine-focuses light rays. When the lens becomes opaque (e.g., in old age), it is a *cataract* (see section J).
4. The *vitreous humor* is the gel that forms the bulk of the interior of the eyeball.

H. *The optic fundus* is the part of the eye seen with an ophthalmoscope. The retina is transparent, and vessels are seen through the retina against the background of the choroid (Fig. 8-6).

1. The *optic disc* is the area of the blindspot. The disc is pink (containing many capillaries) with a white central depression. In optic nerve disease, the entire disc may appear white. The *central retinal artery and vein* are visible in the center of the disc. An increase in intracranial pressure may cause bulging of the disc, or *papilledema*, obscuring the vessels because of edema.
2. *Retinal vessels* are easily seen fanning out from the optic disc. *Importance*: hemorrhages and exudates may be seen in diseases such as arteriosclerosis, hypertension, and diabetes mellitus.
3. The *macula* is a round, dark area situated temporal to the optic disc. Its central area, the *fovea*, can sometimes be seen. The fovea is the area of most acute vision, having an abundance of cones.

I. *Basic physiology of vision.*

1. *General. Stimulation of the rods and cones by light rays initiates the nerve impulse (depolarization)*. The rods and cones synapse with the bipolar cells, and they synapse with the ganglion cells (Fig. 8-4). After crossing, the optic tracts travel to the thalamus, then to the occipital lobe as the optic radiations (Fig. 8-5). *The occipital lobe interprets the nerve impulse as vision.*
2. *Function of the rods and cones.*
   a. The *rods* are responsible for *dim light vision*, and the *cones* for *color vision*. Light rays striking the rods and cones change the structure of pigments within them; this change is responsible for the nerve impulse. The pigment in both is a small molecule, retinene-1 (or retinal), bound to a large protein, opsin. The retinene is the same in both rods and cones, but the opsin is different. The rod pigment is rhodopsin, or visual purple. A cone pigment, iodopsin, appears to be involved with red light. *Vitamin A* is necessary for the manufacture of rhodopsin, and probably is needed in cone pigments as well. Regeneration of rhodopsin is decreased in vitamin A deficiency, resulting in a mild case *of night blindness*.
   b. *Color-blindness* is caused by a recessively inherited gene defect on the X chromosome. Males are affected more frequently because they have only one X chromosome. Females need two for the defect.
3. *Visual acuity means how well one can see,* and is usually measured by the use of a chart of letters (or

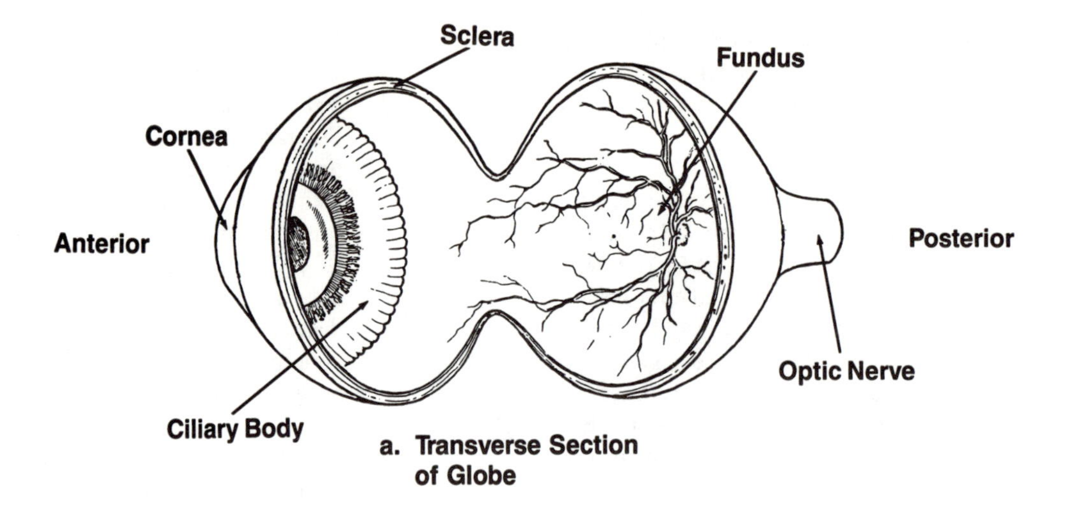

a. Transverse Section of Globe

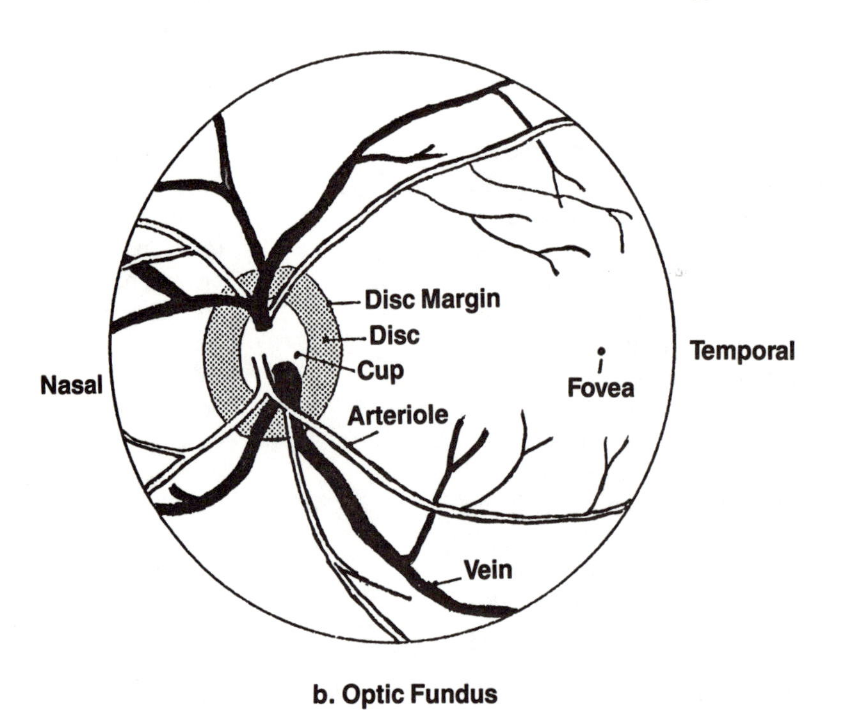

b. Optic Fundus

Fig. 8-6 Transverse Section of Globe and Optic Fundus. (See ref. Fig. 8-2).

pictures, or "E's" for children or the handicapped) in rows at various heights. Commonly, the *Snellen eye chart* is read at 20 feet with each eye. Large letters are in the top row. As the rows move down they get smaller. If a person is able to read the line marked 20, then he has 20/20 vision. If he can read only the 70 line (larger letters) then he has 20/70 vision (meaning a normal person can read at 70 feet what the person being examined reads at 20 feet). *Gross vision* is easily tested by holding up a certain number of fingers and asking the person how many he can see with each eye. It is crucial to test vision on a patient, and never rely on his own statement.

4. *Pupillary reflexes.*
   The pupillary light reflex takes place when light is shined in the eye and the pupil constricts. The other pupil also constricts (*consensual reflex*). Sensory fibers for this reflex travel with optic fibers as far as the midbrain. Other fibers, including those for accommodation, continue to the occipital lobe. Fibers mediating the light and accommodation reflexes synapse in the midbrain. Motor fibers then return to the eye to innervate the iris and ciliary body. Importance: Optic reflexes are important to test in disorders such as head injury, iritis and glaucoma, in which reactions may be depressed. *Pupils equal and reactive to light is often abbreviated as PERL* (and PERLA if the pupillary reaction to accommodation is included).
5. *Accommodation means focusing for near vision.* The *ciliary muscle contracts* and the *lens becomes more convex*. Light can thereby be focused from near. Accommodation is tested by asking the person to read a near vision card. *Importance:* Accommodation decreases with age, primarily because of stiffening of the lens. The *near point of vision* indicates how well the eye can accommodate, and moves farther away with age (presbyopia). This is why the elderly are farsighted and need glasses for near vision. It rests the eyes to look far off because accommodation is not involved.
6. The *visual field is the panorama of vision that the eye can see*. The confrontation method is often used to assess peripheral vision, as follows: with the patient facing you, your hands are outstretched to the sides. With your fingers moving, you draw your hands slowly inward, asking the patient to indicate when he notices your moving fingers. He must be looking straight at you, since the test measures *peripheral vision*.
   Importance: different patterns of visual field loss may occur, depending on where the lesion lies in the visual pathway.
   a. A growing pituitary tumor often impinges on the optic chiasm, causing loss of the lateral sides of the visual field (see Fig. 8-5b).
   b. In pyschoneurosis, a common pattern is loss of peripheral vision ("gunbarrel vision").
7. *Autonomic responses*:
   a. *Parasympathetic drugs* constrict the pupil (miosis), e.g., pilocarpine.
   b. *Sympathetic drugs* dilate the pupil (mydriasis) e.g., neosynephrine.
   c. *Parasympathetic blocking agents* (parasympatholytic agents) inhibit the parasympathetic system, resulting in a sympathetic response, e.g., atropine.

J. *Common disorders of the eye.*
   1. *Amblyopia is decreased vision in the absence of organic disease.* It is seen in those with a deviated eye ("lazy eye"). In children, vision is suppressed in the deviated eye. If the condition is not corrected by the time vision has developed (about first grade), the eye may be become permanently impaired. Correction of the lazy eye involves patching and glasses, and should take place before age 7.
   2. *A "black-eye"* is usually caused by a blow to the eye and/or orbit.

The lids possess a large potential space, and when small vessels are broken, *blood and edema fluid seep into the lid (ecchymosis)*. The orbit may also swell for the same reason, and the extravasation of blood may even extend across the bridge of the nose to involve the non-injured eye. Although the eye may close, vision is usually not affected. However, a black-eye may also accompany significant damage, such as an orbital fracture, rupture of the globe, dislocation of the lens or bleeding into the anterior chamber (hyphema). Primary treatment of an uncomplicated black eye consists of an adequate ice-pack to reduce swelling.

3. A *cataract* is a *thickening (opacity) of the lens* that impairs vision. Usually this occurs over many years. Primary treatment is removal of the lens. Today the implantation of an artificial lens is common.

4. *Corneal injury*. The cornea contains many sensory nerve endings, and a multitude of minor injuries may cause *severe pain and irritation* (a scratch from a branch, contact lenses left in too long, an imbedded foreign body such as a small iron chip, or an abrasion from an inadvertant fingernail). Corneal burns are common from prolonged exposure to a sunlamp or a welding arc. The cornea is one of the fastest healing organs in the body (rapid regeneration of epithelial cells). Even a painful corneal abrasion, treated with an eye patch and an antibacterial ointment, will usually completely heal in 24 hours.

5. *Detached retina*. The retina is loosely attached to a layer of cells, the pigment epithelium, which abuts on the choroid. The tissue-thin retina may become detached from the pigment epithelium and choroid for a variety of reasons: spontaneously, secondary to a blow, or as a complication of cataract surgery. A hole in the retina allows vitreous fluid to accumulate between the retina and the pigment epithelium and detachment progresses. The vision is blurred, like a "shadow" or "curtain" coming across the field of vision. It is sometimes possible to see the detachment with the ophthalmoscope. Primary treatment consists of closing the detachment by the use of a high-enery light beam (laser), or surgery.

6. *Farsightedness (hyperopia)*. The eyeball is shorter than usual and *light rays are focused behind the retina*. Primary treatment: eyeglasses with a convex lens (Fig. 8-7).

7. *Glaucoma is a disease in which the reabsorption of aqueous humor into the venous system is blocked* (Fig. 8-8). Intraocular pressure rises, with damage to the retina. In acute glaucoma, the intraocular pressure is high, the

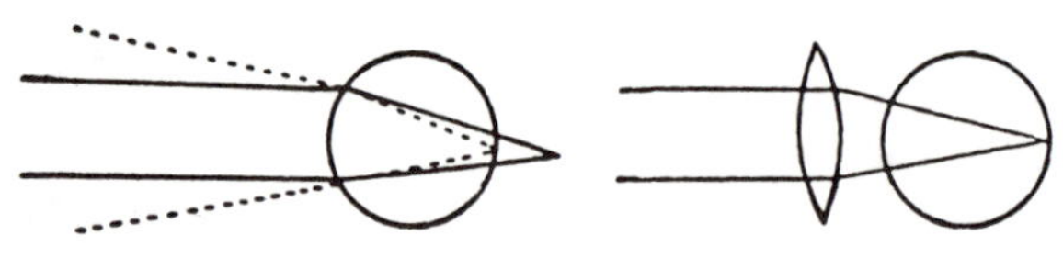

Fig. 8-7 Farsightedness (Hyperopia).

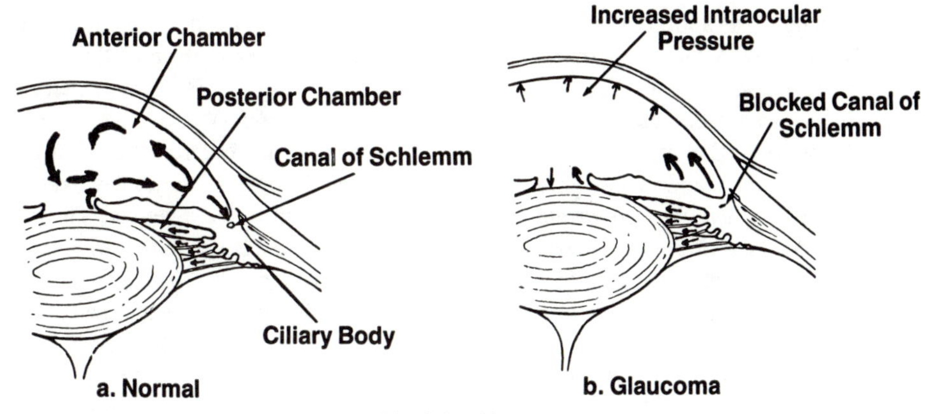

Fig. 8-8 Glaucoma.

a. Normal movement of aqueous humor from posterior chamber to anterior chamber and into canal of Schlemm.
b. Canal of Schlemm blocked resulting in increased ocular pressure.

vision is blurred, pain is severe and is often accompanied by nausea and vomiting. The pupil size is mid-dilated and fixed (will not constrict with light) and the cornea has a "steamy" or hazy appearance. Primary treatment: a drug that constricts the pupil (e.g., pilocarpine) is used. The iris is pulled toward the midline and the drainage system is opened. The definitive treatment is surgery.

8. *Iritis* is *inflammation of the iris*. The cause is often unknown. Eye pain is present, the pupil is constricted and redness is present surrounding the cornea (limbal injection). Frequently both eyes are involved. Primary treatment consists of dilation of the pupil with a parasympathetic blocking agent such as atropine. If an infection is present, an antibiotic is given.
9. *Nearsightedness (myopia)*. The eyeball is longer than normal and *light rays focus in front of the retina*. Primary treatment: eyeglasses with a concave lens (Fig. 8-9).
10. *Nystagmus* is *repetitive jerky movements of the eyes,* seen in disorders of the inner ear, brain stem, cerebellum and in certain drug toxicities.
11. *"Pink-eye" (conjunctivitis)*, the most common eye disease, is an *inflammation of the conjunctiva*. The redness is caused by dilation of the conjunctival vessels. Causes include bacterial infection, viral infection, foreign body in the eye and allergy. Primary treatment is directed at treating the cause. A bacterial infection

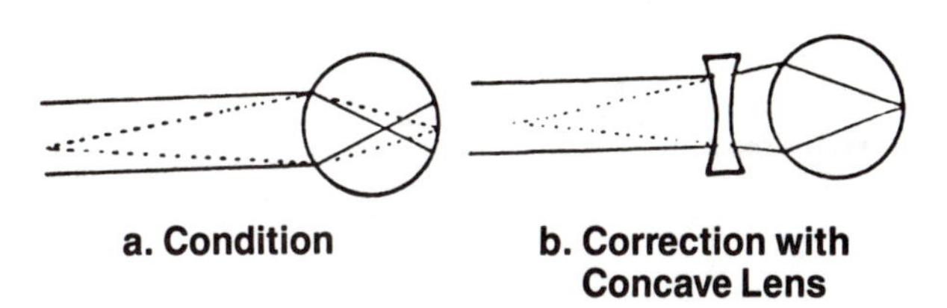

Fig. 8-9 Nearsightedness (Myopia).

responds to an antibiotic.

12. *"Spots" before the eyes (vitreous floaters)*. Dots or filaments sometimes appear and move as the eye or head moves. These are usually benign, and are thought to be *debris of embryologic origin, trapped in the vitreous*. However, a sudden onset of floaters may herald impending retinal detachment.
13. *A sty (hordeolum)* is a *bacterial infection of the eyelid*, characterized by a red, swollen, painful pimple on the eyelid margin. Primary treatment: antibiotic eye drops.

II. *The ear*.

A. The *external ear*.

1. The *outer ear*, or *auricle (pinna)*, consists mostly of cartilage surrounding the external auditory canal. The ear-lobe (lobule) is the fleshy part of the ear (Fig. 8-10)
2. The *external auditory canal* is about 2.5 cm long and terminates at the ear-drum (*tympanic membrane*). Glands in the canal secrete ear-wax (*cerumen*) which captures foreign material and protects the canal epithelium.

B. The *middle ear* or *tympanic cavity* lies in the outer portion of the temporal bone (Fig. 8-11). Three small bones (the *ossicles*) articulate with one another and run from the ear-drum to a small opening, the oval window, the first part of the inner ear. The *Eustachian tube* connects the *tympanic cavity* with the *pharynx*. Its function is to equalize pressure in the cavity. *Importance*: blockage of the Eustachian tube may cause pain, pressure and loss of hearing. It is seen commonly with throat and middle ear infections, although it may occur spontaneously. The tube may be opened by swallowing, yawning, using decongestant nose-drops or oral decongestants. From lateral to medial, the following structures are encountered:

1. *Tympanic membrane*
2. *Ossicles* (malleus, incus and stapes): the malleus attaches to the ear-drum; the stapes inserts in the oval window.

C. The *inner ear (labyrinth)* is a series of fluid-containing channels in the *temporal bones* (Fig. 8-11). When the fluid moves, cells responsible for *hearing* or *equilibrium* are stimulated. The organ of hearing is the *cochlea*, a

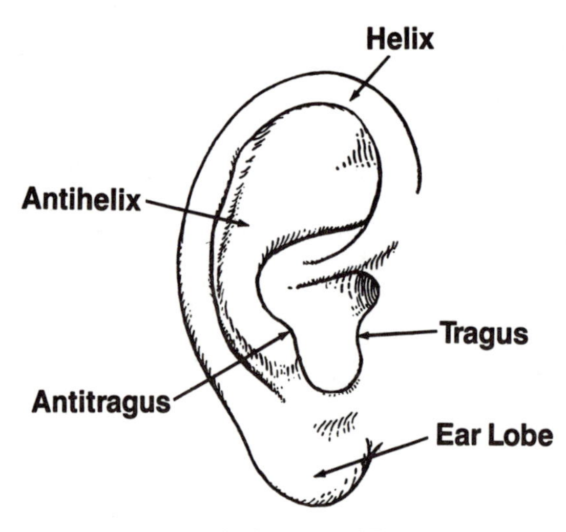

Fig. 8-10 External Ear.

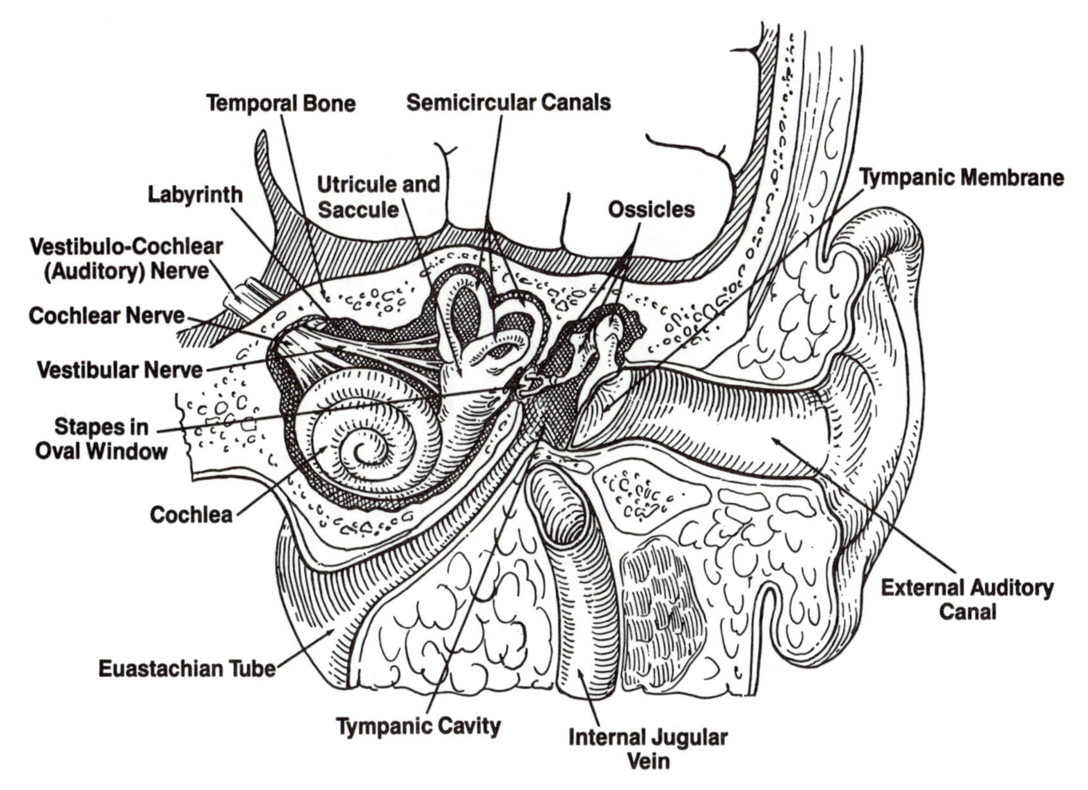

Fig. 8-11 The Auditory Apparatus.

spiral structure running from the oval window to a membrane, the round window. This *spiral organ (of Corti)* contains receptor hair cells that sit on the basilar membrane. The hairs touch the *tectorial membrane*. Hair cells convert fluid waves into nerve impulses. Two bulging areas, the *saccule* and *utricle*, as well as another series of channels, the *semicircular canals*, contain hair cells modified for equilibrium.

D. *Basic physiology of hearing* (Fig. 8-13):
Sound waves stimulate the tympanic membrane to vibrate, setting the ossicles in motion. The stapes creates a fluid wave at the oval window that is transmitted to the round window membrane. The spiral organ is set in motion. Hair cells of the spiral organ are stimulated as they touch the tectorial membrane. They act as tranducers, converting mechanical energy into electrical impulses in adjacent fibers of bipolar cells of the spiral ganglion. Axons of these cells become the cochlear portion of the auditory nerve (vestibulo-cochlear nerve). Nerve fibers pass to the medulla, pons, midbrain, and finally to the *temporal lobes* of the cortex, where the impulses are interpreted as "sound".

E. *Conduction of sound waves* may take place in air (*air conduction*) or through bone (*bone conduction*). Both stimulate the organ of Corti. Air conduction is the normal event of sound waves in air striking the tympanic membrane, setting in motion the ossicles and initiating events in the inner ear. In bone conduction, sound waves move through the skull to the inner ear, bypassing the middle ear. Utilizing principles of bone and air conduction, the tuning fork may be used to distin-

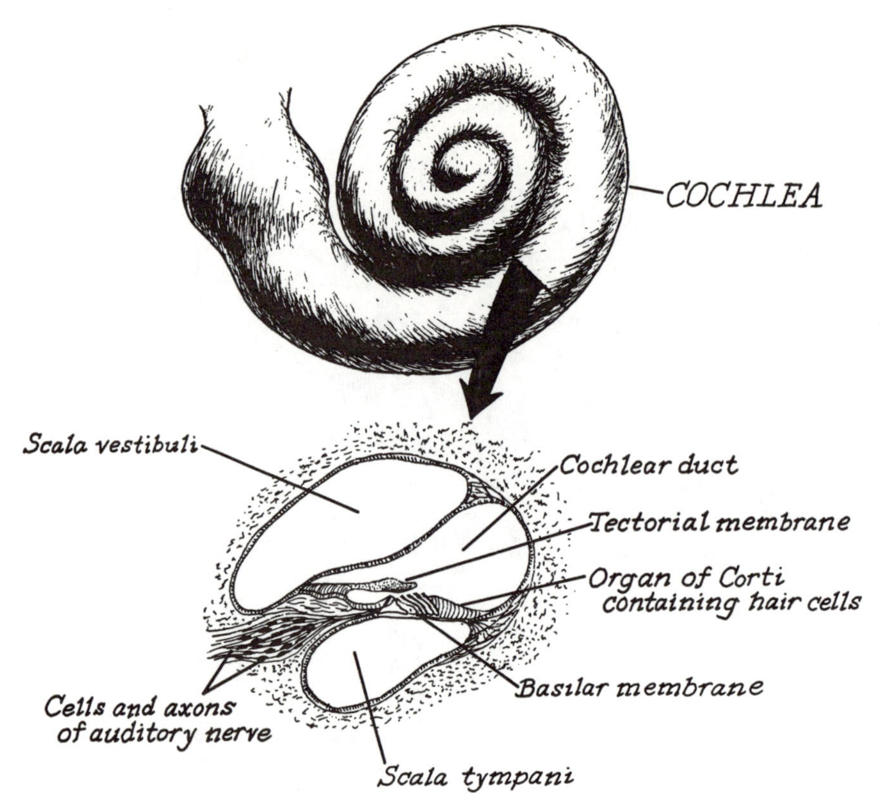

Fig. 8-12 Transverse Section Through Cochlea.

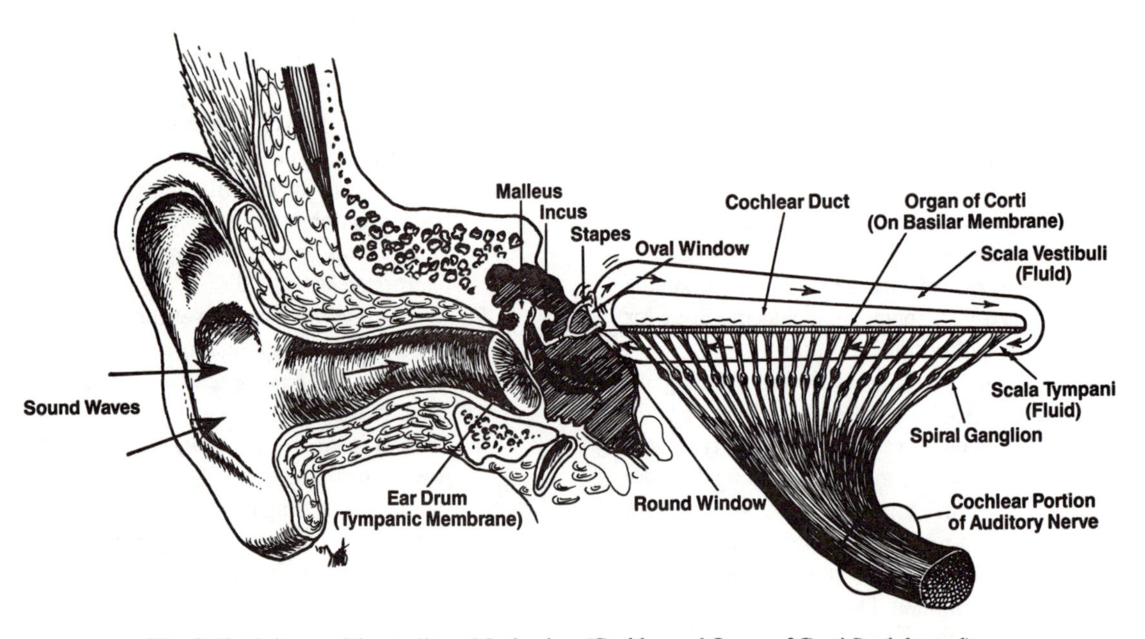

Fig. 8-13 Scheme of the Auditory Mechanism (Cochlea and Organ of Corti Straightened).

guish nerve from conduction deafness (Table 8-1).

F. *Basic physiology of equilibrium.* Rotary movement moves fluid in the semicircular canals, stimulating sensory cells. Hair cells in the utricle and saccule are sensitive to acceleration and deceleration, gravity, the position of the head, and eye movements (Fig. 8-11). Fibers from the three areas unite to form the *vestibular part* of the auditory nerve. Some fibers synapse with ocular and cerebellar neurons. *Importance*: disorders of the inner ear may cause *nystagmus* (back and forth movement of the eyes), *vertigo* (the sensation of rotation in the absence of rotation), *nausea* and *vomiting* due to ocular and *brain stem* connections.

G. *Common disorders of the ear.*

1. *Deafness* can be of two types, conduction or nerve deafness (Table 8-1):

   a. Common causes of *conduction deafness*:

      1) Impacted wax, or a foreign body in the canal.
      2) Otosclerosis: immobility of the stapes in the oval window.
      3) Perforation of the ear-drum.

   b. Common causes of *nerve deafness*:

      1) Drugs such as streptomycin, gentamycin, kanamycin, neomycin.
      2) Congenital
      3) Prolonged exposure to loud noise (e.g., machinery, "hard rock").
      4) Cochlear atrophy in the elderly.

   c. The *hearing aid* is a sound amplifier, and is useful in many cases of nerve deafness and in some cases of conduction deafness. In some instances of nerve deafness (which has usually been considered permanent), *cochlear implants* are successful because it is the hair cells that are faulty, and not afferent fibers of spiral ganglion neurons that touch the hair cells. The implants bypass the hair cells, and stimulate the afferent fibers, which propagate the nerve impulses. A new type of hearing aid that uses *bone conduction* may be implanted in the temporal bone behind the ear.

Table 8-1: Nerve vs. Conduction Deafness

	WEBER TEST: base of tuning fork on top of skull	*RINNE TEST:* base of tuning fork placed on mastoid process until subject no longer hears it, then fork held in air next to ear
NORMAL	Heard equally on both sides	Hears vibration in air after bone conduction is over
CONDUCTION DEAFNESS (one ear)	Sound louder in diseased ear because masking effect of environmental noise is absent on diseased side	Does not hear vibrations in air after bone conduction is over
NERVE DEAFNESS (one ear)	Sound louder in normal ear	Hears vibration in air after bone conduction is over (but a much shorter time than in normal)

2. *Labyrinthitis* is an *inflammation of the inner ear* (the vestibule). The cause may be a viral infection, a drug, or it may be unknown. Some drugs toxic to the inner ear are aspirin (usually reversible), streptomycin and gentamycin (may be permanent).
   a. *Signs and symptoms*.
      1) Vertigo (most common sign)
      2) Ataxia (from cerebellar connections)
      3) Nausea and vomiting
      4) Tinnitus (ringing in the ears) because of involvement of the cochlea.
   b. *Meniere's disease* (paroxysmal labrinthine vertigo, endolymphatic hydrops) consists of intermittent attacks of vertigo, nausea, hearing loss and tinnitus, particularly in older males. The cochlear duct is dilated with fluid. The cause is unknown.
   c. *Primary treatment* for labyrinthitis is symptomatic, including a drug for vertigo such as meclizine (Antivert, Bonine).
3. *Middle ear infection (otitis media)* is an *infection of the tympanic cavity*. It commonly occurs from an untreated throat infection (movement of bacteria up through the Eustachian tube). Signs and symptoms are pain, decreased and usually muffled hearing, and a red or distorted tympanic membrane. The condition is often accompanied by fever. Primary treatment is an antibiotic, an oral decongestant, and decongestant nose drops to open the Eustachian tube.
4. *"Swimmer's Ear" (otitis externa)* is an *infection of the auditory canal*. Predisposing factors are swimming in chlorinated water and cleaning the ears with cotton swabs. Children possess an inventive genius about the variety of objects that can be inserted into the ear. Often there is great pain, particularly with gentle tugging on the earlobe. Primary treatment consists of removal of the foul object, and the instillation of antibiotic ear-drops in all cases.

-NINE-

THE HEART AND VESSELS

The cardiovascular system is a transport system bringing nutrients to and returning waste-products from the tissues. The heart pumps oxygen-rich blood via the arteries, arterioles and capillaries to the tissues. Venules and veins bring oxygen-poor blood back to the heart, which shunts it to the lungs to be oxygenated via the pulmonary circulation. Carbon dioxide is eliminated by the lungs, and other waste products are eliminated by the kidneys. Lymph returns some substances that have leaked from the capillaries into the tissues back into the venous system. The exchange of nutrients and waste-products in tissues involves the capillaries (discussed in next chapter) (Fig. 9-1).

I. *The heart*:

A. *Structure*. The heart is a large muscle pump lying in an oblique position behind the sternum, somewhat to the left (Fig. 9-2). The posterior portion is the *base*, and the lower tip is the *apex*. The small upper chambers are the *atria*, and the large lower chambers are the *ventricles*. Coronary vessels run in grooves between the atria and ventricles and between the two ventricles (Figs. 9-3, 9-4).

1. *The pericardium* is a serous sac surrounding the heart (see Fig. 1-5).

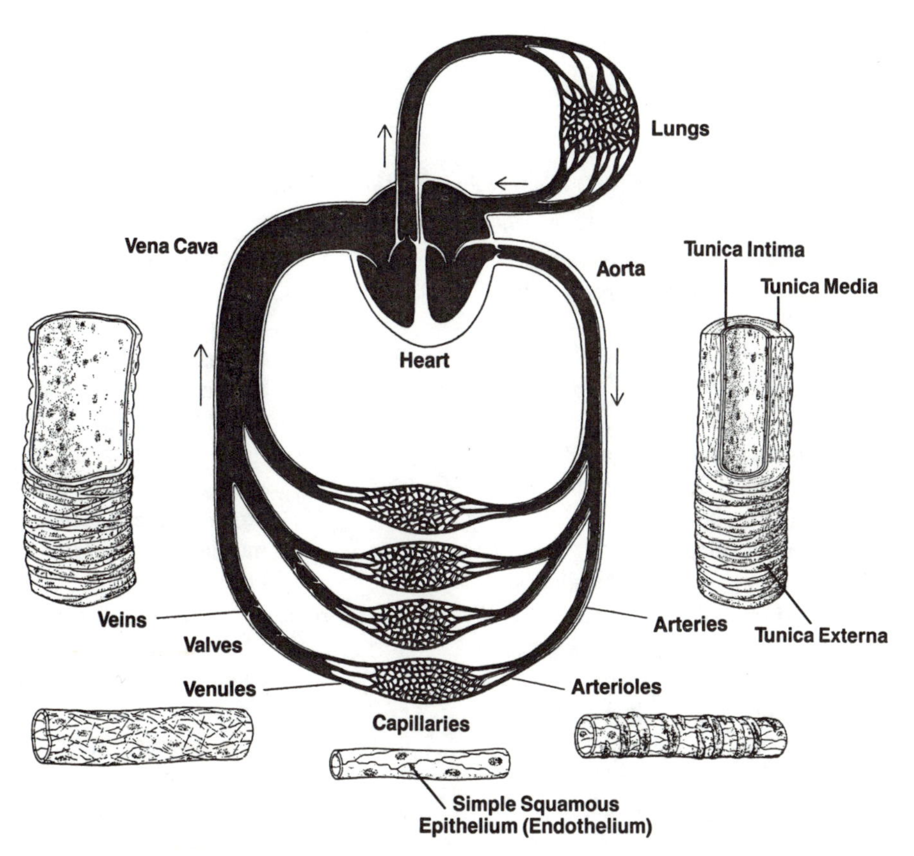

Fig. 9-1 Overview of Circulatory System and Structure of Vessels. (Reproduced and modified, with permission, from Wood, "The Venous System", Scientific American, 1/68).

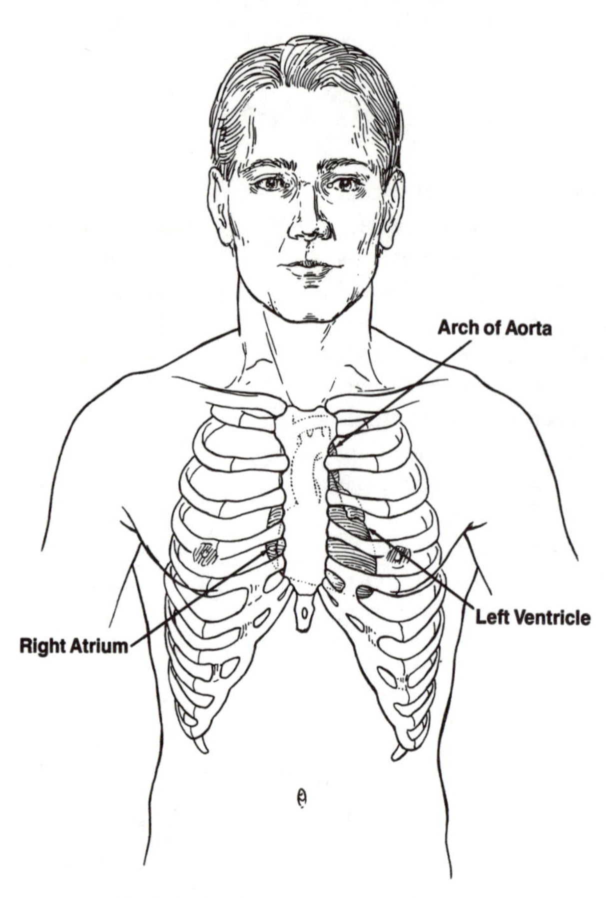

Fig. 9-2 Surface Projection of the Heart.

a. *Visceral pericardium (epicardium)* — inner portion lying directly on the heart
b. *Parietal pericardium* — outer portion

2. *The myocardium,* or heart muscle, forms the bulk of the heart
3. *The endocardium* is the thin inner lining of the heart
4. *Chambers* (Fig. 9-5):

Chambers	
a. *right atrium* b. *left atrium*	thin-walled, separated by a thin *interatrial septum*
c. *right ventricle* d. *left ventricle*	thick-walled, separated by a thick *interventricular septum*

5. *Valves*:
   a. The *semilunar valves (aortic and pulmonary)* prevent the backflow of blood to the ventricles.
   b. *The atrioventricular (AV) valves (tricuspid and mitral)* prevent the backflow of blood into the atria. These valves are under high pressure and are held in place by tendinous cords (*chordae tendinae*) attached to inner musculature (*papillary muscles*).
6. *Vessels* (Figs. 9-3, 9-4):
   a. The "*great vessels*" are those entering or leaving the superior portion of the heart.
      1) Pulmonary trunk — carries oxygen-poor blood to lungs for gas exchange.
      2) Aorta — carries oxygen-rich blood to the general circulation.

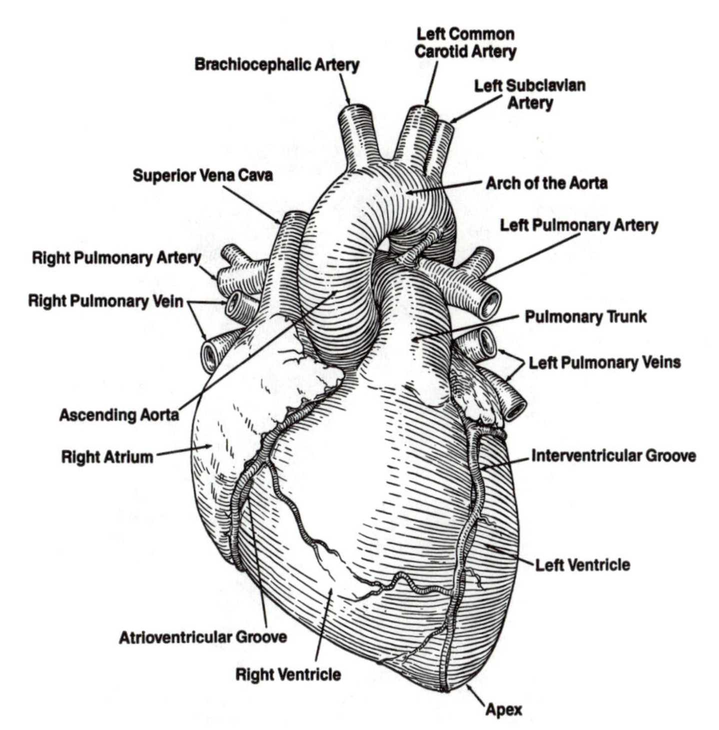

Fig. 9-3 Heart, Anterior Aspect.

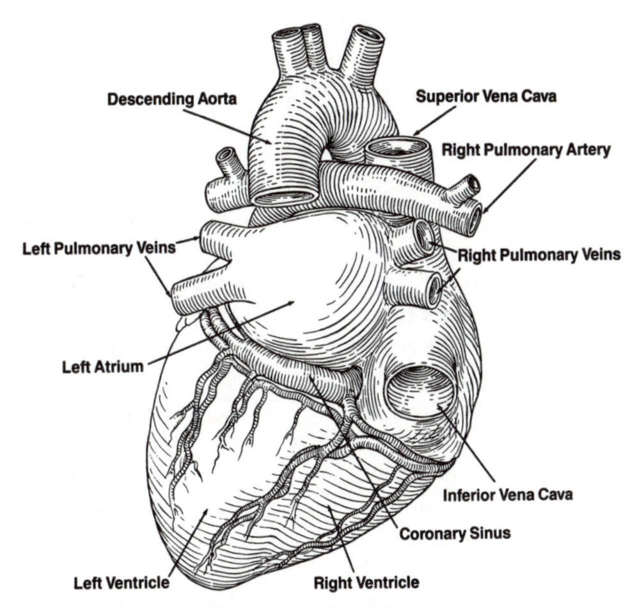

Fig. 9-4 Heart, Posterior Aspect.

3) Superior vena cava — brings oxygen-poor blood to the right atrium from the upper venous circulation.

b. *Other vessels:*

1) Inferior vena cava — brings oxygen-poor blood to the right atrium from the lower

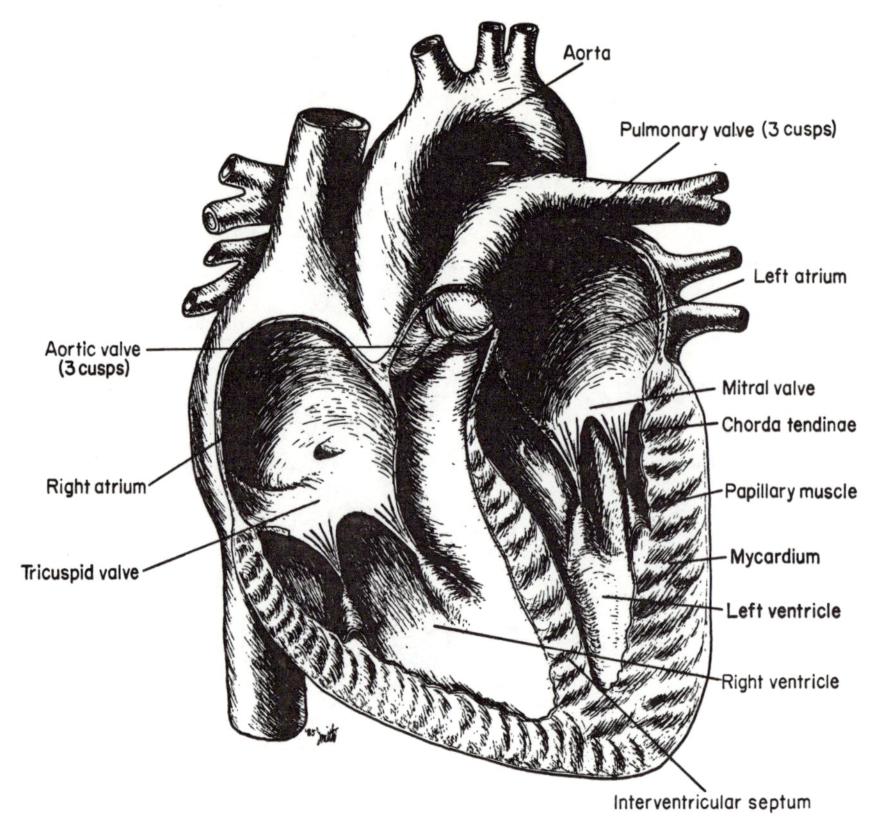

Fig. 9-5 Chambers and Valves of the Heart.

venous circulation.

2) Pulmonary veins — 4 veins lying posteriorly that bring oxygen-rich blood to the left atrium from the lungs.

B. *Circulation of blood in the heart* (Fig. 9-6):

1. *The superior and inferior vena cavae* bring oxygen-poor blood to the *right atrium*.
2. Blood passes from the right atrium through the *tricuspid valve* to the *right ventricle*.
3. From the right ventricle, blood passes through the *pulmonary valve* to the *pulmonary circulation*.
4. The *pulmonary circulation* is the *pulmonary trunk*, (which divides into the *left and right pulmonary arteries* and carries oxygen-poor blood to both lungs), and the *pulmonary veins* which bring oxygen-rich blood to the *left atrium*.
5. Blood passes from the left atrium through the *mitral* (*bicuspid*) valve to the *left ventricle*.
6. From the left ventricle, blood is pumped through the aortic *valve* to the *aorta* and then to the *general circulation*.

C. *Blood supply of the heart*. The *coronary arteries* originate from the *ascending* aorta (Fig. 9-7):

1. The *right coronary artery* curves around the right ventricle to the posterior part of the heart. It supplies much of the *right ventricle*, the *posterior portion of the left ventricle*, the *SA node*, the *AV node*, and the *Bundle of His*.
2. The *left coronary artery* branches quickly into the *circumflex* and *anterior descending* (anterior in-

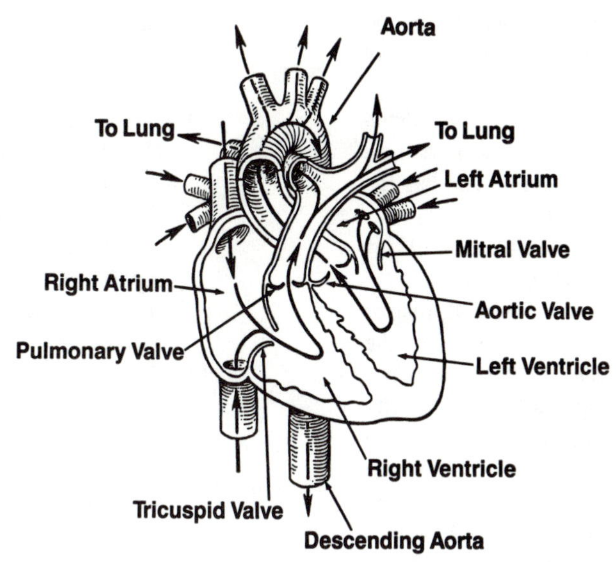

Fig. 9-6 Blood Flow in the Heart.

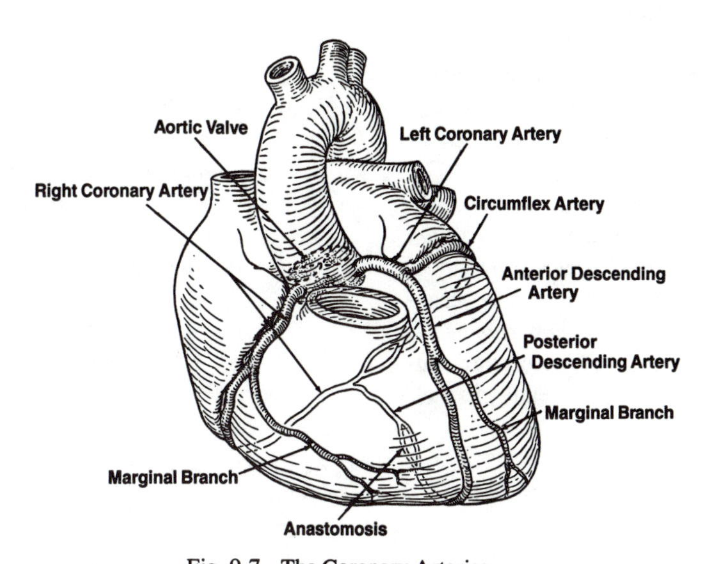

Fig. 9-7 The Coronary Arteries.

terventricular) arteries. The circumflex swings to the left and supplies the *lateral part of the left ventricle*. The anterior descending supplies the *anterior part of the left ventricle* and most of the interventricular *septum*. Most heart attacks involve the anterior or inferior parts of the left ventricle.

D. *Innervation of the heart*. The heart has a *built-in rhythmicity*. It does not require nervous stimulation to contract. If removed from the body, the heart will depolarize and contract for a long time. Sympathetic and parasympathetic nerves alter the rate and force of contraction.

1. *Stimulation of sympathetic nerves* releases norepinephrine which stimulates *beta-1 receptors* on heart muscle. This causes an increase in heart rate (*chronotropic*

effect), and an increase in the force of contraction (*inotropic effect*).

2. *Stimulation of parasympathetic fibers* (from the *vagus nerve*) releases acetylcholine and has the opposite effect.

E. *The conduction system of the heart and the electrocardiogram (EKG):* Certain heart muscle fibers depolarize faster than others and constitute the *conducting system* of the heart. Contraction of the heart takes place shortly after depolarization. Depolarization is recorded by means of electrodes placed on various parts of the body, and the image is shown on an oscilloscope, or recorded on paper (the electrocardiogram).

1. *The electrical impulse generated in heart muscle* (Fig. 9-8):
   a. During *depolarization, sodium ions rapidly enter cells* through channels termed "*fast channels*".
   b. During *repolarization* (as potassium leaves the cell), *calcium ions enter cells slower* by way of channels termed "slow channels", or "*calcium channels*". The SA and AV nodes are activated by the calcium channel. *Calcium ions then initiate contraction of heart muscle* (see also Figs. 1-11, 5-2).

2. *The conducting system* (Fig. 9-9):
   a. *The sino-atrial node (SA node)* is an area of muscle in the upper right atrium, and is the first portion of the conducting system to depolarize, setting the heart-rate at about 70 beats per minute (the *pacemaker*).
   b. The wave of depolarization spreads from the SA node throughout the atria (*atrial depolarization*), which is the first wave of the EKG, the *Polar wave.*
   c. The *atrioventricular node (AV node),* which lies in the right posterior portion of the interatrial septum, depolarizes and the wave spreads down the interventricular septum (*AV bundle* or bundle of His, and the *right and left bundle branches*) to the *Purkinje system* in the ventricles.
   d. The rest of the ventricle depolarizes (the *QRS wave complex*) and contraction takes place.
   e. The *T-wave* on the EKG repre-

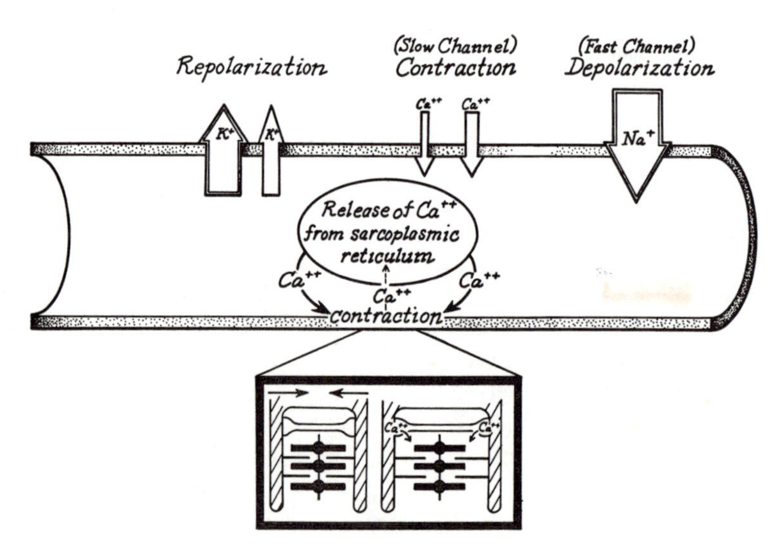

Fig. 9-8 The Heart Muscle Impulse.

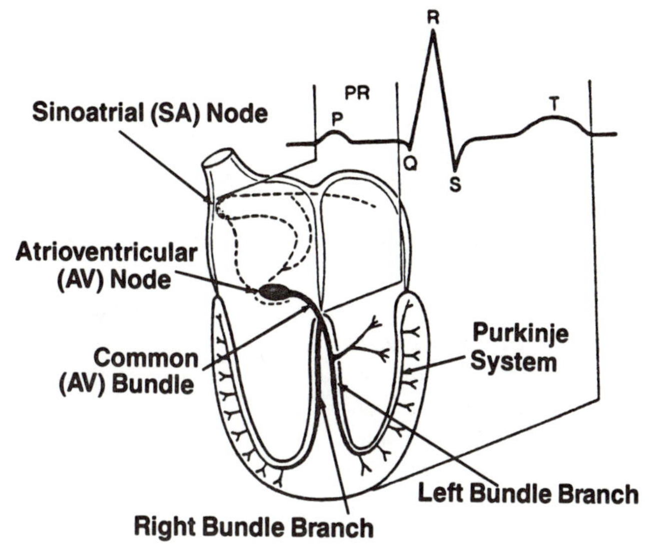

Fig. 9-9 The Conducting System of the Heart. (Reproduced, with permission, from Abridged Textbook of Advanced Cardiac Life Support, American Heart Association, 1983).

sents *ventricular repolarization*. The wave for atrial repolarization is masked by the QRS complex.

F. *The heart as a pump*. The right ventricle pumps blood to the lungs via the *pulmonary arteries*, and the left ventricle pumps blood to the general circulation via the *aorta*. The thin-walled atria pump blood into the ventricles.

1. *The cardiac cycle* is the sequence of events in one heart-beat:
   a. *Diastole*: relaxation of the ventricles during filling.
   b. *Atrial systole*: contraction of the atria, moving blood into the ventricles.
   c. *Ventricular systole* (or *systole*): contraction of the ventricles.
2. *Heart sounds* are heard with a stethoscope. Phonetically, the sound is "lub"-"dup". The *first sound* (lub) is *closure of the atrioventricular (AV) valves* at ventricular systole. The *second sound* (dup) is the *closure of the semilunar valves* at the beginning of diastole (Fig. 9-10).
3. There are about *70 cardiac cycles per minute (the heartrate)* in the average person.
4. The *cardiac output* is the amount of blood ejected by the left ventricle per minute, and represents the

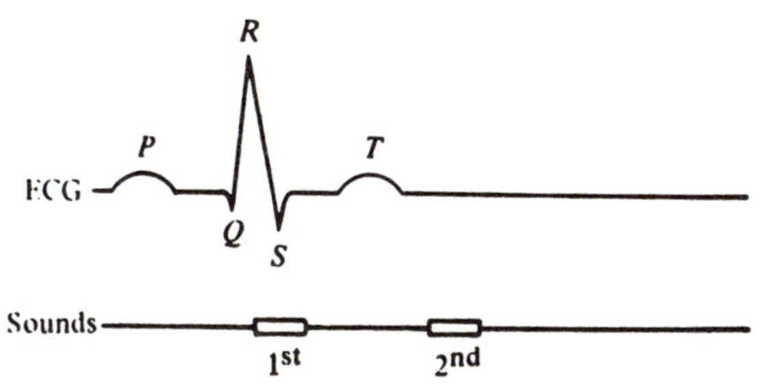

Fig. 9-10 Heart Sounds and the Electrocardiogram. (Reproduced, with permission, from Wiley, Human Physiology, Certified Medical Representatives Institute, 1969).

stroke volume (the amount of blood ejected per beat) times the number of beats per minute (the *heart-rate*). *Example*: if the stroke volume is 90 ml, and the heart-rate 70, what is the cardiac output? *Answer*: 6300 ml, or 6.3 liters per minute. Under normal conditions, 5-6 liters is an average output. During exercise it may rise to 20 or more liters. Since cardiac output depends on both stroke volume and heart-rate, a change in either can affect the body. In a "fit" person, *long term exercise enlarges the heart and the stroke volume increases, thus decreasing the number of beats necessary for normal cardiac output*. Many athletes have heart-rates of 45 to 50. This represents an efficient machine. In an "unfit" person, an increase in heart-rate attempts to compensate for the low stroke volume. Consequently, a fatal arrhythmia may occur during sustained physical activity.

G. *Auscultation (listening with the stethoscope) of the heart*:
   1. The *point of maximum impulse (PMI)*, is usually heard best at the left *5th interspace* (mitral area). The PMI is sometimes called the *apical impulse* (impulse at the apex of the heart) (Fig. 9-11).
   2. A *murmur* is the sound of blood flowing through an *abnormal opening*. It is heard in *valves that are damaged and do not open properly (narrowing, or stenosis)*, and in valves that *do not close well*, letting blood back up through the valve (*regurgitation, or insufficiency*). A congenital abnormal opening in children also produces a murmur. Murmurs that occur after the first heart sound (closure of the AV valves, S-1) are *systolic*. Those occurring after the second heart sound (closure of the semi-lunar

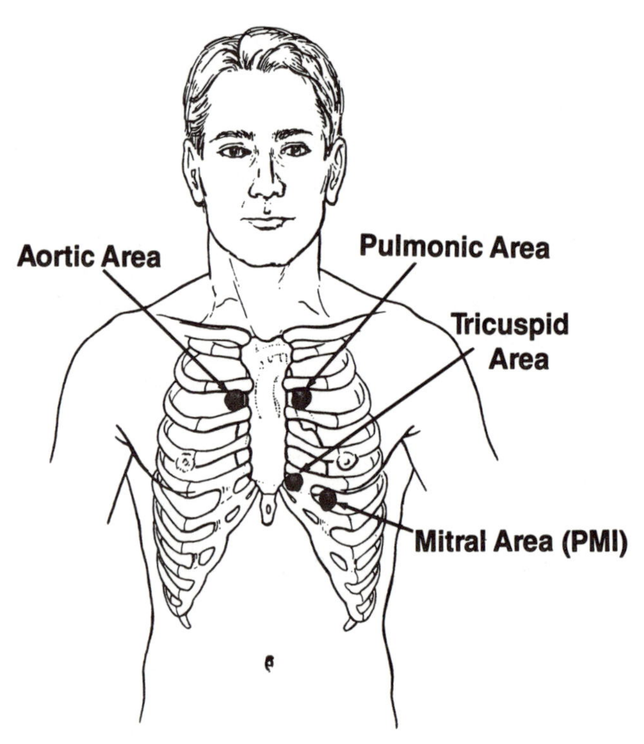

Fig. 9-11 Location of Valve Sounds on Chest Wall.

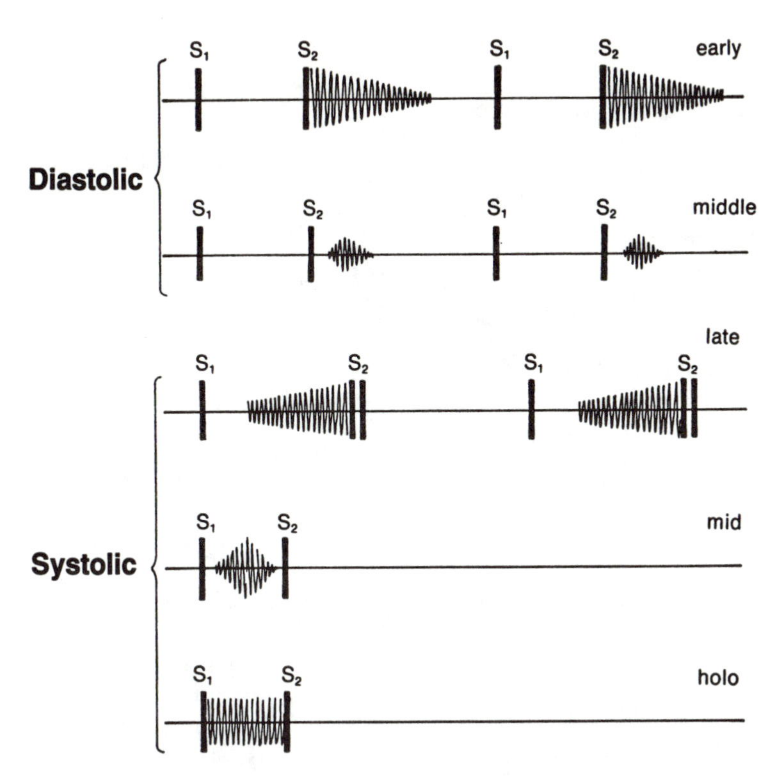

Fig. 9-12 Heart Murmurs. (Reproduced, with permission, from Tilkian, Understanding Heart Sounds and Murmurs, Saunders, 1979).

valves, S-2) are *diastolic* (Fig. 9-12).

II. *Arteries* transport oxygen-rich blood from the heart to the tissues. Arteries are thick-walled to tolerate the high pressure from heart contraction. Smaller vessels are arterioles. The extensive network of *arteriolar branching* is important because *sympathetic nerves* innervating the smooth muscle layer control the calibre of the vessels, thus the *blood pressure*. When the vessel constricts, the pressure goes up; when it relaxes, the pressure goes down. The structure of an artery and vein is shown in Fig. 9-1.

A. *Tunica externa* (adventitia) is the outer *connective-tissue* layer.
B. *Tunica media* is the middle *smooth-muscle* layer, innervated by sympathetic nerves.
C. *Tunica intima* is the inner layer of *simple squamous epithelium* (endothelium).
D. *Main arteries of the head and neck* (Figs. 9-13, 9-14):
The *arch of the aorta* gives rise to three arteries (from right to left): the brachiocephalic, the left common carotid and the left subclavian. The subclavians supply the upper extremities. The arteries of the brain are discussed in chapter 6.
1. The *brachiocephalic* (innominate) artery, a short artery, becomes the *right common carotid* and the *right subclavian.*
2. The *common carotid arteries* branch at the level of the upper part of the thyroid cartilage to become the *external and internal carotid arteries*. *Importance*: The common carotid is an important *pulse-taking artery,* and when damaged may cause a *transient ischemic attack*.
3. The *internal carotid artery* sup-

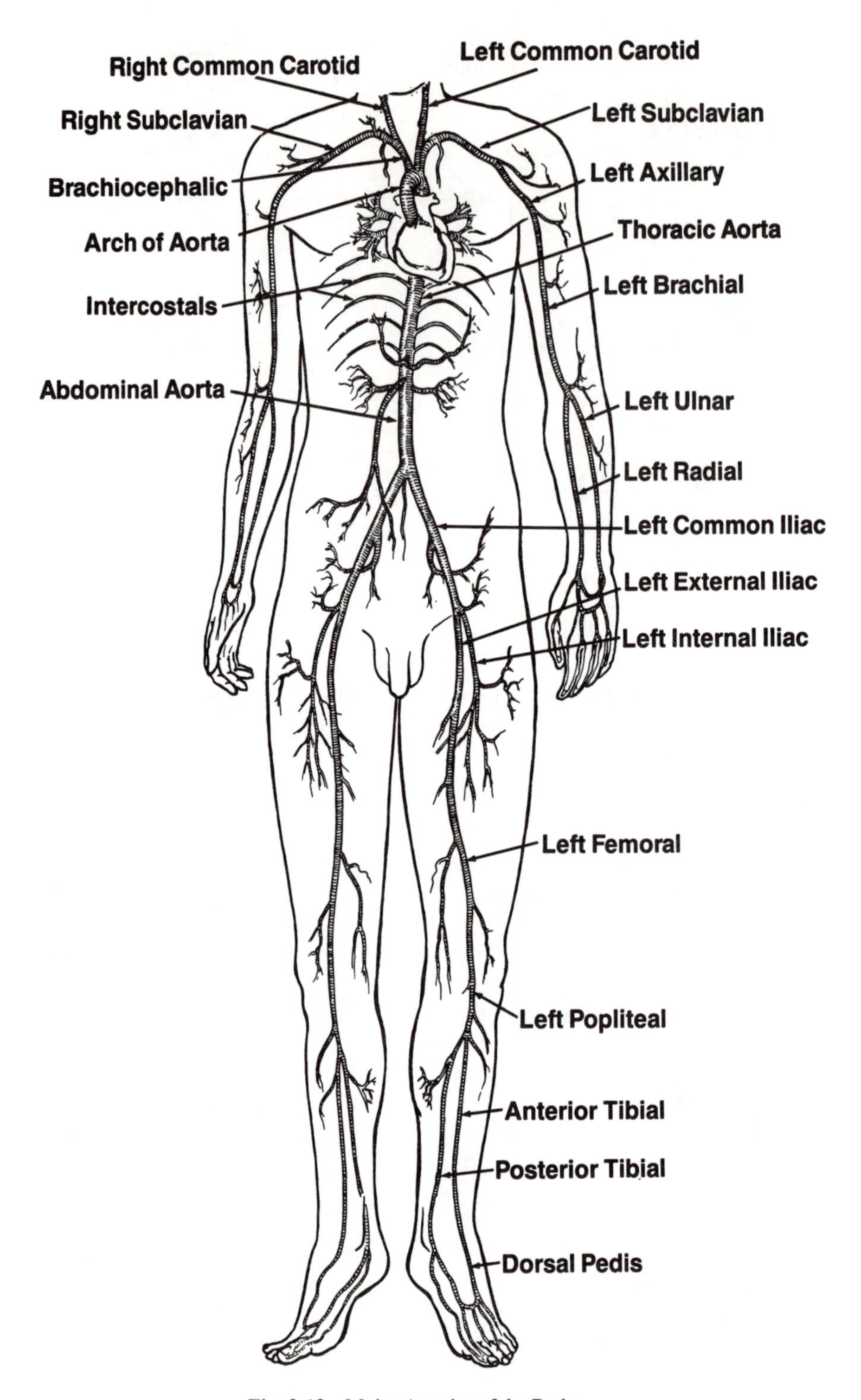

Fig. 9-13 Major Arteries of the Body.

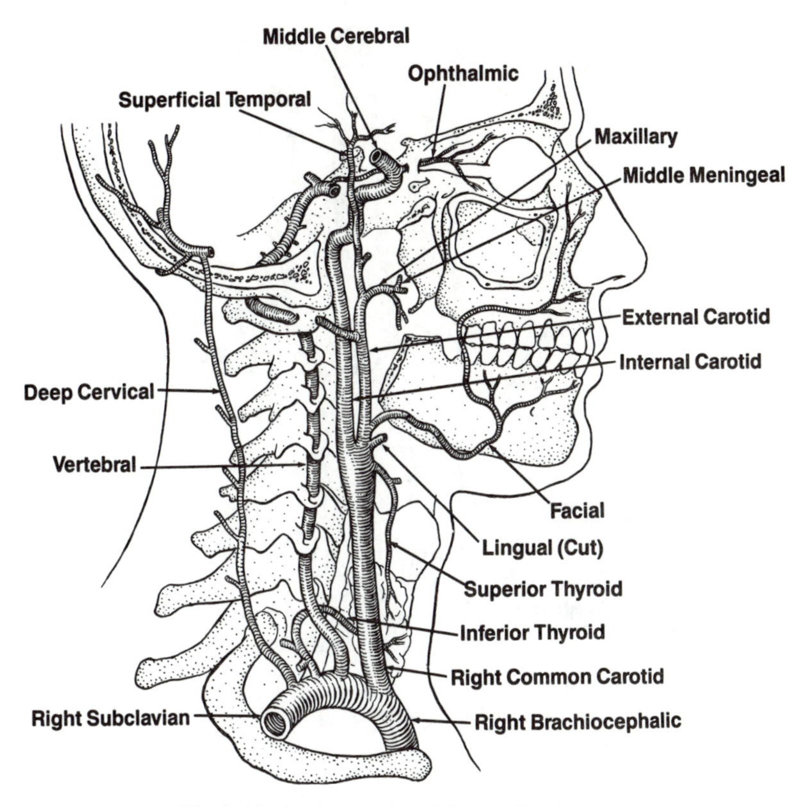

Fig. 9-14 Main Arteries of the Head and Neck.

plies the brain.

4. The *external carotid artery* supplies the face, head and neck. The *superficial temporal artery* is the cranial termination of the external carotid. *Importance*: The superficial temporal is a *pulse-taking artery* (superior and anterior to the ear).
5. The two *vertebral arteries* become the *basilar artery*, which help supply the brain.

E. *Main arteries of the upper extremity* (Fig. 9-13):

1. The *subclavian artery* becomes the *axillary artery* at the clavicle.
2. Near the head of the humerus, the axillary artery becomes the *brachial artery*. *Importance*: the brachial artery is the *main blood-pressure taking artery*. It is also a *pulse-taking artery*.
3. The brachial artery divides at the elbow region into the radial and ulnar arteries.
4. The *ulnar artery* lies deep and medial.
5. The *radial artery* lies more superficial and lateral. Both arteries communicate in the hand by two deep anastomoses, a *superficial and deep palmar arch* (Fig. 9-15). *Importance*: the radial artery is a common *pulse-taking artery* at the wrist (thumb side), and is used in *arterial puncture for blood-gas analysis*.

F. *Main arteries of the trunk* (Fig. 9-16):

1. *The chest (thorax)* (Fig. 9-17):
After supplying the head, neck and upper extremity, the aorta descends posteriorly as the *thoracic aorta*, sending branches to the intercostal muscles (right and left *intercostal*

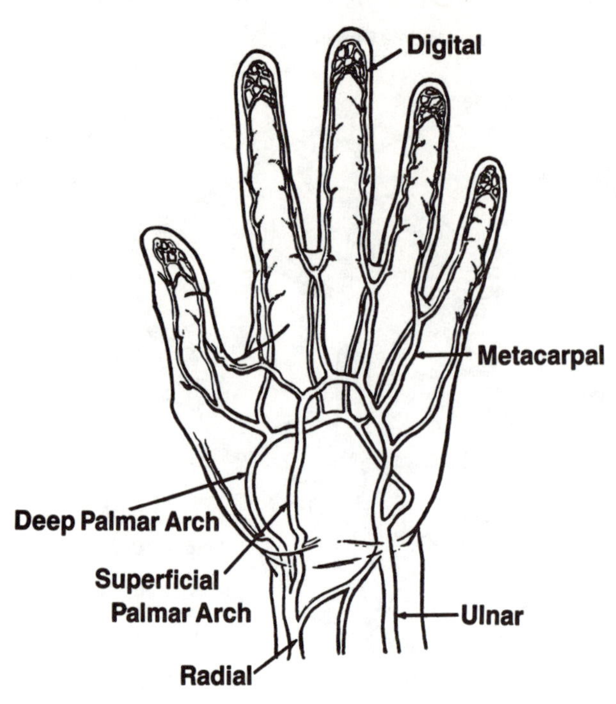

Fig. 9-15 Arteries of the Hand.

arteries). The intercostal arteries anastomose anteriorly with the left and right *internal thoracic arteries*. *Importance*: if the aorta is damaged, the intercostal muscles (breathing muscles) may still receive a blood supply by way of the internal thoracic arteries.

2. *The abdomen* (Fig. 9-16):
   When the thoracic aorta penetrates the diaphragm, it is the *abdominal aorta*, and supplies the abdominal organs. The *main branches of the abdominal aorta* (from cranial to caudal) are:
   a. *The celiac trunk* supplies the stomach, spleen and liver (left gastric, splenic and hepatic arteries).
   b. *The superior mesenteric artery* supplies the small intestine and half of the colon (ascending and first half of the transverse).
   c. The two *renal arteries* supply the kidneys.
   d. The two *testicular (or ovarian) arteries* supply the gonads.
   e. The *inferior mesenteric artery* supplies the remaining half of the transverse, and all of the descending and sigmoid colon.
   f. The abdominal aorta then divides into the left and right *common iliac arteries*. The common iliac divides into:
      1) The *internal iliac artery* supplying the pelvic organs, and
      2) the *external iliac artery*

G. *Main arteries of the lower extremity* (Fig. 9-13):
   1. The *external iliac artery* becomes the *femoral artery* after passing under the inguinal ligament. The femoral artery lies superficially at the femoral triangle, then descends posteriorly through the adductor muscles (adductor canal) (see Fig. 16-25). *Importance*: the femoral artery is an important *pulse-taking artery*. It is used extensively for

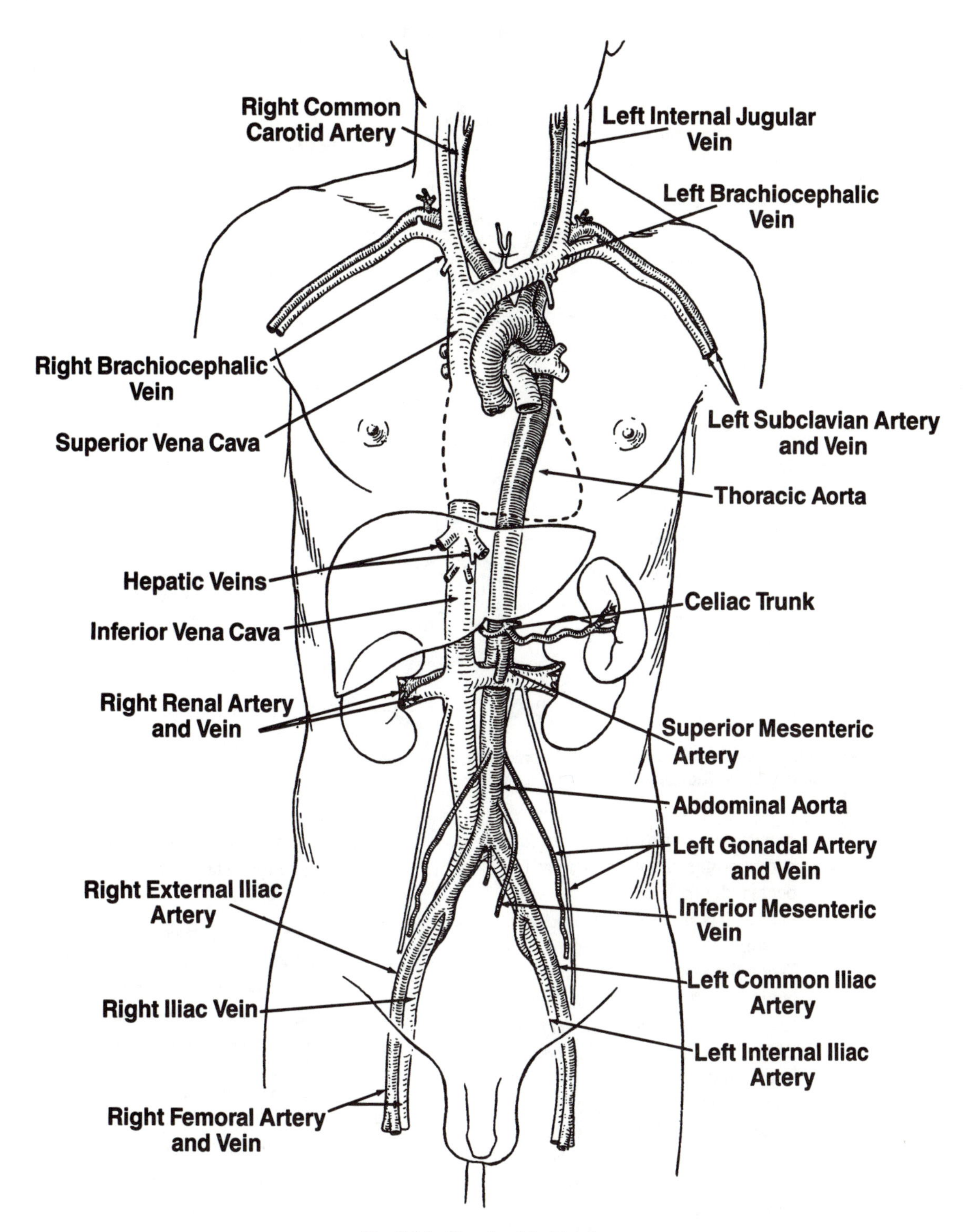

Fig. 9-16 Vessels of the Trunk.

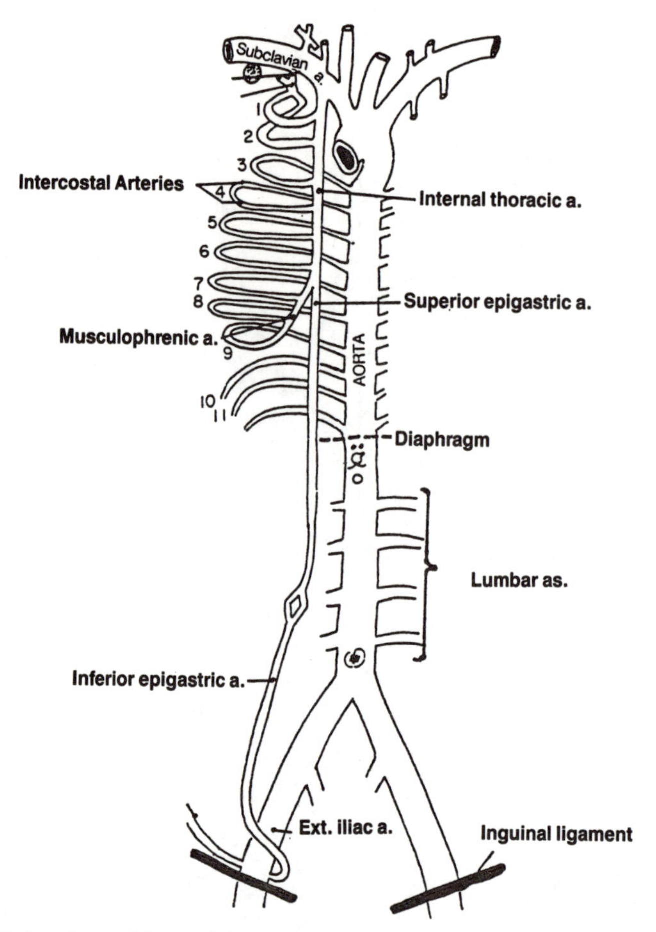

Fig. 9-17 Intercostal Arteries and Internal Thoracic Artery. (Reproduced, with permission, from Goldberg, Clinical Anatomy Made Ridiculously Simple, MedMaster, 1984).

monitoring a patient in *cardiac arrest*, since the carotid may be occluded, particularly in the elderly. It is also used for *blood-gas determinations*.

2. When the *femoral artery* emerges behind the knee at the popliteal region, it is the *popliteal artery* (p. 256). This is also a pulse-taking artery, but a poor one. A strong femoral pulse, and a good foot (pedal) pulse makes it unnecessary to search for the popliteal. *Importance:* if the femoral artery is blocked, a segment of saphenous vein may be transplanted between the femoral and popliteal arteries (femoro-popliteal bypass) to supply blood to the leg.

3. The *popliteal artery* divides to become the *anterior and posterior tibial arteries*.

   a. The *anterior tibial artery* becomes the *dorsalis pedis artery* on the dorsal aspect of the foot. *Importance*: this is an important *pulse-taking artery*. Marking the spot of the pulse with an "X" on the foot enables one to find it again without extensive searching (Fig. 9-13).

   b. The *posterior tibial artery* descends behind the medial malleolus. *Importance*: This is also a pulse-taking artery, but usually more difficult to find than the dorsalis pedis.

III. *Veins*, as shown in Fig. 9-1, are thinner and less circular than arteries. *Venules* emerge from capillaries and become veins. Veins transport *oxygen-poor* blood from the tissues back to the heart. 75% of the blood of the body is in the venous system. Veins possess *valves* to prevent the backflow of blood. The *superficial* (*peripheral*) veins are often visible underneath the skin (e.g. veins of the arm and hand) and empty into the *deep veins*, which usually accompany arteries. The *deep veins of the neck and trunk are* the *central veins*. The main central veins are the *subclavian, internal jugular* and *femoral*. These are often sites for emergency catheter insertions.

A. *Major veins of the head and neck* (Fig. 9-18):

1. *Superficial*:
   a. The right and left *external jugular veins* drain blood from the *face, head and neck*, and are visible in the neck. *Importance*: these are prominent if the *heart fails* and blood backs up in the venous system (*jugular venous distention, or JVD*).
   b. Each external jugular vein empties into a *subclavian vein*.
2. *Deep*:
   a. Venous drainage from the *brain* is by way of the *internal jugular veins* (see Fig. 6-7).
   b. Each internal jugular vein joins a *subclavian vein* to form a *brachiocephalic vein*.

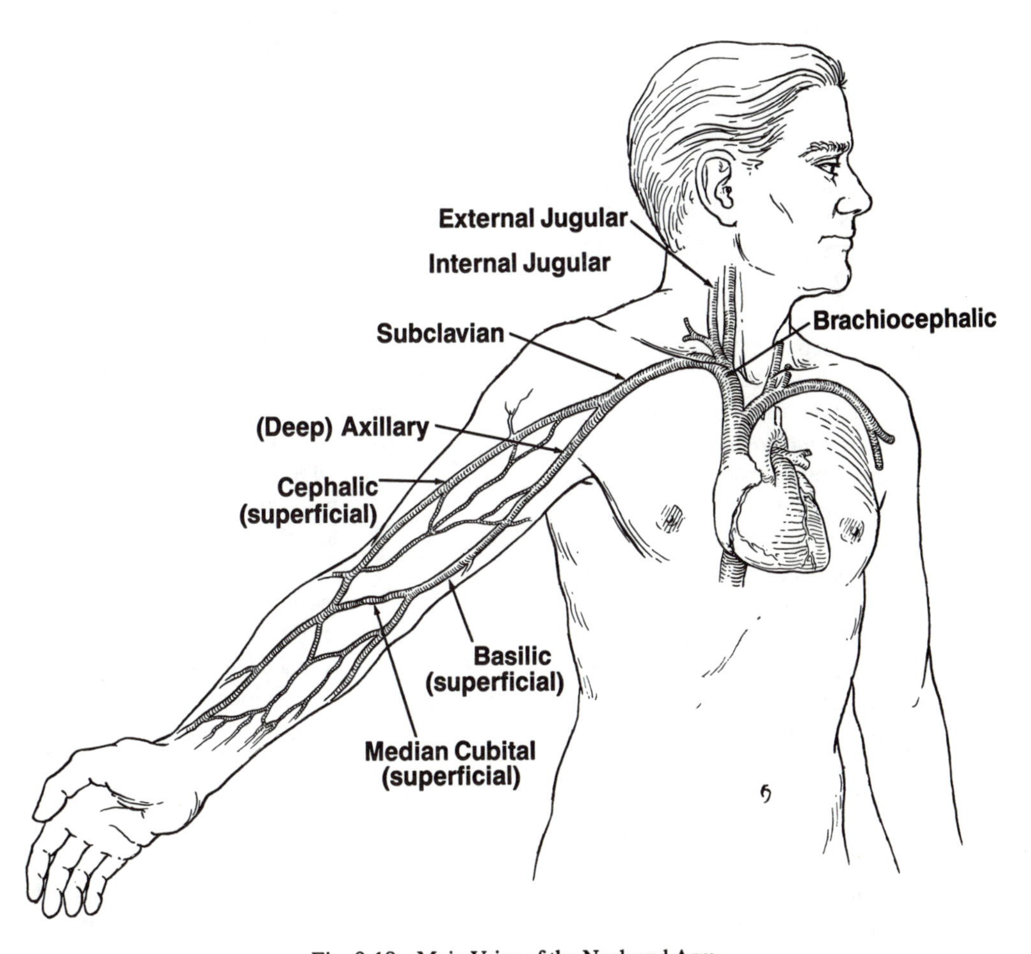

Fig. 9-18 Main Veins of the Neck and Arm.

B. *Main veins of the upper extremity* Fig. 9-18):
   1. Superficial:
      a. Superficial veins originate on the dorsal side of the hand as a *dorsal venous plexus*. They curve around the wrist to the ventral side as the:
      b. *Cephalic vein* which runs along the lateral aspect of forearm and arm, going deep at the deltoid muscle, and the:
      c. *Basilic vein* which runs along the medial aspect of the forearm and arm and goes deep at the biceps muscle.
      d. The *median cubital vein* is an anastomosis between the basilic and cephalic veins.
      e. *Importance*: all are common *venipuncture sites*. Hand veins are preferred over arm veins for intravenous infusions, since the entire arm is thus not immobilized. Catheters are often threaded through superficial veins of the arm into the right atrium (see also Figs. 16-11, 16-12, 16-15).
   2. *Deep*:
      a. The deep veins are formed from twigs in the hand and forearm. Although there is sometimes a short *brachial vein*, essentially the first main deep vein is the *axillary vein*.
      b. The *axillary vein* becomes the *subclavian vein* when it passes under the clavicle.

C. *Main veins of the trunk* (Fig. 9-16):
   1. The *subclavian vein* joins the *internal jugular vein* to become the *brachiocephalic vein*. *Importance*: the subclavian is an important central vein for *intravenous infusion* (IV), central venous pressure measurement, feeding by vein (hyperalimentation) and temporary pacemaker electrode insertion.
   2. Two *brachiocephalic veins* join to become the *superior vena cava* that empties into the right atrium.
   3. The *azygous system*, which lies on the posterior body wall, drains the *intercostal veins*. The azygous vein empties into the *superior vena cava* (Fig. 9-19).
   4. The *inferior vena cava* drains blood from the abdominal viscera into the right atrium (the digestive organs and spleen first drain into the *portal vein*, which empties into the liver. The portal vein and its tributaries are discussed in the section on the liver in chapter 13). *Branches of the inferior vena cava* (from cranial to caudal) (Fig. 9-16) are:
      a. *Hepatic veins* from the liver.
      b. Right and left *renal veins* from the kidneys.
      c. Right and left *testicular (or ovarian) veins* from the gonads.
      d. Two *common iliac veins* form the inferior vena cava.
         1) The *internal iliac veins* drain the pelvic structures.
         2) The *external iliac veins* are the continuations of the femoral veins.

D. *Main veins of the lower extremity* (Fig. 9-20):
   1. *Superficial*:
      a. Superficial veins of the leg begin as a *dorsal venous arch* on top of the foot.
      b. The *greater saphenous vein* ascends medially from the foot up the leg to the thigh and drains into the femoral vein at the femoral region. *Importance*: the greater saphenous veins may become chronically dilated in certain people (*varicose veins*). They may then become inflammed and form blood clots (*thrombophlebitis*). The greater saphenous is a commonly used

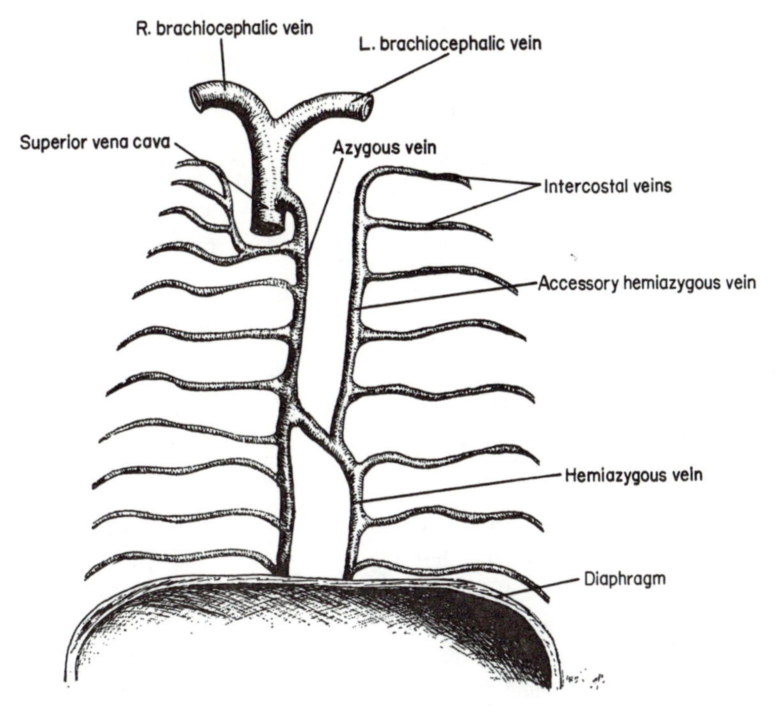

Fig. 9-19 Azygous System. (Posterior Chest Wall).

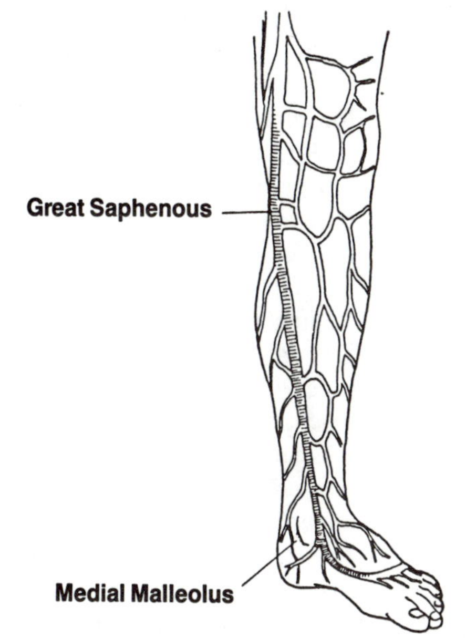

Fig. 9-20 Superficial Veins of the Lower Extremity. (See ref. Fig. 9-9).

vessel in heart bypass surgery.

c. The *lesser saphenous vein* runs laterally from the foot along the gastrocnemius muscle, and drains into the popliteal vein.

2. *Deep*:
   a. The *anterial tibial vein* and
   b. The *posterior tibial vein* drain into the
   c. *Popliteal vein.*
   d. The popliteal vein becomes the *femoral vein*. *Importance*: the deep veins of the leg may also become inflamed (*deep vein thrombosis*). This is a more serious condition than superficial thrombophlebitis. The clot may break off and travel to the heart, then lodge in the lung (*pulmonary embolism*).

IV. *Cardiovascular regulatory mechanisms*:

A. *Blood volume and flow*. The average blood volume is from five to six liters. Blood flow is:
   1. *Fastest in arteries*
   2. *Moderately fast in veins*
   3. *Slowest in capillaries* (because of exchange of nutrients and waste

products between tissues and blood)

B. *The pulse rate — vital sign #2.*
Blood forced into the aorta during systole sets up a pressure wave that travels down the arteries. The wave expands the arterial wall and the expansion is palpable by pressing the artery against tissue, which enables one to measure the *pulse rate.* The rate may be *regular or irregular,* and *strong or weak.* Irregular pulses are commonly found in atrial fibrillation and premature ventricular contractions. A strong pulse is seen in hyperthyroidism, a weak one in shock and myocardial infarction. A heart rate *over 100 is tachycardia; less than 60 is bradycardia* Figs. 9-29, 9-30). The pulse rate at birth is about 120, decreasing to 90 at about age 10. The heart rate rises as the stroke volume decreases in order to maintain normal cardiac output. For instance, in shock, the blood pressure declines and the heart rate rises.

C. *Blood pressure — vital sign #4*:
Pressure is regulated by *sympathetic nerves* to the *arterioles.* Normally, arterioles are in a state of partial constriction (arteriolar tone). *Stimulation of the sympathetic system* causes *arteriolar constriction* and an *increase in blood pressure. Non-stimulation* results in a *decrease in blood pressure.* In hypertension, the sympathetic system is in a state of continuous stimulation, resulting in constant high blood pressure.

1. *Blood pressure* is measured with a *sphygmomanometer,* a cloth-covered rubber bag, wrapped around the arm over the *brachial artery.* A tube connects the inside of the bag with a manometer, containing *air (anaeroid)* or *Mercury.* Blood pressure is highest during contraction of the heart (systole) — the systolic blood pressure. It is lowest when the heart is relaxing (diastole) — the diastolic pressure. Pressure is blocked in the brachial artery by inflating the cuff to about 20 or 30 mm Hg above the point at which the *palpable radial pulse* disappears. As the cuff is deflated with the stethoscope on the brachial artery (the brachial artery is located medial to the biceps tendon on the arm, not in the antecubital fossa!), sounds are heard (the *sounds of Korotkoff). This is the systolic pressure. The diastolic pressure is recorded when the sounds are no longer heard.* The figure is expressed as *systolic over diastolic* in mm Hg (Mercury) (e.g., 140/80). As the vessels become more remote from the heart, the systolic and diastolic pressures merge into one pressure. When the vessels change from arteries to arterioles to capillaries to venules to veins, the pressure gets lower and lower, until in the large veins there may be no pressure at all, or a negative pressure. This is why, when drawing venous blood, one has to pull back on the syringe.
2. Blood pressure in various vessels:
   a. *Highest in arteries: e.g. 120/80 mm Hg*
   b. *Arterioles: about 65 mm Hg*
   c. *Capillaries: about 30 mm Hg*
   d. *Veins: may be 0,* or a negative pressure
3. *The pulse pressure* is the difference between the systolic and diastolic pressures. Example: blood pressure of 120/80. The pulse pressure is 40 mm Hg. If this is multiplied by 1.7 (a conversion factor), this is the *stroke volume.* The *cardiac output* may then be calculated (section I-F-4).
4. *Normal blood pressure range:*
The blood pressure depends on the size of the person. On an average, below 100/60 is usually considered hypotension; above 140/90 hypertension. The average newborn has a blood pressure of 90/

60. At age 15 the average blood pressure is about 120/60. The young adult has a blood pressure of 120/80.

5. *Blood pressure changes under various conditions:*
   a. A *systolic pressure increase* (without diastolic increase) is seen in transient conditions such as anxiety and exercise.
   b. A *widened pulse pressure* may be also seen in hypertension, aortic regurgitation, hyperthyroidism and congenital heart disease. In some cases of hyperthyroidism and aortic regurgitation, the sounds of Korotkoff may persist to O. In this case, the *muffling* of sounds should be used as the diastolic pressure.
   c. *Hypertension* (see end of chapter).
   e. *Hypotension* is a *fall in the systolic and diastolic blood pressure. Profound hypotension results in shock* (see vii, h). *Hypotension exists when the systolic pressure is less than about 90 mm Hg.* It is seen in hemorrhage, heart attack and heart failure. Slow bleeding, as with a peptic ulcer, may cause hypotension. In this case, sudden standing from a sitting or lying position may cause the blood pressure to fall, and fainting may occur (orthostatic hypotension).

D. *Factors affecting venous return:*
Venous return is important because it insures a proper amount of blood in the heart for normal cardiac output. Some factors affecting return are:
   1. *Gravity:* Standing in one position for a long period of time is not good. Blood in the veins below the heart must constantly overcome gravity to get to the heart.
   2. *Respiratory pump:* As one breathes in and out, movement of the diaphragm and intercostal muscles help move blood through the venous system.
   3. *Muscular pump:* Exercise (contraction of muscles) helps push venous blood to the heart.
   4. *Cardiac pump:* The heart functions as a pump. An efficient pump is seen in a "fit" individual (large stroke volume and low heart-rate).
   5. *Valves:* If valves are incompetent, blood backs up in the venous system (e.g., varicose veins).

E. *Brain centers and baroreceptors*:
   1. The *medulla* is the center for the regulation of the heart rate (*cardioinhibitory center*), blood pressure (*vasomotor center*) and respiration (*respiratory center*).
   2. *Baroreceptors* are *stretch receptors* located near the *carotid bifurcation* and in the *arch of the aorta* (Fig. 9-21). They are sensitive to pressure, or stretch. From the carotid area, afferent fibers run via the 9th cranial nerve and from the aortic arch via the 10th cranial nerve to the medulla. When blood pressure increases, the *depressor portion of the vasomotor center* and the *cardioinhibitory center* of the medulla, where the vagus nerve originates, are stimulated. The heart rate slows, sympathetic activity is blocked and vasodilation of arterioles occurs, resulting in a fall in blood pressure. Blood pressure and heart rate are increased if the baroreceptors are not stimulated. *Importance*: these are regulatory mechanisms designed to *stabilize the blood pressure under various physiological circumstances*. They fail in many cases of hypertension, heart and vessel disease. Clinically, baroreceptor stimulation is applied in the emergency setting as therapy for paroxysmal atrial tachycardia (PAT): manually mas-

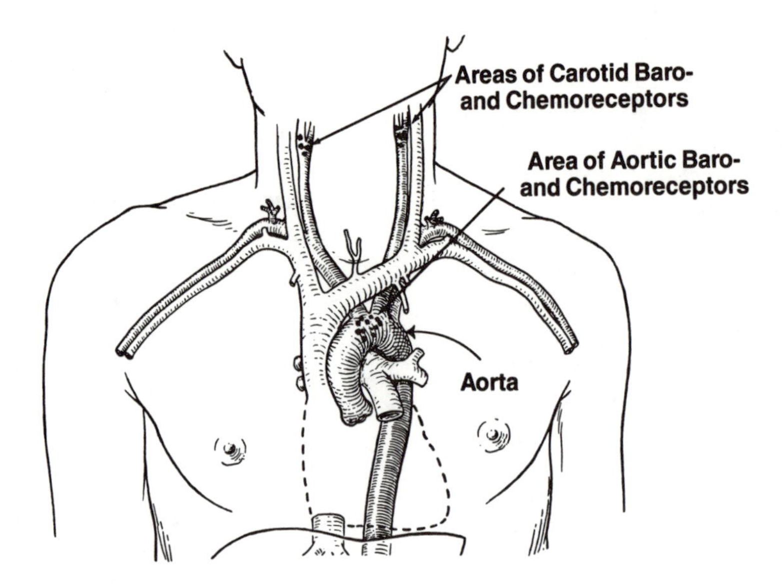

Fig. 9-21 Baroreceptor and Chemoreceptor Areas.

saging the carotid artery on one side near the jaw angle decreases the heart rate.

V. *Arteriosclerosis is "hardening of the arteries".* Although several causes exist for the hardening, the most common and important is the deposition of fatty plaques (atheromas) in medium and large arteries, "*atherosclerosis*". Often the two terms are used synonymously. Atheromas are often found near bifurcations, where high pressure and flow relationships exist. Smaller arteries, arterioles and veins are usually spared. Increased arteriosclerosis is seen in countries with high fat diets, particularly saturated fatty acids and cholesterol. The "*lipid theory*" is the most widely accepted theory on the origin of arteriosclerosis.

A. *Lesion formation* (Fig. 9-22).
Lipids enter endothelial cells of the *tunica intima*. Dead cells coalesce, and create masses of debris. Cholesterol and calcium precipitate and crystalize, forming an atheromatous plaque. Damage may extend into the *tunica media*. Hard plaques prevent the muscle layer from dilating or con-

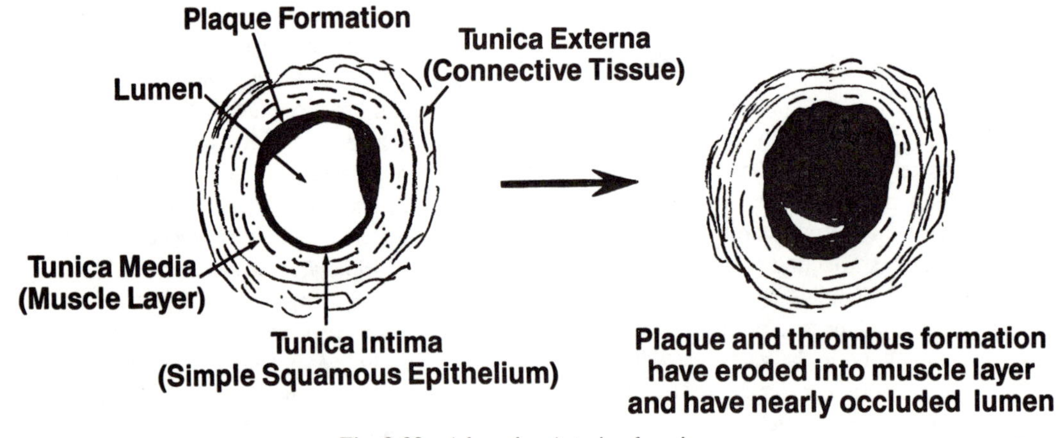

Fig. 9-22 Advancing Arteriosclerosis.

stricting. The vessel thus becomes rigid rather than elastic. Since blood coagulates when exposed to a foreign surface, it may clot around the plaque (*thrombus formation*). The vessel becomes more narrowed, with a cycle of plaque and thrombus formation. Finally a completely occluded lumen results in death of tissue supplied by the vessel (*infarction*). The weak wall may also balloon out (aneurysm) and rupture.

B. *Arteriosclerosis causes angina, myocardial infarction (MI), TIA, stroke, aneurysm and peripheral vascular disease (PVD).* Important arteries involved, in order of frequency, are:
   1. Lower aorta and iliacs (aneurysm, PVD)
   2. Proximal coronaries (angina, MI)
   3. Femorals, popliteal and thoracic aorta (PVD, aneurysm)
   4. Internal carotids (TIA, stroke)
   5. Cerebral vessels (TIA, stroke)

C. *Aerobic conditioning is frequently able to*:
   1. *Enlarge the artery*, thus enabling blood to pass. In some cases, the arteries may increase 2 to 3 times the normal size.
   2. *Develop collateral circulation* around the occlusion. Small new vessels are created (see chapter 5).

VI. *Beta-blocking and calcium-channel blocking agents slow the heart rate* (negative chronotropic effect) and *depress myocardial contractility* (negative inotropic effect). In addition, *beta-blockers lower the blood pressure*, and *calcium-channel blockers dilate the coronary arteries*.

A. *Beta-blockers*. As discussed in chapter 7, beta-l blockers block adrenergic receptors in heart muscle. Drugs that perform this function are used in the treatment of *coronary heart disease* (angina, myocardial infarction), *supraventricular tachycardia* and *hypertension*. Some are non-selective; that is, they also block some beta-2 receptors. They thus may precipitate bronchospasm in a susceptible individual: i.e. propranolol (Inderal), nadolol (Corgard), timolol (Blocadren) and pindolol (Visken). Some are more beta-1 (cardio-) selective: i.e. metoprolol (Lopressor) and atenolol (Tenormin).

B. *Calcium-channel blockers reduce the influx of calcium.* The electrical impulse and contraction of the heart depend on calcium flow into the cell via the slow calcium channel. Calcium blockers also depress the SA and AV nodes. Drugs that perform this function are used in the treatment of *coronary heart disease*, *supraventricular tachycardias* and *atrial arrhythmias*. Verapamil (Isoptin, Calan) is used in the treatment of supraventricular tachycardia. Nifedipine (Procardia) is used in the treatment of coronary artery disease.

VII. *Common disorders of the cardiovascular system:*

A. *Incidence of main heart diseases in the US:*
   1. *Arteriosclerotic heart disease* — 80%
   2. *Hypertensive heart disease* — 9%
   3. *Rheumatic heart disease* — 2%
   4. *Other* — 9%

B. *Terms:*
   1. An *aneurysm* is an *abnormal dilation of a part of a vessel.* Usually the result of arteriosclerosis, it may also be congenital. The most common site is the aorta below the renal arteries. The second most common site is the thoracic aorta (Fig. 9-23).
   2. An *embolus* is a *mass traveling in the bloodstream.* It may be a blood clot (thrombus), air, fat, tumor cells or clumps of bacteria.
   3. An *infarct* is *an area of dead tissue caused by a loss of blood supply.*
   4. *Ischemia* means a *temporary deficiency of blood supply* to a tissue.

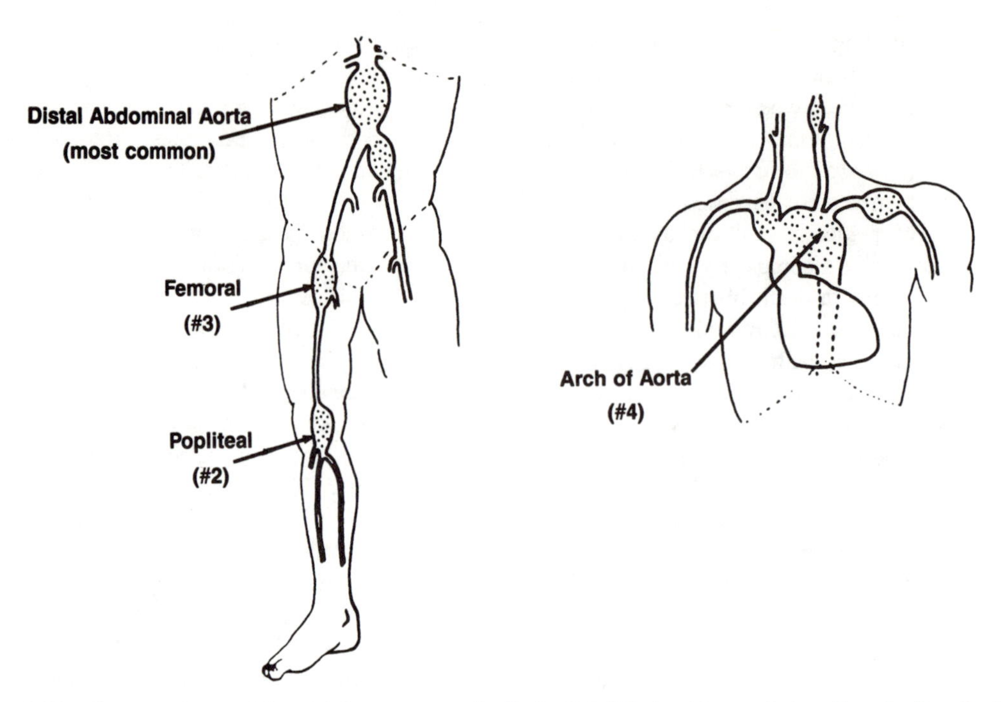

Fig. 9-23 Common Areas of Arterial Aneurysms in the Body. (Modified, with permission, from Ludbrook, An Introduction to Surgery, Academic Press, NY, 1971).

5. An *occlusion* is the *blockage of a vessel*.
6. A *thrombus* is *an intravascular clot*.

C. *Cardiac arrest* may occur from many conditions. The most common is a *heart attack* (myocardial infarction). The heart may stop completely (*asystole*), it may fibrillate (*ventricular fibrillation*), or there may be an EKG tracing without a pulse (*electromechanical dissociation, EMD*). Untreated, all three conditions are rapidly fatal. The majority of arrests are ventricular fibrillation (Fig. 9-24).

1. *In the hospital setting* a cardiac monitor will show the rhythm. The inotropic and chronotropic drug *epinephrine* is given during *cardiopulmonary resuscitation (CPR)*. If the person is in ventricular fibrillation, he is defibrillated. The antiarrhythmic drugs bretylium or lidocaine are given. Asystole may respond to the parasympathetic blocking agent atropine, and both asystole and EMD may respond to the sympathetic agent isoproterenol (Isuprel).
2. *In the non-hospital setting*, a cardiac arrest may be confused with choking. However, if the person can make a sound, he is not choking. In the *un-witnessed arrest* — finding someone with no heartbeat or respirations — there is a problem. Brain damage occurs in 4-5 minutes. CPR may be successful, but with brain death. It may be proper under these circumstances not to resuscitate. In a *witnessed arrest, CPR should be begun* in the following manner:
   a. Tell one or two people to call the paramedics or fire department.
   b. Tilt the head far backward.
   c. If the person is not breathing, clear the mouth of vomitus, dentures, etc.
   d. Blow air into the mouth, holding the nose closed. Inflate the lungs 3-5 times.

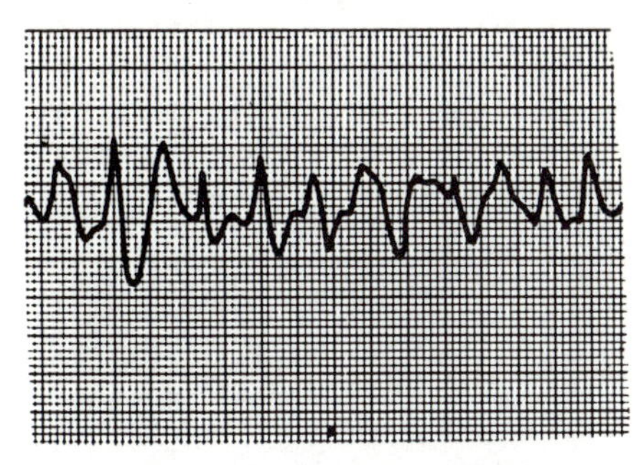

a. Ventricular Fibrillation

(Reproduced, by permission, of Merck Sharp & Dohm, Division of Merck & Co., Inc.).

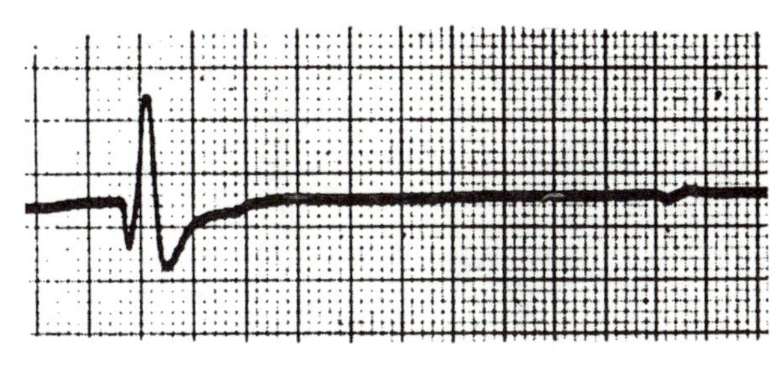

b. Asystole

(Reproduced, with permission, from Conway, *A Pocket Atlas of Arrhythmias,* Year Book Medical Publishers, 1974).

Fig. 9-24 Fatal Arrhythmias.

e. Check for a carotid pulse. If a pulse is absent, then:

f. Give a short, quick blow of the fist in the mid-sternum. This generates a small electrical impulse and may be effective in restoring a beat. Check if it is.

g. If there is no pulse, CPR is begun, depressing the lower third of the sternum with the heel of the hand about two inches each second (being careful not to depress the xyphoid and thus the abdomen, which does no good) (Fig. 9-25).

h. After 15 sternal compressions, give 2 quick, deep lung inflations. If another person who knows CPR is available, have him blow air into the mouth after every 5th sternal compression.

i. Trade off at some point if you are tiring. If you are by yourself, do not stop until the paramedics arrive.

D. *Coronary artery disease* is most often caused by *arteriosclerosis and superimposed plaque and thrombus formation in one or more of the coronary arteries. Partial occlusion* causes the transient chest and arm pain of *angina. Total occlusion* causes the crushing or squeezing pain of *infarction.* The six most common sites of occlusion are shown in Fig. 9-26. Table 9-1 lists some differences between angina and infarction (the table represents average circumstances; e.g., exertion may trigger angina, infarction, or both).

1. *Angina pectoris* (angina: to suffocate; pectoris: chest) is *chest pain, or discomfort, brought on by exertion, emotion or cold weather.* It is usually relieved by rest or nitroglycerin (NTG). The pain often radiates to the arms, neck or jaw. During an attack, the EKG typically shows S-T depressions or *T-wave inversions (ischemia).* The condition may be replicated by a *treadmill stress test.* Confirmation of occlusion is accomplished by injecting a dye into the coronary vessels and determining the amount of blockage (*coronary angiography*).

   a. *Medical treatment* consists of the use of *nitrate vasodilators* such as NTG and skin patches or ointments. They decrease blood flow to the heart and improve coronary blood flow. *Beta-blocking* and *calcium-channel blocking agents* are useful (see

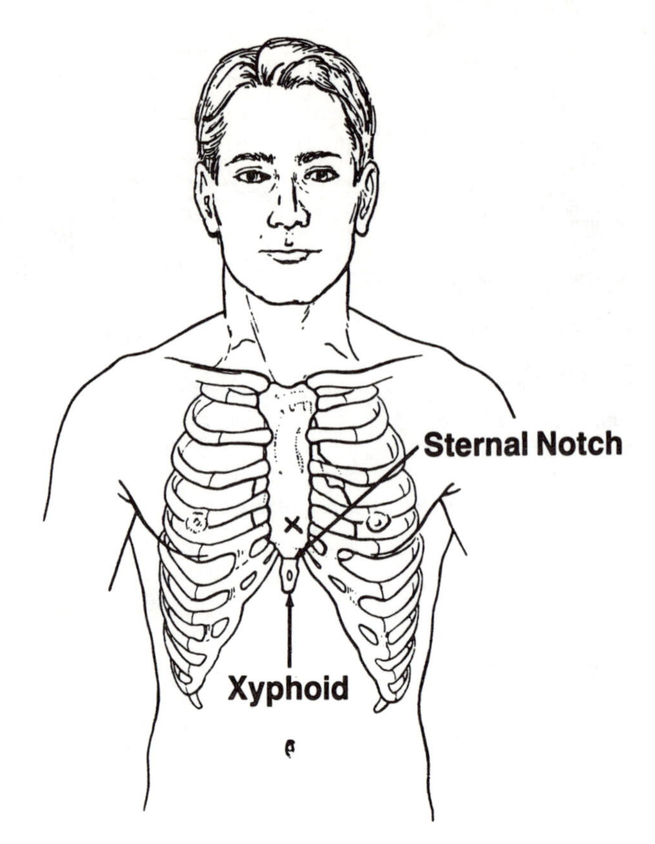

Fig. 9-25 Hand Placement in CPR at "x".

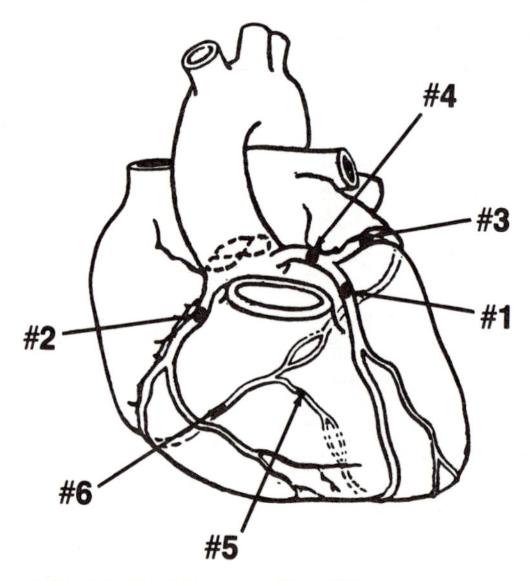

Fig. 9-26 Six Most Common Sites of Coronary Occlusion.

Table 9-1: Acute Myocardial Infarction (MI) vs. Angina Pectoris

	MI	*Angina*
Signs and Symptoms:	sudden squeezing or crushing retrosternal chest pain at rest; symptoms may mimic indigestion	chest pain, or jaw or arm pain brought on by exercise
	unrelieved by vasodilators such as nitroglycerin (NTG), or resting	relieved by NTG, or resting
Lab:	elevated cardiac enzymes (CK, LDH)	normal cardiac enzymes
EKG:	abnormal EKG with ST segment elevations, T wave inversions, or Q waves	usually normal, but may have ST changes during attack
Treatment:	always coronary care unit of hospital, because major complication is arrhythmia and shock	hospitalization seldom necessary, stress test useful, possibly coronary angiography and bypass surgery
	In both cases slow but progressive cardiovascular exercise (jogging, cycling or swimming) is essential	

section VI). Combinations of the three agents are often used.

b. *Surgical treatment* consists of bypassing one or more of the obstructed arteries with grafts from a saphenous vein (e.g., "double", "triple" *coronary bypass*). Another procedure is to maneuver a balloon-tipped catheter through the skin into a coronary artery. The balloon is inflated and dilates the narrowed area (percutaneous transluminal *coronary angioplasty,* PTCA).

2. *Heart attack* (myocardial infarction, MI) is the #1 killer of people in the US and Europe. The hallmark is *substernal crushing or squeezing chest pain,* often accompanied by weakness and cold sweats.
   a. *Laboratory findings.*
      1) When an organ is injured, *enzymes are released* into the bloodstream. The enzymes in the table (*CK, LDH*) are abbreviations for creatine kinase and lactic dehydrogenase, found in several organs, including *muscle.* It is possible to accurately determine whether skeletal or *heart muscle* is involved by the use of *isoenzymes,* more specific chemical forms.
      2) The *EKG* in patients with infarction commonly shows *S-T segment elevations* (acute *injury* to heart muscle), *Q-waves (infarction),* or *T-wave inversions,* as in angina. Q-waves may not be present initially but develop later (Fig. 9-27).
   b. *Primary treatment.* Hospitalization in a *coronary care unit (CCU)* is mandatory. Complications of an infarct include *heart failure, shock and arrhythmias.* Oxygen, morphine and antiarrhythmic drugs are given as needed. In some instances, a *fibrinolytic agent* may be useful (see chapter 10). Beta-blockers and anti-platelet drugs (e.g., aspirin) are sometimes used. Rehabilitation by means of a gradual, progressive exercise program is beneficial.

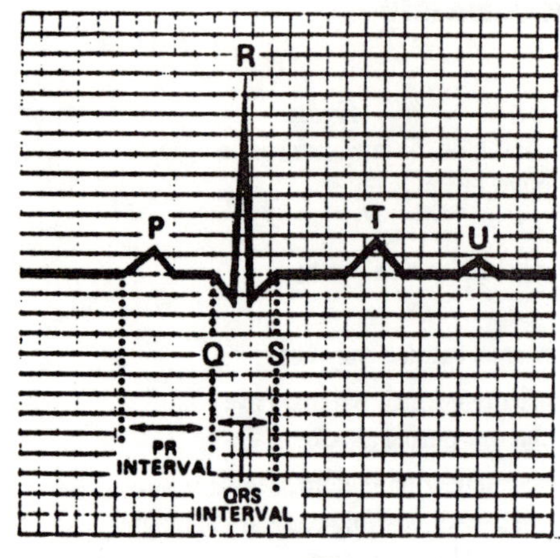

a. The Normal EKG. (See ref., Fig. 9-9).

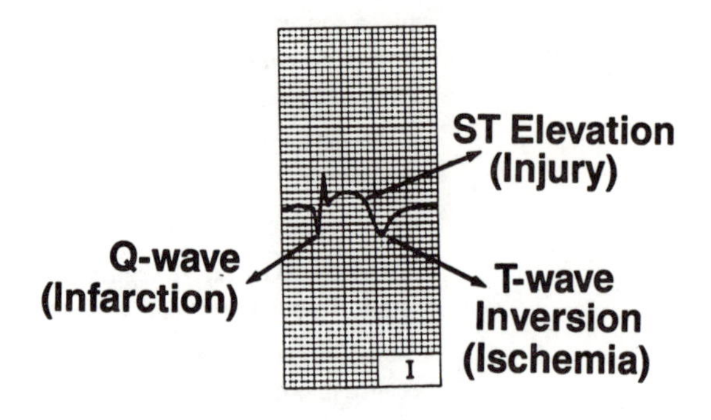

b. The EKG in Myocardial Infarction. (See ref., Fig. 9-24a).

Fig. 9-27 The EKG in a Heart Attack.

E. *Heart failure (congestive heart failure, CHF)* may result from a variety of heart disorders, including myocardial infarction, hypertension or rheumatic heart disease. *When the heart fails as a pump, blood backs up in the vessels returning it to the heart*: the superior and inferior *vena cavae* and the *pulmonary veins*. The backup of blood is *congestion*. Fluid from the bloodstream leaks from the capillaries into the tissues (*edema*). The left or right ventricle may fail, but usually both are involved (Table 9-2). Primary treatment for acute heart failure consists of:
   1. Oxygen
   2. An upright position in bed decreases venous return and blood pressure in the upper portion of the lungs
   3. A diuretic such as furosemide (Lasix) results in fluid loss and a decrease in blood pressure.
   4. Digoxin (an inotropic drug) increases the force of cardiac contraction.
   5. Aminophylline is a bronchodilator.
   6. Morphine decreases venous return by pooling blood in the splanchnic (internal organs) bed.
   7. Nitroglycerine (NTG) sublingually is a vasodilator.

F. *Hypertensive heart disease*. Sustained hypertension leads to hypertensive heart disease. The cause of *continuous sympathetic stimulation and constriction of arterioles is unknown* (*essential*

Table 9-2: Heart Failure

Left-sided:

Pulmonary congestion and edema, resulting in:

1. Dyspnea on exertion (DOE)
2. Fatigue and weakness
3. Cough
4. Rales
5. Orthopnea, and paroxysmal nocturnal dyspnea (PND)
6. Occasional coughing up of blood (hemoptysis)

Right-sided:

General venous congestion and edema, resulting in:

1. Prominent superficial veins. The external jugular vein is distended when the patient is sitting up in bed at about a 45° angle: jugular venous distention (JVD).
2. Dependent edema (pitting edema, particularly in ankles)
3. Liver tenderness and enlargement (hepatomegaly)
4. Splenomegaly

hypertension), although kidney disease and arteriosclerosis play significant roles. The heart enlarges because of the increased work of the left ventricle against arteriolar resistance, and may go into failure. *Hemorrhagic stroke* is an important complication of untreated hypertension. On an average, hypertension exists with a blood pressure above 140/90. The goal of therapy is to try to maintain a normal resting blood pressure below 140/90. A *"stepped-care" regimen* is usually followed, adding the next agent as needed.

1. *Salt restriction* decreases fluid, thus lowering blood pressure.
2. *Weight loss* decreases the resistance against which the heart must pump.
3. Step 1: a diuretic, such as hydrochlorothiazide (HydroDiuril, Moduretic) decreases fluid and lowers blood pressure.
4. Step 2: beta-blockers (see section VI) are widely used in the treatment of hypertension and angina. Sometimes they are used as Step 1 agents. Methyldopa (Aldomet) and clonidine (Catapres) suppress CNS sympathetic outflow. Prazosin (Minipress) is an alpha-blocker.
5. Step 3: hydralazine is a potent vasodilating agent.
6. Step 4: guanethidine (Ismelin) blocks norepinephrine release.

G. *Rheumatic fever is a sequel in some children to an untreated streptococcal throat infection two weeks earlier.* The cause is thought to be an immunological response to streptococcal substances remaining in the body. Common signs and symptoms include fever, joint pain and heart problems. Rheumatic fever may progress to *rheumatic heart disease,* in which the heart valves (particularly the mitral) become deformed. Rheumatic heart disease is the most common cause of valve replacement. A complication of rheumatic valvular disease is the seeding of bacteria on the deformed valve cusps (*bacterial endocarditis*), with high fever, heart murmur and bacteria in the bloodstream. *Primary treatment* is an appropriate antibiotic.

H. *Shock* is an *inadequate blood supply to vital organs*. The main clinical manifestations are low blood pressure — usually below 90 mm Hg systolic *(hypotension),* and an altered mental status (decreased oxygen to the brain). Glucose is poorly utilized, and products of anaerobic metabolism, such as lactic acid, accumulate (see chapter 13).The main types of shock are *hypovolemic, cardiogenic, septic* and *anaphylactic*. Compensatory mechanisms go into effect, such as vasoconstriction of kidney vessels to conserve fluid (reducing urine formation), thirst (indicating the need for fluid), a thready fast pulse (to try to restore cardiac output by increasing the heart rate), and rapid respirations (attempting to increase oxygen in the blood). Obtundation (decreased mental status from decreased oxygen to the brain) is sometimes present and the skin is cool, pale, and clammy. Certain treatment measures are applicable to all forms of shock, such as respiratory support with oxygen, and the infusion of intravenous fluid.

1. In *hypovolemic shock*, the infusion of lactated Ringer's is the solution of choice. It is then replaced by whole blood. Military Anti-Shock Trousers (MAST suit) may be applied to the legs and abdomen to increase blood pressure to more vital areas (Fig. 9-28).
2. The cautious administration of fluid is indicated in *cardiogenic shock. Dopamine* (Intropin) may be added to sustain blood pressure. Dopamine is the catecholamine precursor of norepinephrine, and

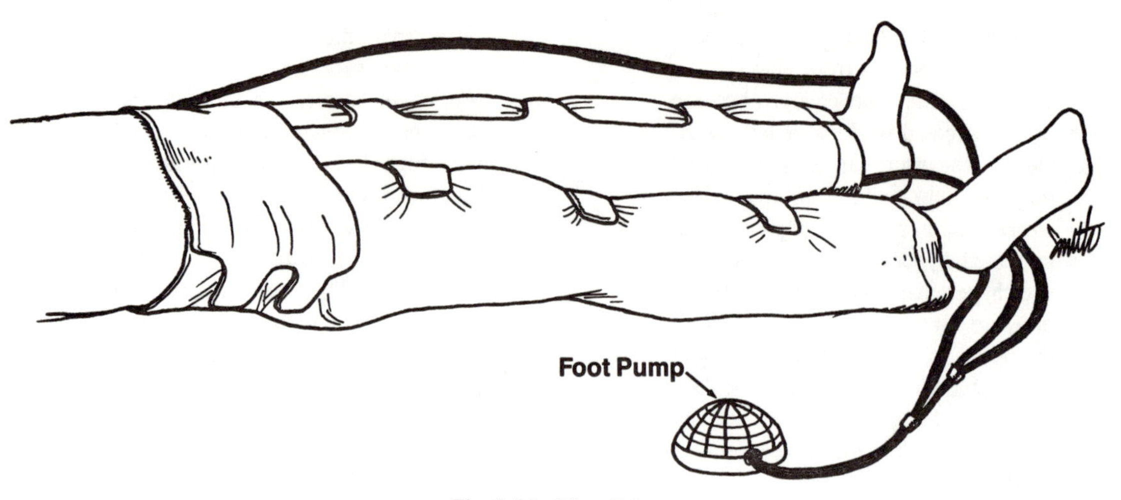

Fig. 9-28 Mast Suit.

exerts a positive inotropic effect on the heart, with little chronotropic effect.

3. *Septic shock* is caused by *endotoxins of gram-negative bacteria.* The toxin causes a decrease in venous tone and permits movement of large quantities of fluid from the blood to the tissues. Treatment of septic shock includes measures as for other forms of shock, but also an appropriate antibiotic.
4. *Anaphylactic shock* is discussed in chapter 2.

I. *Some disturbances of heart rate and rhythm*:

1. In *bradycardia* (heart rate less than 60), primary treatment, when necessary, is the administration of atropine, a parasympathetic blocking agent (Fig. 9-29). Athletes often have heart rates in the 50's.
2. In *tachycardia,* usually no treatment is necessary (Fig. 9-30). In *paroxysmal atrial (supraventricular) tachycardia (PAT),* seen in otherwise healthy young people, the rate ranges from 150 to 250 per minute (Fig. 9-31). Primary treatment involves several important anatomic and physiological principles:
   a. *Stimulation of the vagus nerve slows the heart rate.* The vagal center is located in the medulla (the cardioinhibitory center). Several vagal maneuvers are possible:
      1) Pressure on the eyeballs stimulates the vagus via the 3rd cranial nerve.
      2) Massaging the carotid baroreceptor area (*carotid massage*) on one side of the

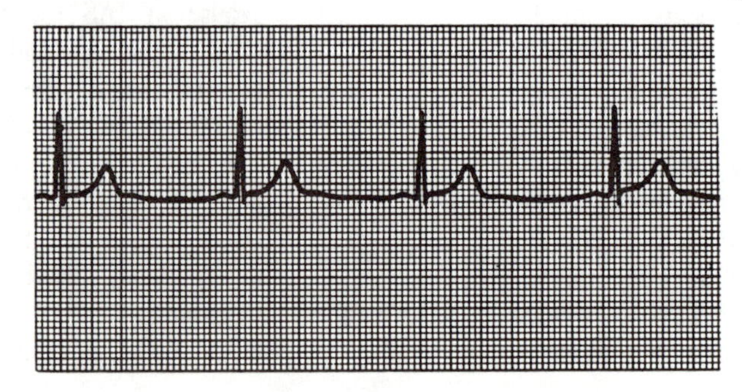

Fig. 9-29 Bradycardia. (See ref., Fig. 9-24a).

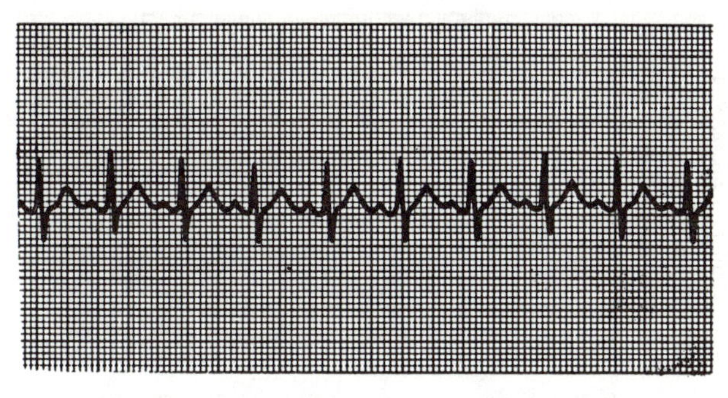

Fig. 9-30 Tachycardia. (See ref., Fig. 9-24a).

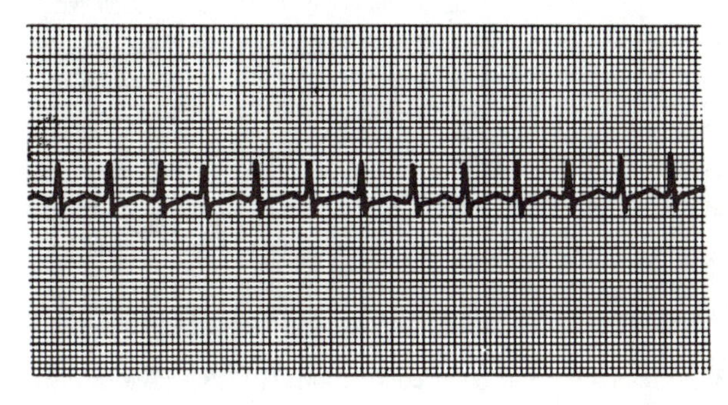

Fig. 9-31 Paroxysmal atrial tachycardia (PAT). (See ref. Fig. 9-24a).

neck at the jaw angle stimulates the vagal center in the medulla via the 9th cranial nerve.

3) Emersion of the face in ice water stimulates the vagal center via the ophthalmic branch of the 5th cranial nerve.

4) The *valsalva maneuver* (straining against a closed glottis) stimulates carotid and aortic baroreceptors.

b. *Propranolol* (Inderal) is a sympathetic blocking agent (beta blocker).

c. *Digoxin* decreases SA node depolarization by way of the vagus (negative chronotropic action). In addition, digoxin increases the force of contraction (positive inotropic action).

d. *Verapamil* (Isoptin, Calan) is a calcium-channel blocker that slows AV node conduction, thus decreasing the heart rate (negative chronotropic effect).

3. *Atrial fibrillation* is a common arrhythmia in the elderly and is treated if the person is symptomatic or if the ventricular rate is greater than 100 (Fig. 9-32). The arrhythmia responds to digoxin, propranolol or verapamil.

4. *Premature ventricular contractions (PVC's),* also common in older people, are often not treated unless associated with significant heart disease (Fig. 9-33). Lidocaine (Xylocaine) may be given.

5. *Ventricular tachycardia,* a life-threatening arrhythmia indicating significant heart disease, often evolves into ventricular fibrillation (Fig. 9-34). Either lidocaine or bretylium is given. If there is no

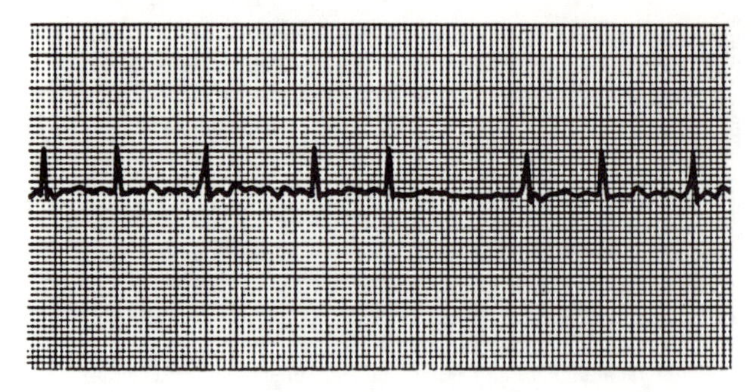

Fig. 9-32 Atrial fibrillation. (See ref. Fig. 9-24a).

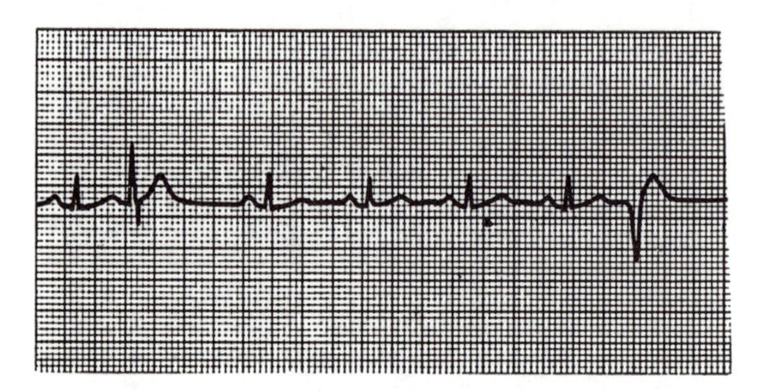

Fig. 9-33 Premature ventricular contractions (PVC's). (See ref. Fig. 9-24a).

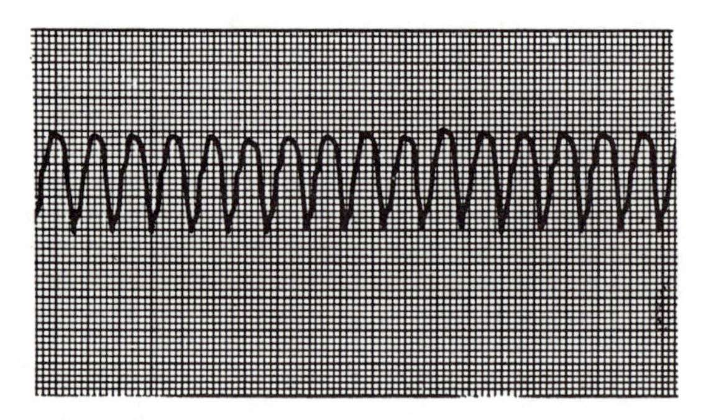

Fig. 9-34 Ventricular tachycardia. (See ref. Fig. 9-24a).

pulse or respiration, CPR is begun.

J. *Thrombophlebitis* is *obstruction of a vein by a thrombus, with inflammation of the wall.* Veins commonly affected are those of the leg (see earlier sections). *Primary treatment* for superficial thrombophlebitis is elevation of the leg, not moist packs and an anti-inflammatory agent. Deep vein thrombosis is treated in the hospital with an anti-coagulant. *Heparin* is first given, followed by an oral agent. *Thrombolytic agents* are also useful (see Ch. 10, 11).

-TEN-

BLOOD, ELECTROLYTES AND BODY FLUIDS

I. *Whole blood* consists of two portions, the *formed elements* and the liquid *plasma*.

A. *Plasma* constitutes about 55% and contains many dissolved substances. If blood is allowed to clot, and the clot is removed, the resulting fluid is *serum*.

B. The *formed elements* are *red blood cells*, *white blood cells* and *platelets*. The *hematocrit* is the *percentage of red cells* after centrifugation in a tube (Fig. 10-1). The white cells form a thin line at the top of the red cells ("buffy coat"). In males the average hematocrit is 47, with a range of 40-52. In females the average is 42, with a range of 37-47. A hematocrit below these figures indicates possible *anemia*. A hematocrit above these figures indicates possible *polycythemia*.

II. *Blood cell formation*:

In the adult, blood cells are formed mainly in the marrow of the bones of the chest, vertebrae and pelvis. In times of need, yellow marrow can convert to red marrow. The stages of development of blood cells in red marrow is *hemopoiesis* (Fig. 10-2). Blood cells originate from a common precursor cell, the *stem cell*. Immature forms are *blast cells*. When the cells are mature (below the line in Fig. 10-2), they move into the bloodstream. Blast cells may be seen in peripheral blood in *leukemia*.

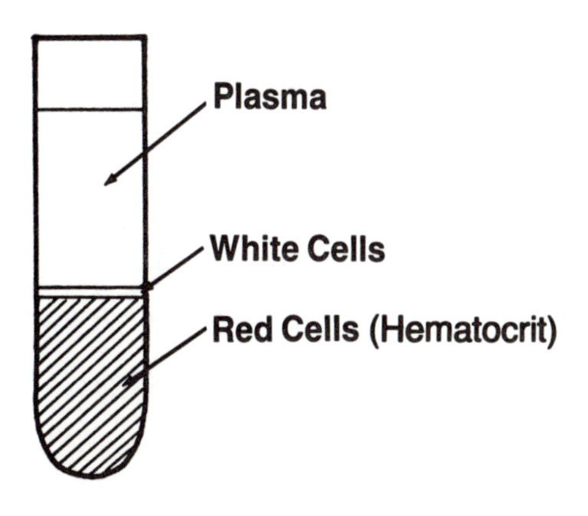

Fig. 10-1 The Hematocrit.

III. *Blood cells*:

A. *Red cells (erythrocytes)*:

1. *Characteristics*: The red cell loses its nucleus during development and is the only functional cell *without a nucleus*. The life span of the red cell is about *120 days*. *The normal red cell count is from 4.0 to 6.0 million per cubic millimeter*. Below this suggests *anemia*; above this suggests *polycythemia*. Approximately 200 billion red cells are produced (and die) each day. The red cell requires both *vitamin B-12 and folic acid* for normal development. A lack of oxygen (*hypoxia*), or *anemia*, stimulates the production of cells from the bone marrow (*erythropoiesis*).

2. *Functions*:

   a. The *main function* of the red cell is to *carry oxygen to the tissues*. Oxygen is carried *bound to iron* in the red pigment *hemoglobin* (Hb) as oxyhemoglobin (HbO_2). Hemoglobin is a large molecule with several *heme subunits* bound to the protein *globin* (Fig. 10-3). Oxygen is released to the tissues after transport. *Normal values for Hb: males 14 to 18* grams per 100 ml; females *12 to 16 grams*. Values below these ranges are suggestive of *anemia*.

   b. A second function of the red cell is the processing of CO_2 from tissues. A small amount of CO_2 is transported in plasma. The majority, however, is converted to HCO_3^- in the red cell, diffuses into the plasma, and is transported in this form. About 1/4 of

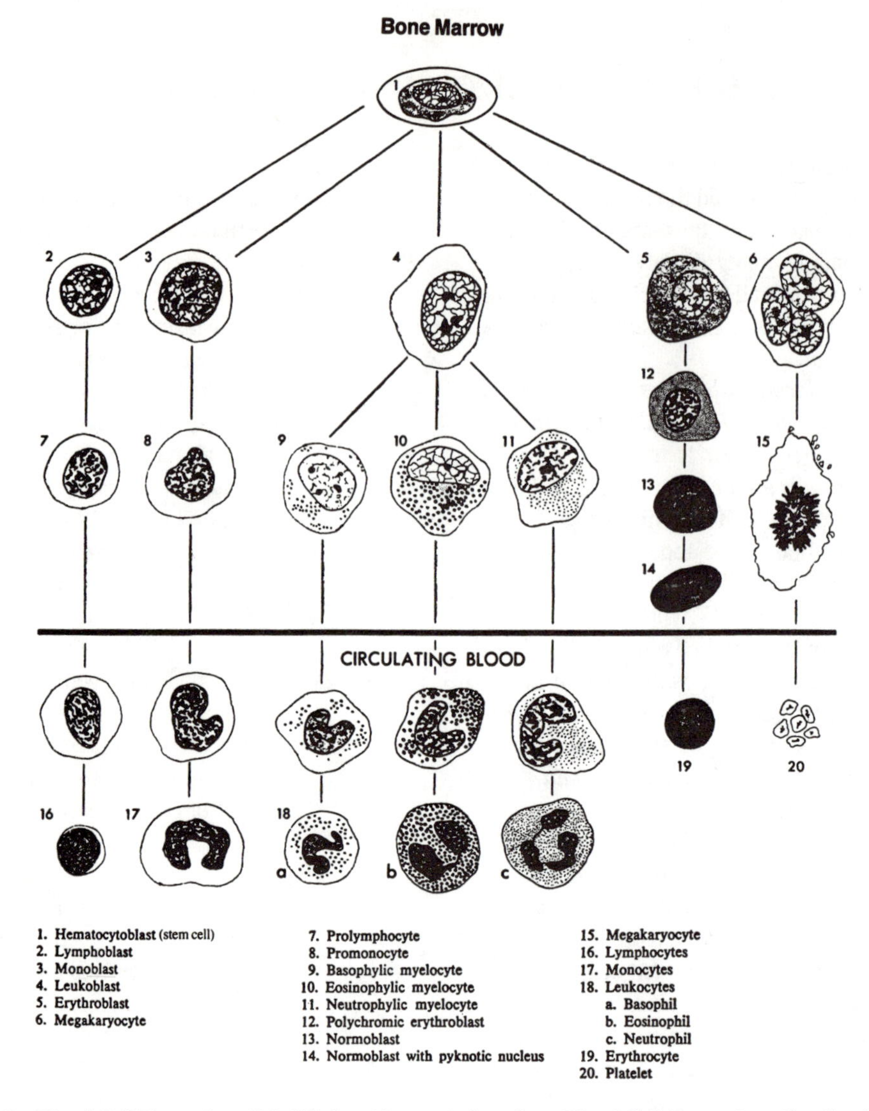

Fig. 10-2 Blood Cell Formation. (Modified, with permission, from Blood Cell Formation, Carolina Biological, 1966).

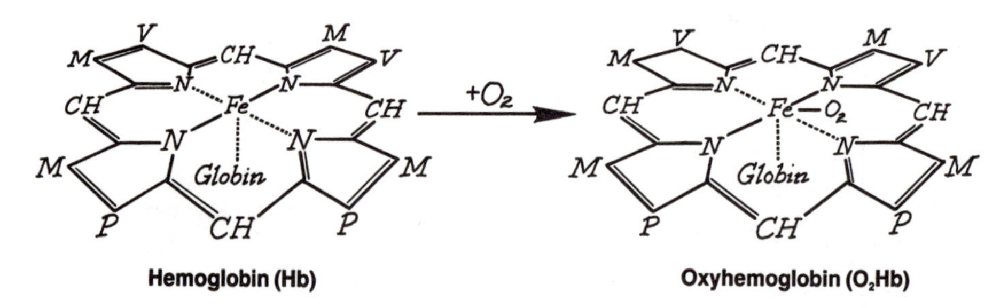

Fig. 10-3 Hemoglobin and Oxyhemoglobin. (M, P and V are abbreviations for methyl, proprionyl and vinyl groups)

the CO_2 is transported bound to Hb (carbamino-hemoglobin).

c. Hb is an important *acid-base buffer* (70% of the buffering of blood; see chapter 11).

d. The *blood-group antigens* are present on the red cell membrane (see section IV).

3. *Reticulocytes*, newly formed red cells which contain small fragments of a nucleus, constitute about 1% of the total red cell count. The *reticulocyte count* is an important index to production. It increases as the body responds to blood loss by producing more cells.

4. *Breakdown of old red cells*.

   a. After 120 days, aging red cells are destroyed by phagocytes, collectively called the *mononuclear phagocyte system* (*MPS*, tissue macrophage system, reticuloendothelial system, RES) (the earlier term "reticuloendothelial", used to describe phagocytic activity of endothelial cells, particularly in the spleen, liver, bone marrow, adrenal gland, pituitary and involving reticular fibers, has been discarded, since these cells are modified monocytes, widely distributed throughout many tissues of the body). Each of these monouclear phagocytes has receptor sites for antibodies on the cell membrane. *The spleen is the primary MPS organ.*

   b. *Iron is reused in the manufacture of new heme*. Old heme is split from globin, the porphyrin ring opens, and the green bile pigment *biliverdin* is formed. Biliverdin is quickly reduced to the yellow pigment *bilirubin* (Fig. 10-4). Bilirubin combines with albumin and is *transported to the liver*. The liver excretes bilirubin, as a constituent of bile, to the gall bladder. When the liver is damaged or diseased, or if red cells break (lysis) as occurs in certain of the hemolytic anemias, bilirubin may accumulate in the bloodstream, giving the skin a yellow color (*jaundice*).

B. *White cells (leukocytes)* (Fig. 10-2):

1. *Characteristics*: The *normal white count is 5 to 10 thousand* per cubic millimeter of blood. *Under 5 thousand is leukopenia*, which is seen in some viral infections, overwhelming bacterial infections (neutrophils are literally used up), radiation poisoning, malnutrition, hypersplenism and bone marrow failure. *Above 10 thousand is leukocytosis*, commonly seen in bacterial infections. *The main function of the white cell is to combat infection*.

2. *Five cell types* are recognized, and are categorized by the presence or absence of granules in the cytoplasm.

   a. *Granulocytes* (polymorphonuclear leukocytes, "polys", "segs") have nuclei that appear segmented. They live several hours. Types of granulocytes are:

      1) The *neutrophil* contains neutral-staining granules. It is the most abundant of the white cells: 50 to 70 percent. Juvenile forms, newly released from bone marrow, show nuclei which are band-shaped rather than segmented and are known as "bands" or stabs" (the German for band is "stab"). Bands make up about 3 to 5% of neutrophils. *The neutrophil is the main phagocyte in acute bacterial infections*.

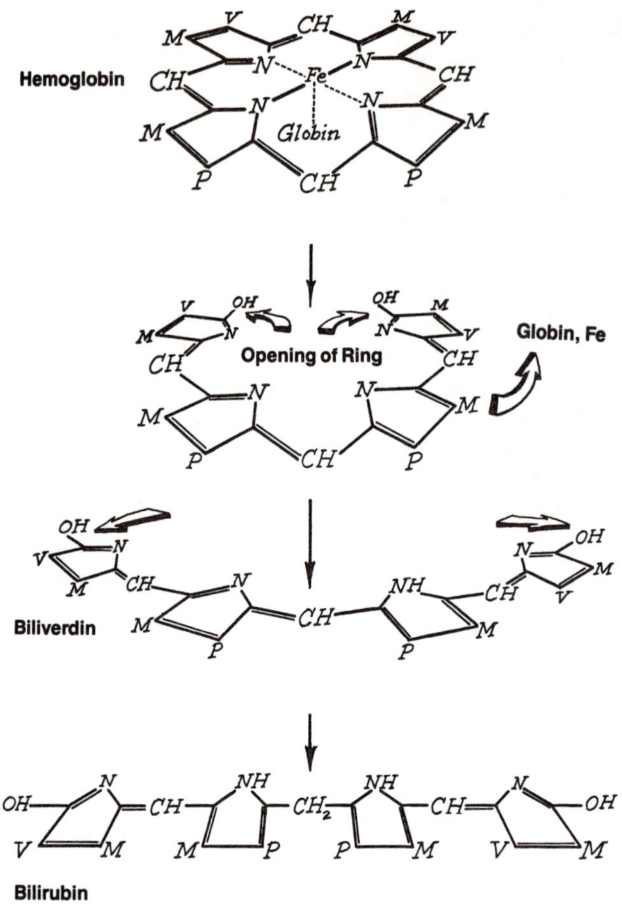

Fig. 10-4 Breakdown of Hemoglobin.

2) The *eosinophil* contains acidic-staining (eosin, or red) granules in the cytoplasm, and constitutes 0-5% of white cells. The eosinophil *phagocytizes antigen-antibody complexes*. Its number increases in parasitic and allergic conditions, such as asthma.
3) The *basophil* contains basic-staining (purple) granules in the cytoplasm, and constitutes 0-1% of white cells. The basophil contains both heparin and histamine, and its structure is similar to the *mast cell*, found in connective tissue. Its exact physiological function is unknown.

b. *Agranulocytes* (no granules in the cytoplasm):
   1) The *lymphocyte* is slightly larger than the red cell. Its nucleus is usually dark-staining and large compared to the cytoplasm, which stains light blue. It comprises 20 to 40% of the white cells. B and T-lymphocytes play an essential role in the *immune reaction* (see chapter 2). There is often a lymphocytosis in *viral infections*.
   2) The *monocyte* is the largest

white cell. Both nucleus and cytoplasm stain faintly. Monocytes make up 0-7% of white cells. They are part of the reticulo-endothelial system, and in tissues are known as *macrophages*. These are *second-line phagocytes*, and their numbers increase in chronic diseases such as tuberculosis, the recovery stage of acute infections, and in some rickettsial and protozoal diseases.

C. *Platelets (thrombocytes) are small cell fragments originating from giant cells in the bone marrow*. In a blood smear, they appear as tiny bodies on high-power, sometimes in clumps. The normal platelet count is 150 to 300 thousand per cubic millimeter. The life of platelets is about 10 days. One third is in the spleen. Platelets are *necessary for normal blood clotting.* Bleeding occurs when the platelet count is low (thrombocytopenia — see end of chapter):

1. When a vessel is injured, platelets adhere to the wound site. Adhesion causes *platelet activation.*
2. Activated platelets release ADP and the vasoconstrictor thromboxane A_2, derived from arachidonic acid and related to the prostaglandins. This causes more *platelets to aggregate* and become activated, increasing the size of the plug (see also chapters 12 and 13).

IV. *Blood Types.*

Naturally occurring antigens exist in the body, such as those located on the *red cell*, and comprise the various *blood groups* (Table 10-1). The more important are the *ABO and the Rh systems* (Rh stands for *rh*esus from early monkey experiments). The A and B antigens are strong. The O antigen is weak and is usually considered as being absent. The same applies to the Rh factor. The more powerful Rh antigens are C, D and E, with D being the strongest. It is usual to consider persons with either c, d or e as not having the antigen. *Naturally occurring antibodies* are also present in the plasma.

A. *The ABO system.*

1. *One cannot possess an antigen with an incompatible antibody in his plasma*. For instance, a person with A antigen on his red cells has anti-B antibodies in his plasma. The same applies for type B. He has the B antigen and anti-A antibodies in his plasma. Type AB, having both antigens on the red cell, has no antibodies in the plasma. Type O, possessing a weak antigen, has both antibodies in the plasma (Table 10-2).
2. *If antigens are placed with incompatible antibodies, clumping*

Table 10-1: The ABO and Rh blood group systems

ANTIGEN (AGGLUTINOGEN) ON RED CELL	ANTIBODY (AGGLUTININ) IN PLASMA	
	Anti-A antibody	*Anti-B antibody*
O	+	+
A	–	+
B	+	–
AB	–	–
	Rh (Anti-D) antibody in plasma	
Rh^+ (D antigen)	–	
Rh^-	–	

Table 10-2: Distribution of ABO and Rh blood types in the United States and Europe

	PERCENTAGE
O	45
A	42
B	10
AB	3
Rh^+	85
Rh^-	15

(agglutination) of red cells results. This is seen when a person of one blood type gives blood to an incompatible type; for instance, type A giving to type B. *The donor's cells are agglutinated by the receiver's plasma.* Thus, in the above example, the anti-A antibodies in blood type B's plasma would agglutinate the cells of type A. Or, A antigens would react with anti-A antibodies. Type B donors, having B antigens, would react with anti-B antibodies of A recipients. Type AB, having both antigens, could give to no one except his own type. However, having no antibodies in his plasma (as they would react with his own antigens), he can receive blood from *all* types. He is thus the *universal receiver*. Blood type O, possessing weak antigens, may give to all types (but receive only from O) and is the *universal donor*.

3. When an antigen-antibody response results in agglutination, the *antibodies* are *agglutinins*, and the *antigens* are *agglutinogens*.
4. In practice, *cross-matching* is done between cells and plasma of both donor and recipient to detect other antigens and antibodies, as well as previous sensitizations. Type A should only give to type A, O to O, etc. Only in emergencies should O be the universal donor and AB the universal receiver. It is often better to give saline or a plasma expander in emergencies rather than type O negative blood and risk a possible transfusion reaction.

B. *The Rh system.*

1. The *D antigen* has *no naturally occurring antibody*. 85% of people have the D antigen and are Rh positive. 15% do not have the antigen and are Rh negative. An Rh negative person can give blood to an Rh positive, since his cells do not possess the antigen. If an Rh positive gives to an Rh negative, the negative will develop antibodies to the foreign antigen, and agglutination occurs.
2. *If an Rh negative mother is carrying an Rh positive fetus*, the hazard exists of immature fetal cells crossing the placenta (mature blood cells do not cross the placental membrane) and stimulating maternal production of anti-D antibodies, similar to a transfusion reaction. Antibodies *may* diffuse back across the placental membrane to the fetus and agglutinate and destroy fetal red cells. Destruction of fetal red cells releases hemoglobin (hemolysis), causing anemia, jaundice (from the increase in bilirubin from heme), enlargement of the liver and spleen, and generalized edema in the newborn. This is *erythroblastosis fetalis*.
3. *Rh negative individuals may receive one Rh positive transfusion without difficulty*, since it takes time to develop antibodies. However, subsequent transfusions, even years later, may prove fatal, since a higher titer of Rh antibodies may have developed. The same applies to the Rh negative mother. She can have the first Rh positive baby without difficulty. During delivery, however, there is a flood of fetal cells across the placental membrane, and antibodies develop

in the mother. This creates a problem with *subsequent pregnancies*. If, however, during the first pregnancy when she has not had time to develop Rh antibodies, she receives an injection of artificial antibodies *immediately after birth*, the artificial antibodies react with fetal cells and the mother will not produce antibodies. *Human immune globulin (Rho-gam)* (artificial antibodies, or gamma globulin against the Rh factor) is usually injected to suppress the mother's immune response. However, each time the mother is pregnant with an Rh positive baby, *immune globulin* (Rho-gam) must be given.

V. *Hemostasis* means the arrest of bleeding:
   A. *Vessel injury*: After a vessel is injured, the following events take place:
      1. *Local vasoconstriction* of the vessel.
      2. *Platelet aggregation* at the site of injury. Platelets plug the injured vessel wall.
      3. *Blood coagulates at the site of injury (Fig. 10-5a).*
         a. *Stage one* of the clotting scheme involves two mechanisms.
            1) Damaged tissue releases a substance with an enzyme-effect, *thromboplastin*. Thromboplastin activates factor 7, which in turn activates factor 10, which *initiates clotting in 12 to 15 seconds*. The clotting is only partial, since not enough factor 10 has been formed. This is the *extrinsic system*.
            2) Exposure of blood to damaged vessel endothelium also initiates a series of cascade reactions leading to the *final clot in 5 to 10 minutes*. This is the *intrinsic system*. Each factor is a globulin in blood that is activated by the previously activated factor above. The intrinsic system activates enough factor 10 to produce the definitive clot (fibrin).
         b. *Stage two* is the conversion of the inactive plasma protein *prothrombin* to the active *thrombin*, under the influence of factor 10 and calcium. *Vitamin K* catalyses the formation of prothrombin in the liver.
         c. *Stage three* is the conversion of the plasma protein *fibrinogen* to the *fibrin* clot, under the influence of thrombin.
   B. *The fibrinolytic (clot-dissolving,*

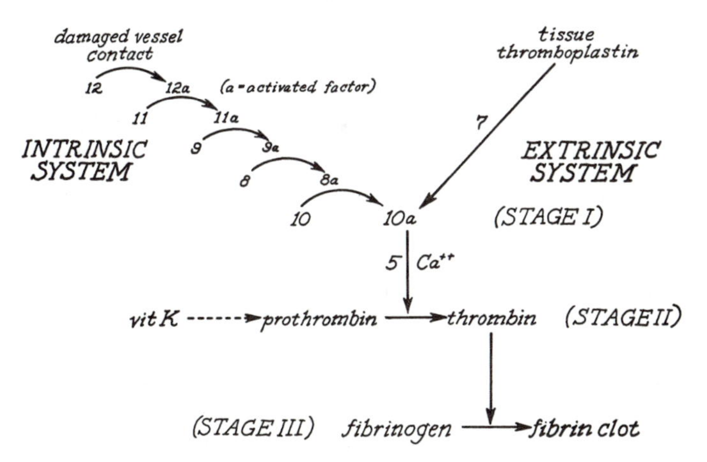

Fig. 10-5a The Clotting Mechanism.

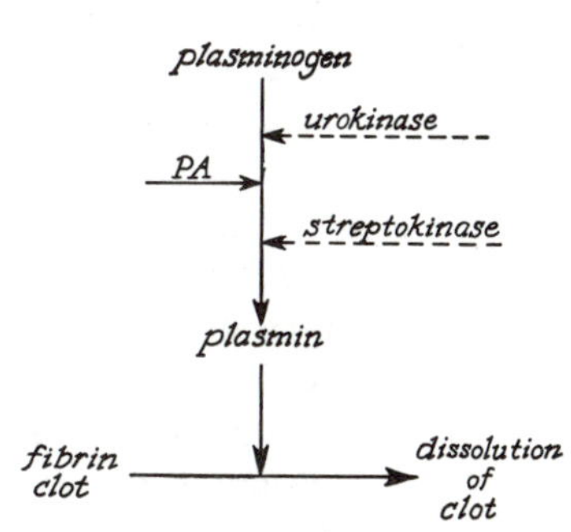

Fig. 10-5b Fibrinolysis

thrombolytic) system dissolves fibrin (Fig. 10-5b). The same factors that initiate the coagulation scheme (contact activation and tissue damage) activate this system.

1. *Plasminogen-activator* (PA) is released from endothelial cells and binds to fibrin.
2. *Plasminogen* in the bloodstream is converted to plasmin by PA, dissolving the clot.
3. *Commercial thrombolytic agents:*
   a. *Urokinase,* derived from human fetal kidney cells, is used in the treatment of *pulmonary embolism.*
   b. *Streptokinase,* derived from the streptococcus bacterium, is useful in pulmonary embolism, *deep vein thrombosis,* and *myocardial infarction.* Urokinase and streptokinase unfortunately cause bleeding.
   c. *Tissue plasminogen activator* (TPA) is used in the treatment of *coronary occlusion and infarction.* It acts locally, thus avoiding a bleeding tendency.
   d. *Pro-UK,* a form of urokinase, acts as a back-up thrombolytic agent, usually to TPA. It dissolves clots anywhere in the body. TPA and pro-UK are important agents in the treatment of *coronary artery disease.*
   e. At this time, it appears that the combination of TPA, and immediately afterwards, coronary angioplasty, is optimum therapy for many cases of MI.

C. *Some causes of bleeding.*
   1. A *genetic defect* in any of the clotting factors (usually factor 8 or 9) is *hemophilia.*
   2. Bleeding may take place if vessel walls are fragile, as in Vitamin C deficiency, or in infections.
   3. Faulty or low platelets in *thrombocytopenia.*
   4. *Liver disease* (e.g., cirrhosis in alcoholics) causes bleeding because Vitamin K is not able to contribute to the formation of *prothrombin.*
   5. *Disseminated intravascular coagulation (DIC) is a condition in which the clotting factors are depleted because of excessive intravascular clotting.* It is seen following an overwhelming gram negative bacterial infection, certain snake bites, surgery, cancer, and transfusion reactions. Primary treatment is directed at remedying the underlying cause. Platelets, plasma and fibrinogen are useful. Sometimes the administration of heparin is necessary to prevent further clotting.
   6. *A common cause of bleeding is seen in people taking certain drugs.* The *oral anticoagulant medicines* such as bishydroxycoumarin and warfarin are discussed in section D. Another drug which causes bleeding is *aspirin,* which inhibits platelet aggregation. Because of this trait, it is useful for the prevention of strokes. Primary treatment for excess aspirin is stopping the drug. As a rule of thumb, more than about eight aspirin tablets (325 mg per tablet) per day may cause bleeding.

D. *Simple screening tests for bleeding disorders*:
   1. The simplest test is the complete blood count (CBC). The blood smear detects a *low platelet count.*
   2. *The partial thromboplastin time (PTT)* is a test in which a platelet factor and calcium are added to plasma (for standardization) and the time noted until clotting occurs. The PTT has replaced the older more time-consuming whole-blood-coagulation-time (Lee-White), where one waited for blood to clot in a tube (5 to 10 minutes). The normal value for the PTT is 60-

100 seconds. Many labs use an "activated-PTT", where the normal value is 30-45 seconds. If the prothrombin time (PT) is normal (see below), the PTT measures defects in the *intrinsic system*, such as *hemophilia*. It is widely used to *monitor patients on heparin* in the hospital.

3. *The prothrombin time (PT)* is a test in which tissue thromboplastin and calcium are added to plasma (for standardization), and the time noted until clotting takes place (from 12 to 15 seconds). It measures defects in the *extrinsic system*. Commonly, it is used to evaluate people with *liver disease* (inability to utilize vitamin K in the formation of prothrombin), and to monitor those on *oral anticoagulants*.

E. *Anticoagulants*:

1. *Heparin*: The main action of heparin is the *inhibition of thrombin*. It is given intravenously in the hospital as a constant infusion or in intermittent doses. Heparin is *monitored by the PTT*. As a rule of thumb, the therapeutic range of heparin should increase the PTT to about *2 1/2 times normal*. Heparin is used in venous thrombosis, pulmonary embolism, DIC, heart and blood vessel surgery, and arterial and coronary occlusion. Primary treatment for bleeding: *stopping heparin* infusion is usually sufficient, since 4 hours after cessation no heparin will be found in the bloodstream. In an emergency situation, protamine may be given.

2. *Oral anticoagulants* such as warfarin (Coumadin) *block the formation by vitamin K of prothrombin in the liver*. They are *monitored by the prothrombin time (PT)*. The *therapeutic range is also about 2 1/2 times normal*. Indications for oral anticoagulants are the same as those for heparin (except DIC). In the hospital setting, heparin is begun for a rapid effect and the patient is gradually transfered to the oral anticoagulants. Primary treatment for bleeding from oral anticoagulants is *vitamin K*, although sometimes prothrombin is required. *Note*: oral anticoagulants are among the most dangerous drugs, since they are taken at home, sometimes with little follow-up. Profound bleeding may occur, sometimes necessitating a transfusion.

VI. *Body fluids*:

A. *Definitions and units of measurement*:

1. *Diffusion* means the *movement of a substance from an area of a higher to an area of a lower concentration* (e.g., water, NaCl, O_2, CO_2).

2. *Filtration* occurs when *fluid is forced through a membrane from high pressure on one side* (e.g., capillary, nephron).

3. *Osmosis is the movement of fluid (solvent) through a semipermeable membrane separating different concentrations of substances (solute)*. The fluid moves from the low to the high solute region (e.g., capillary, cell).

4. *Electrolytes* are substances which, *when placed in a solution, are decomposed and conduct an electric current*. Common electrolytes are *salts*, such as sodium chloride (NaCl: table salt). When these substances separate in solution they carry an *electrical charge* and are *ions*. Ions with a positive charge are *cations* (e.g. Na^+, K^+, Ca^{++}, NH_4^+, Mg^{++}); those with a negative charge are *anions* (e.g., Cl^-, HCO_3^-, $HPO_4^=$, $SO_4^=$).

5. *The milliliter (ml) or cubic centimeter (cc)* is the unit of measurement for *fluid volume*. 1000 ml equals one liter. Sometimes values are expressed in deciliters (dL).

100 ml equals one deciliter.

6. *The milliequivalent (mEq)* is the chemical *combining power of an ion*, based on the number of charges (valence) of the ion (e.g., Na^+ has one, Ca^{++} has two). 1000 mEq equals 1 equivalent. Many substances in blood are expressed as mEq per liter (e.g. Na^+, K^+, Cl^-).
7. *The milliosmol (mosmol)* is the *unit of measurement of a substance with osmotic activity*, such as ions (Na^+, Cl^-) or non-ions such as glucose or urea.
8. The *milligram* (mg) is the *unit of weight of a substance*. Many substances in plasma are expressed as mg/dL (e.g., glucose, uric acid).

B. *Distribution of water and ions in body fluids*. Various fluid regions are sometimes referred to as *compartments* (e.g., the intracellular compartment) (Fig. 10-6).

1. *Intracellular fluid* — 30 liters
   a. *main cation* — K^+
   b. *main anion* — $PO4^=$
2. *Extracelluar fluid* — 15 liters (interstitital or tissue fluid — 12 liters; plasma — 3 liters)
   a. *main cation* — Na^+
   b. *main anion* — Cl^-

C. *Osmosis* can be thought of as the "drawing-power" of osmotically active substances for water. The more substances, the more water will be drawn. *Semipermeable membranes* in the biological system include the *cell membrane* (red and white blood cells, tissue cells) and the *capillary membrane*. *Sodium constitutes the main osmotic force in the plasma and tissue*

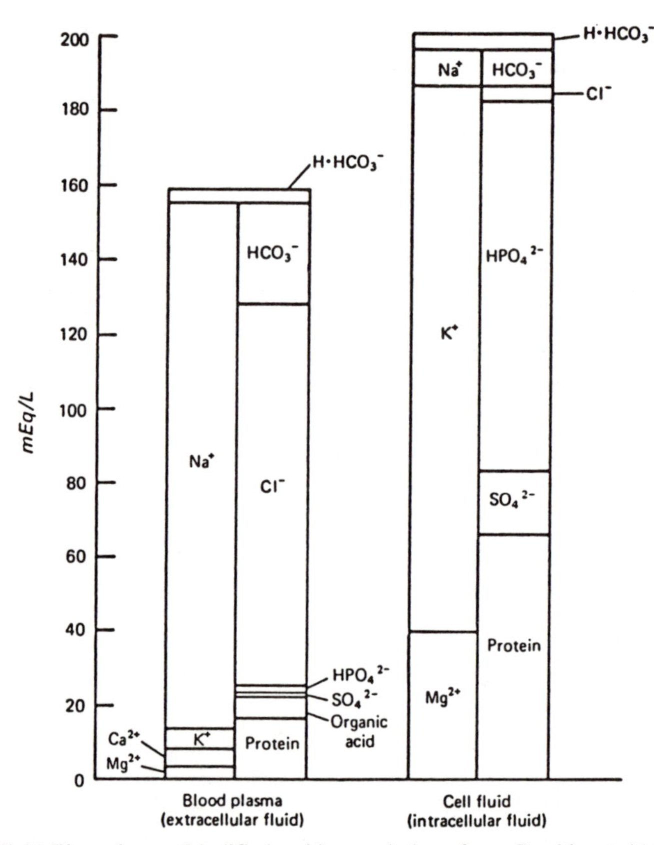

Fig. 10-6 Body Fluid Electrolytes. (Modified, with permission, from Gamble, Acid-Base Physiology, John Hopkins Univ. Press, 1982).

fluid (the extracellular compartment). Potassium is high inside the cell and provides most of the osmolality of the intracellular compartment. *Sodium* and *potassium* are in equilibrium with each other on either side of the cell membrane.

1. *An illustration of how osmosis works* is seen when one observes red blood cells in different concentrations of plasma. Normally, the osmotic pressure in plasma and red cells are in equilibrium (an *isotonic solution*). However, if sodium is removed from plasma, fewer particles are in plasma compared to the red cell (a *hypotonic solution*) and fluid moves into the cells (more osmotically active particles compared to plasma). The red cell swells and burst (*hemolysis*). If sodium is added to plasma, now there are more osmotically active particles (a *hypertonic solution*) and fluid moves from the red cells into the plasma. In this case cells will shrink (*crenation*) (Fig. 10-7).
2. *Examples of osmotic disorders in the living system.*
   a. In *dehydration,* low fluid intake, causes a relative increase of sodium in the extracellular fluid (e.g., plasma). This creates a hypertonic solution. Fluid is drawn from the cells into the plasma. Primary treatment consists of restoration of fluid, sometimes intravenously.
   b. Loss of osmotically active particles such as glucose through the urine in *diabetes* results in water loss (since water accompanies glucose).
   c. In *cerebral edema*, the cells of the brain swell and fluid accumulates. This is seen after a *head injury*, and sometimes in non-traumatic conditions such as *Reye's syndrome* (a rare condition seen in children following the flu or chickenpox, consisting of sudden vomiting and disturbed brain function). The use of an osmotically active substance such as *mannitol* will draw fluid from inside the cells to the plasma, preventing serious brain damage.
   d. In *liver and kidney disease*, plasma proteins (osmotically important substances) are lowered in the body, creating a hypotonic plasma. The result is the movement of fluid into the tissue spaces (see Edema, end of chapter).
   e. The *capillary exchange mechanism* (see below).
   f. *Diabetes insipidus* (see chapters 12, 14).
3. *The following solutions are isotonic with plasma*:

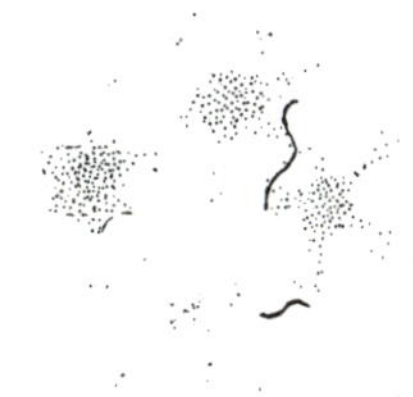

Fig. 10-7 Osmosis.

a. *0.9% NaCl* is physiological, or *normal, saline*.
b. *Ringer's solution* contains many ions and most closely resembles plasma. The term "Ringer's" means a solution containing most of the electrolytes for that species. For instance, there is "frog Ringer's", "dog Ringer's", etc.
c. A *5% glucose (dextrose)* solution is written "D_5W", meaning 5% dextrose in water. "D_5¼NS" is .25% normal saline in 5% dextrose.

D. *The "third-space"*. In addition to the intracellular and extracellular compartments, fluid may accumulate in *physiologically non-reactive areas* such as the intestines (in bowel obstruction), the peritoneal cavity (ascites in cirrhosis) or the skin (in burns) and be *unavailable to the body.* This fluid is gradually mobilized after the restoration of extracellular fluid and ionic deficits, and excreted.

E. *The capillary exchange mechanism.* The exchange of nutrients and waste products between blood and tissue cells is an important part of bodily functioning and utilizes the three basics: *diffusion, filtration* and *osmosis.* Nutrients flow through the capillary membrane into the tissue spaces and from there to the cells. Waste products pass in the reverse direction. Oxygen and glucose move into the tissues by diffusion, and carbon dioxide moves out in the same manner. Plasma proteins are too large to pass through the capillary endothelium. They thus create an osmotic pressure gradient that pulls fluid into the plasma. The osmotic pressure exerted by the plasma proteins (*oncotic pressure*) is about 25 mm Hg (towards the bloodstream). There is usually a small *interstitial fluid pressure* of about 2 mm Hg, which acts to move fluid into the plasma (Fig. 10-8).

1. At the *arteriole end* of the capillary, the *blood pressure (hydrostatic pressure)* is about 37 mm Hg. The *net pressure* (37 minus 25 and 2) is *10 mm Hg into the tissues.*
2. As blood moves down the capillary, the hydrostatic pressure declines (the oncotic and interstitial fluid pressures remain the same).
3. At the *venule end* of the capillary, the hydrostatic pressure is about 17 mm Hg. The *net pressure* (25 and 2 minus 17) is thus *10 mm Hg back into the bloodstream.*
4. In summary, nutrients move from the bloodstream into the tissues at the arteriole end of the capillary, and waste products move from the tissues into the bloodstream at the venule end.

VII. *Important electrolytes* are sodium, chlo-

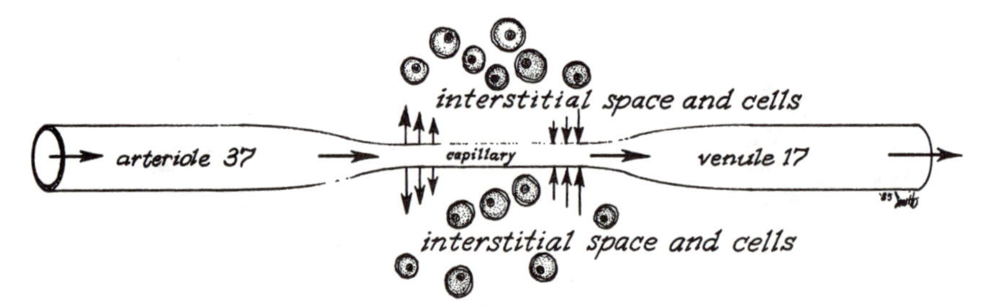

interstitial pressure: 2
oncotic pressure: 25

Fig. 10-8 The Capillary Exchange Mechanism.

ride, potassium, calcium and bicarbonate. Isolated chloride problems are comparatively uncommon, and will not be discussed. Bicarbonate is discussed with acid-base balance in chapter 11. Signs and symptoms of electrolyte disturbances are similar: lassitude, weakness, confusion, fatigue, sometimes coma.

A. *Sodium (German: Natrium: symbol Na)* (Normal Range 135-145 mEq):
   1. *Function*: Regulates the extracellular fluid volume, is necessary for transmission of the nerve and muscle impulses, and is involved with the secretion of ADH. Sodium constitutes the *main osmotically active solute in the extracellular fluid*. When sodium moves, water accompanies it. Water loss always accompanies a loss of sodium (or glucose, or any other osmotically active substance) from the bloodstream. An increase in sodium (as in hyperaldosteronism) causes water gain (and high blood pressure).
   2. *Excess (hypernatremia)*: Sodium is increased in water loss (dehydration) and in adrenal excess. *The most common cause of dehydration is failure to drink in debilitated and older patients*. Primary treatment: water by mouth or intravenously.
   3. *Deficiency (hyponatremia)*: Sodium is decreased in adrenal insufficiency. *The most common causes are excessive sweating, GI losses and the use of potent diuretics*. Primary treatment: restricting fluid intake and the administration of high sodium foods.

B. *Potassium (German: Kalium; symbol K)* (Normal Range 3.5-5 mEq):
   1. *Function*: Necessary for *transmission of the nerve and muscle impulses*. It is particularly *important in heart muscle contraction*, and accompanies glucose into the muscle cell in the presence of insulin.
   2. *Excess (hyperkalemia)*: *This is the most serious electrolyte disorder,* and may be life-threatening, causing ventricular fibrillation or asystole. The main cause is *kidney disease*. It is also seen in adrenal insufficiency. An important laboratory sign of hyperkalemia is an *elevation of the T-wave on the EKG*. Primary treatment: calcium, which antagonizes the toxicity of potassium on heart muscle. The administration of sodium bicarbonate causes a rapid movement of potassium into the cells. Glucose may also be given, which stimulates insulin, causing the movement of both glucose and potassium into the cells. In both treatments, potassium is decreased in the bloodstream.
   3. *Deficiency (hypokalemia)*: The main cause is from *diuretics*. It is also seen in diarrhea. An important sign is a flattened *T-wave on the EKG*. Primary treatment is directed at replacing *potassium* (KCl) orally or IV.

C. *Calcium* (see also chapters 11 and 12). (Normal Range 9-10.5 mg or 4.5-5.5 mEq):
   1. *Function*: Calcium is absorbed in the duodenum and jejunum by active transport, and is regulated by a metabolite of vitamin D. It is present in the body in several forms. Calcium combines with phosphate to form crystalline *salts of bone* (hydroxyapatites). Calcium in blood regulates *parathyroid hormone release*, is necessary for *blood clotting*, *transmission of the muscle and nerve impulse*, *muscle contraction* and several enzyme reactions. Most plasma calcium is in the ionized form (Ca^{++}). Some is bound to albumin. A small amount is bound to globulin. Increased calcium decreases, and a low calcium increases, the

permeability of the neuron membrane to sodium.

2. *Excess (hypercalcemia)*: Two main causes are *excessive milk-drinking* and *excessive vitamin D intake* by health food faddists. Hypercalcemia is also seen in hyperparathyroidism and kidney disease. Another cause is from *cancer* (see chapter 12). Primary treatment: treating the initial cause, and in the emergency situation giving NaCl (antagonizes calcium) and the diuretic *furosemide* to excrete calcium in the urine.
3. *Deficiency (hypocalcemia)*: The main cause of a slight decrease in blood calcium is a *decrease in plasma protein* (some calcium binds to albumin and globulin). In this case, a low calcium reflects low plasma protein, and not a calcium deficit. Hypocalcemia is seen after *parathyroid removal* in thyroid surgery. It is also present if *vitamin D is lacking*, and in *renal failure*. When Ca^{++} is low, depolarization of motor nerves begins to occur spontaneously and the muscles go into spasm (tetany). This is particularly evident in the hand and feet muscles (carpopedal spasm). Primary treatment is the *administration of calcium or vitamin D*. An important and common cause of transient hypocalcemia is *respiratory alkalosis from hyperventilation*. The alkalosis causes an increase in binding of calcium to plasma protein, thereby decreasing the amount of ionized calcium. The result is *transient hypocalcemic tetany*, with carpopedal spasm. It is treated by rebreathing $C0_2$, reversing the alkalosis (see chapter 11).

D. *Solutions containing electrolytes*:
Common intravenous solutions are listed in section VI-C-3. The glucose in a *D_5W* solution is used for *energy*, and D_5W is a common initial solution used in the hospital setting. *Lactated Ringer's* (Ringer's lactate) *contains most of the electrolytes of plasma* (lactate is converted to bicarbonate in the liver) and is commonly used in a long hospital stay. *Normal (isotonic) saline* (0.9% NaCl solution) is the usual solution used in *shock*, since the *NaCl restores osmotically active particles*, helping to raise the blood pressure. Saline is also used in *diabetes mellitus* (since a D_5W solution would raise the glucose level). *Maintenance* parenteral (intravenous) therapy for most patients with normal kidney function is two to three liters of 5% glucose per day, plus:
1. Sodium: 30 mEq per liter
2. Chloride: 20 mEq per liter
3. Potassium: 20 mEq per liter

In practice, normal NaCl is often added to a liter of fluid (D_5W), which supplies about 35 mEq of salt per liter to the patient. An order written "D_5.25NS at 100 ml/hr" means a 5% dextrose (glucose) solution with ¼ normal saline to run at a rate of 2.4 liters per day. 20mEq of potassium is usually added to every other liter.

VIII. *The lymphatic system* (Fig. 10-9):

A. *Lymph is tissue fluid (interstitial fluid) that bathes the cells*. Lymphatic drainage begins in blind channels in tissue spaces. Larger and larger vessels form, which eventually become two main ducts: the *right lymphatic duct* drains the upper right half of the body and empties into the *right subclavian vein;* the *thoracic duct* drains the rest of the body and empties into the *left subclavian vein*. When infection is present, a lymph channel under the skin may show a red streak. This is *lymphangitis* (the spread of infection by the lymphatic vessels, commonly called "blood-poisoning"). Cancer frequently travels this way.

B. Interspersed within the lymphatic circulation are *lymph nodes*, containing macrophages, B and T lymphocytes

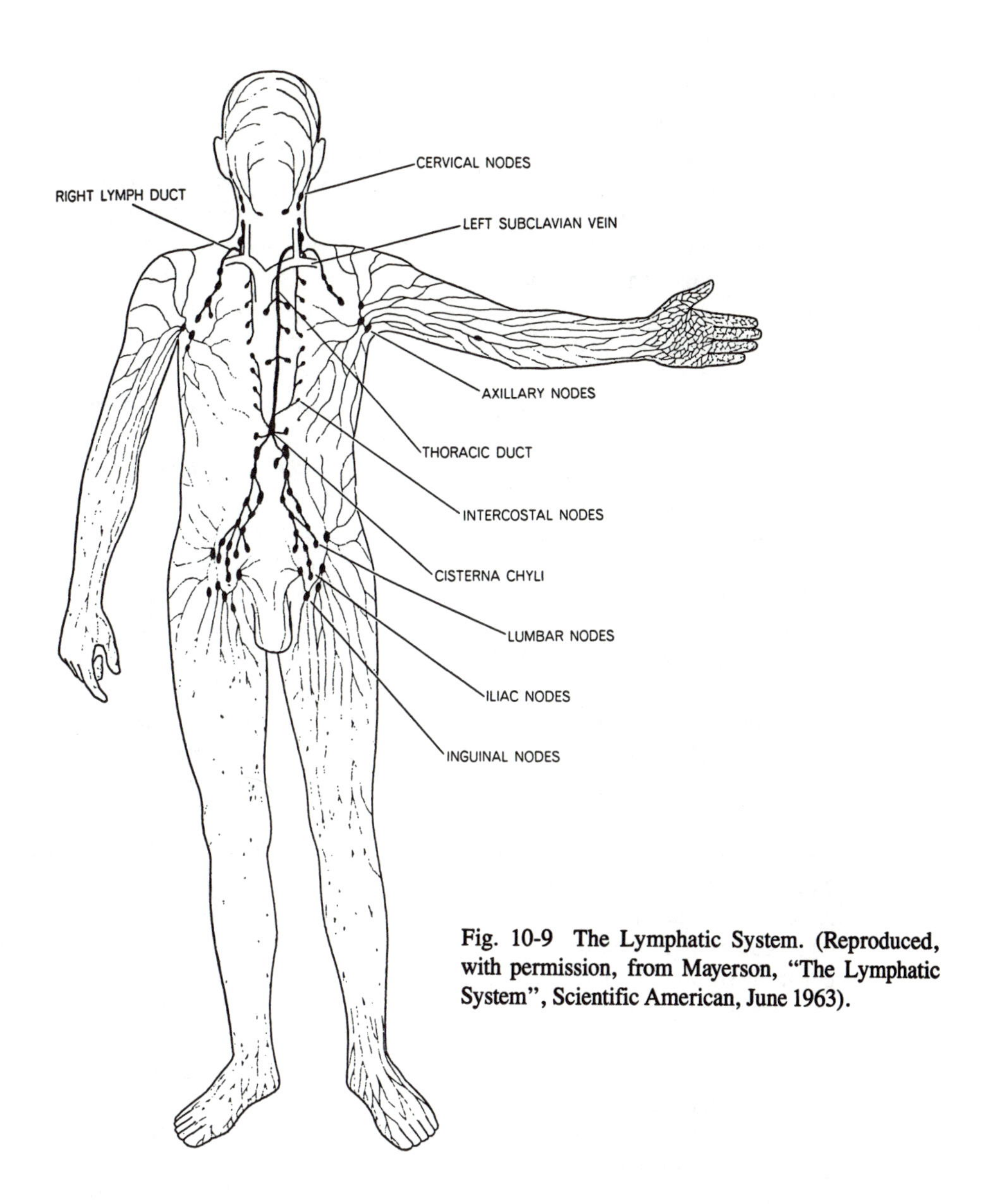

Fig. 10-9 The Lymphatic System. (Reproduced, with permission, from Mayerson, "The Lymphatic System", Scientific American, June 1963).

and plasma cells. During infection, a heightened activity in the nodes takes place. They enlarge (lymphadenopathy) and are often painful. Their main function is to prevent bacteria and viruses from gaining access to the bloodstream; they can be thought of as filter stations for lymph. Main lymph node areas of the body are shown in the above diagram. The gastrointestinal tract has an abundance of small lymph nodules.

C. *Lymph organs* are the spleen, thymus and tonsils.

1. *The spleen* lies in the left upper quadrant of the abdomen, posterior and superior to the stomach. The arterial supply is the splenic artery, a branch of the celiac trunk. Venous drainage is to the splenic vein, which helps form the portal vein. *Functions*: the spleen is the largest lymph organ in the body. Although not essential for life, it appears to protect against certain infections (pneumoccocal pneumonia and sepsis), particularly in infants. It is the main organ of the MPS. If the

spleen is removed (splenectomy), other phagocytes of the MPS assume this responsibility. In animals, the spleen stores and releases blood in times of increased physical activity. In man, this appears to be a minor function. *Main functions:*

a. *Destruction of old red cells* by phagocytes of the MPS.
b. *Immunological defense.* The spleen is abundantly supplied with B and T lymphocytes, plasma cells and macrophages.
c. Serves as a *reservoir for platelets.* About 40% of the total blood platelets is sequestered in the spleen.
d. Produces blood in early fetal life.

Importance: The spleen is frequently injured in trauma to the abdomen, and sometimes must be removed to prevent a fatal hemorrhage. It may rupture as a complication of infectious mononucleosis. The spleen may enlarge (splenomegaly) in liver disease or congestive heart failure, because of the back-up of blood in the portal system.

2. *The thymus* is a small lymphoid remnant in the adult, lying below the thyroid gland against the trachea and above the superior vena cava. It is large and active in the newborn, and atrophies after puberty. Its function is to program a colony of lymphocytes to become *T-cells*. T-cells, and thus an intact thymus, are necessary for *cellular immunity*. *Importance*: a loss of cellular immunity either from an atrophic thymus or a lack of development causes impaired *protection against infection*. The thymus is also involved in myasthenia gravis. Over 60% of the cases of myasthenia show an enlarged thymus.
3. *The tonsils* are described in chapter 12.

D. *Function of lymph*:
Lymph differs from blood in that it has a lower protein content, contains no red blood cells and has fewer white cells. A main function is to *restore plasma proteins to the bloodstream* that have leaked through the capillary walls. A second function is to *transport absorbed fats from the gastrointestinal system to the bloodstream.*

IX. *Common Disorders of Blood and Body Fluids.*

A. *Anemia is a lack of red cells, or red cell substances such as hemoglobin or iron*. The anemias may be classified according to whether there is an increased loss, or a decreased production, of red cells.
1. *An increased loss of red cells*:
a. *Hemorrhage*
b. *Hemolytic* (cell-destruction) (7% of all anemias).
1) Sickle-cell anemia is a genetic condition mainly affecting blacks, and is the most prevalent hemolytic anemia. The incidence of the *trait* in black Americans is about 10%, and the *disease* is about 0.15%. Faulty amino acid substitution on the hemoglobin molecule results in some red cells collapsing and assuming "sickle" shapes (Fig. 10-10). These cells do not carry oxygen. In some localities in South Africa, 25% of the population have the trait. Evolutionarily, *the trait protects against invasion of the red cell by the amebic parasite, P. Falciparium, which causes malaria*. Those with the trait usually have no symptoms. Those with the disease do not survive much

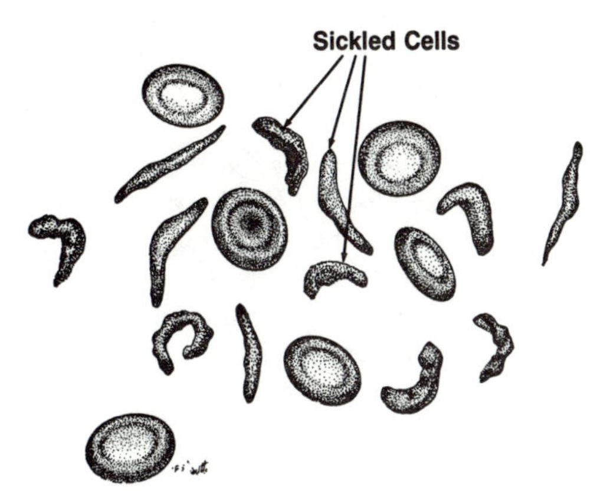

Fig. 10-10 Sickle Cell Anemia.

beyond age 30. Sickle cells block small blood vessels, causing multiple infarctions throughout the body. Common signs and symptoms are jaundice, decreased growth and development and pain in the arms, legs and abdomen. Death is usually caused by infection or a cerebrovascular accident. *Primary treatment* is symptomatic, and includes oxygen, blood transfusions and analgesics. Prevention: genetic counseling. If both parents have the trait, there is a 25% chance that the baby will have the *disease,* and a 50% chance that he will carry the *trait.*

2) Erythroblastosis fetalis (see earlier section, this chapter).

2. *A decreased production of red cells (or their constituents).*
   a. *Nutritional*:
      1) Iron deficiency (70%)
      2) Pernicious anemia (2%) is caused by a lack of *vitamin B-12, folic acid*, or "*intrinsic-factor*" from the stomach (see chapter 12).
   b. *Bone-marrow suppression* (20%): This type of anemia is rising. It is seen in those taking chemotherapeutic agents for cancer, in some taking certain antibiotics, and in patients with chronic diseases. It is also a complication of radiation therapy. Primary treatment is to remove the source of the problem. In some cases, an androgen (male hormone) will re-stimulate the bone-marrow. In severe cases transfusions must be given, along with a bone-marrow transplant.

B. *Edema* is the accumulation of abnormal amounts of fluid in tissue spaces. It often accompanies congestion, or an increased volume of blood in dilated vessels. Common causes of edema are:
   1. *Heart failure*: (see chapter 9)
   2. *Kidney disease*: a *loss of plasma proteins* leads to a decreased osmotic pressure and a loss of fluid from plasma into the tissue spaces.
   3. *Liver disease*: a *decreased production of plasma proteins* results in decreased osmotic pressure and fluid movement from plasma into the tissue spaces (as in 2).
   4. *Localized edema* is seen in inflammation and lymphatic obstruction.
      a. In *inflammation* there is an increased hydrostatic pressure and endothelial permeability from the release of histamine and kinins.
      b. *Lymphedema* is an increase in tissue fluid caused by the obstruction of lymph flow. For example, following a radical mastectomy which removes axillary lymph channels, arm drainage is often partially blocked and the arm swells.
   5. Primary treatment for generalized edema is the cautious use of *diuretics* to remove fluid.

C. *Hodgkin's disease is cancer of lymph tissue (lymphoma)*. There are several variations, but classical Hodgkin's involves a *T-cell deficiency.* Lymph nodes of the neck and groin are often palpable, but non-tender. A lymph node biopsy reveals the lesion. With the proper combinations of radiation and chemotherapy, the cure rate in the initial stages is high — one of the highest for any cancer — in some cases 100%.

D. *Leukemia* is *cancer of blood-forming tissue (i.e. bone marrow) resulting in an increase of mature and immature white cells in blood.* It may be rapidly progressing (acute) or slowly progressing (chronic). Generally, the acute forms show many immature, or "blast" cells, and the chronic forms show mostly mature types. Three categories are: *lymphocytic*, *monocytic* and *granulocytic* (plasma cell leukemia is *multiple myeloma*). The suffixes "cytic" and "blastic" refer to mature versus immature; chronic versus acute types. Sometimes, the term "myelogenous" is used, referring to a bone-marrow origin rather than a lymphoid one. The terms "myelocytic" and "myeloblastic" refer to precursor cells of the granulocytes (see Fig. 10-2). In most cases, not only are certain white cells abnormal, but other marrow cells are injured. This results in faulty red cell and platelet production, causing *anemia*, *thrombocytopenia* and *infection*. Brain hemorrhage and infection are the main causes of death. *Common leukemias* are:
   1. *Acute lymphoblastic* (acute lymphocytic) affects children, with a peak age of 5 years. It is often curative with chemotherapy. Complete remissions are frequent.
   2. *Acute myeloblastic* (myelocytic) affects adults, with no cure at present.
   3. *Chronic lymphocytic* affects older people. Often no therapy is required unless symptoms are evident, which is rare.
   4. *Chronic myelocytic* is found in young adults, with occasional remissions. Chemotherapy may help.

E. *Infectious mononucleosis* is a *viral disease affecting lymphocytes,* resulting in abnormal-appearing cells and lymphocytosis. Primarily a disease of young adults, it is transmitted by respiratory droplets ("kissing disease"). Common signs and symptoms are fever, sore throat, enlarged cervical lymph nodes and occasionally a rash. A blood smear shows atypical lymphocytes. *Complications* are usually the result of *increased activity* and include *ruptured spleen* and *hepatitis.* Primary treatment is bedrest (for up to one month).

F. *Polycythemia* is an *increase in red cells.* This may be a normal occurrence for people living at high altitudes where oxygen is low (more red cells are required to carry what little oxygen is available). *Polycythemia vera* is a condition in which not only the red cells but also white cells and platelets are increased. The red cell count is above 7 million per cubic millimeter, the hemoglobin is above 18 grams and the hematocrit is high. The person is fatigued and weak, with a red face. Complications are hemorrhage and thrombosis. Some may become leukemic. Primary treatment: the removal of about a liter of blood from the person per week (phlebotomy) and radioactive phosphorous therapy.

G. *Thrombocytopenia* is a *decrease in platelets.* Common causes are a low count after surgery, aspirin toxicity, uremia, multiple myeloma, infection, disseminated intravascular coagulation (DIC) and leukemia. *Idiopathic thrombocytopenic purpura (ITP)* is an autoimmune disease (antiplatelet antibodies are present). Common manifestations are easy bruising, nosebleeds (epistaxis), bleeding gums and blood in the urine (hematuria). The platelet count is usually below 100,000 per cubic millimeter. Complications include cerebral hemorrhage and bleeding into nerve tissue, causing paralysis. *Primary treatment*: removal of the spleen (splenectomy), since the spleen destroys platelets coated with antiplatelet antibodies.

-ELEVEN-

RESPIRATION

In terms of vital functions, the respiratory system is the most important of the various systems, since the heart and brain require a continuous supply of oxygen in order to function. Not breathing (apnea) for more than three or four minutes usually causes irreversible brain damage. The *respiratory rate* is one of the *four vital signs*. Respiration is an integral part of *cardiopulmonary resuscitation (CPR)*. This chapter deals with *external respiration*: the movement of oxygen into the lungs, through the bloodstream, to the tissues; and the movement of carbon dioxide from the tissues, through the bloodstream and out the lungs (*internal respiration* is the metabolism of oxygen and carbon dioxide *inside the cell*, and is briefly discussed in chapter 5). The *upper respiratory tract* consists of the nasal cavity, sinuses, pharynx and larynx; the *lower tract* is the trachea, bronchi and alveoli.

I. *Organs of the respiratory system.*

A. *The nose and nasal cavity* (Fig. 11-1):

1. The external *nose*, shown above, is composed mostly of *cartilage*, with two small hard *nasal bones* forming the upper part, or *bridge*. The tip of the nose is the *apex*, and the nostrils are the *nares*.
2. The *nasal cavity* consists of two parts, the *nasal fossae*, separated by a cartilage and bony wall (*septum*). The *roof* is formed by a small portion of the frontal bone, the ethmoid and sphenoid bones; the *floor* by the hard palate, consisting of the maxillae and palatine bones. The portion of the nasal cavity just inside the nares, the *vestibule*, contains stiff hairs (*vibrissae*) which prevent foreign material from entering. Three thin curled bones, the *turbinates* (conchae), project inward from the two outer walls. The spaces above and below each turbinate are the *meatuses*. The lower one receives the *nasolacrimal duct* which drains tears from the eye (Fig. 11-2).
3. A *mucous membrane (mucosa)* consisting of *epithelium*, *glands* and *connective tissue*, lines the nasal cavity and much of the respiratory system (as well as all of the gastrointestinal tract) (Fig. 11-3). Respiratory glands secrete a slimy substance called *mucous*, which, along with small, fine, undulating, hairlike processes (*cilia*) in the pharynx, larynx, trachea, bronchi and bronchioles trap particles such as dust and remove them. The cilia of the trachea and bronchi move upward, in contrast to those of the pharynx which move downward towards the esophagus, where foreign material is removed by swallowing.
4. Nerve endings for the *olfactory nerve* lie on the upper third of both sides of the nasal septum, the *olfactory areas*. They are stimulated by changes in gas consistency. Olfactory nerve fibers pass through small holes in the ethmoid bone to the olfactory bulb, and then to the *cortex*, where the impulses are interpreted as smell.
5. The *function of the nasal cavity* is to *warm*, *filter* and *moisten* incoming air. If this function is bypassed,

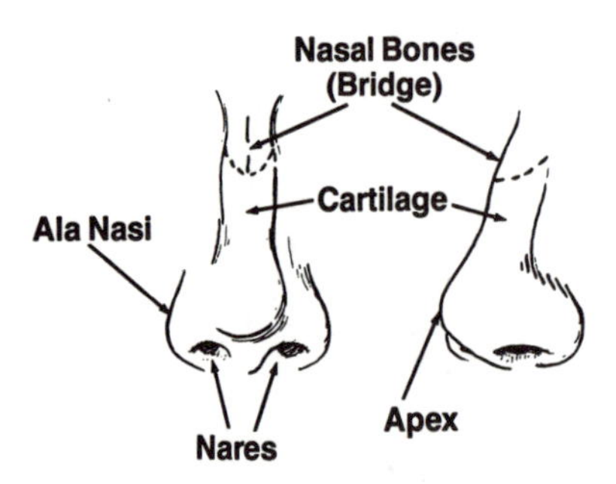

Fig. 11-1 External Nose.

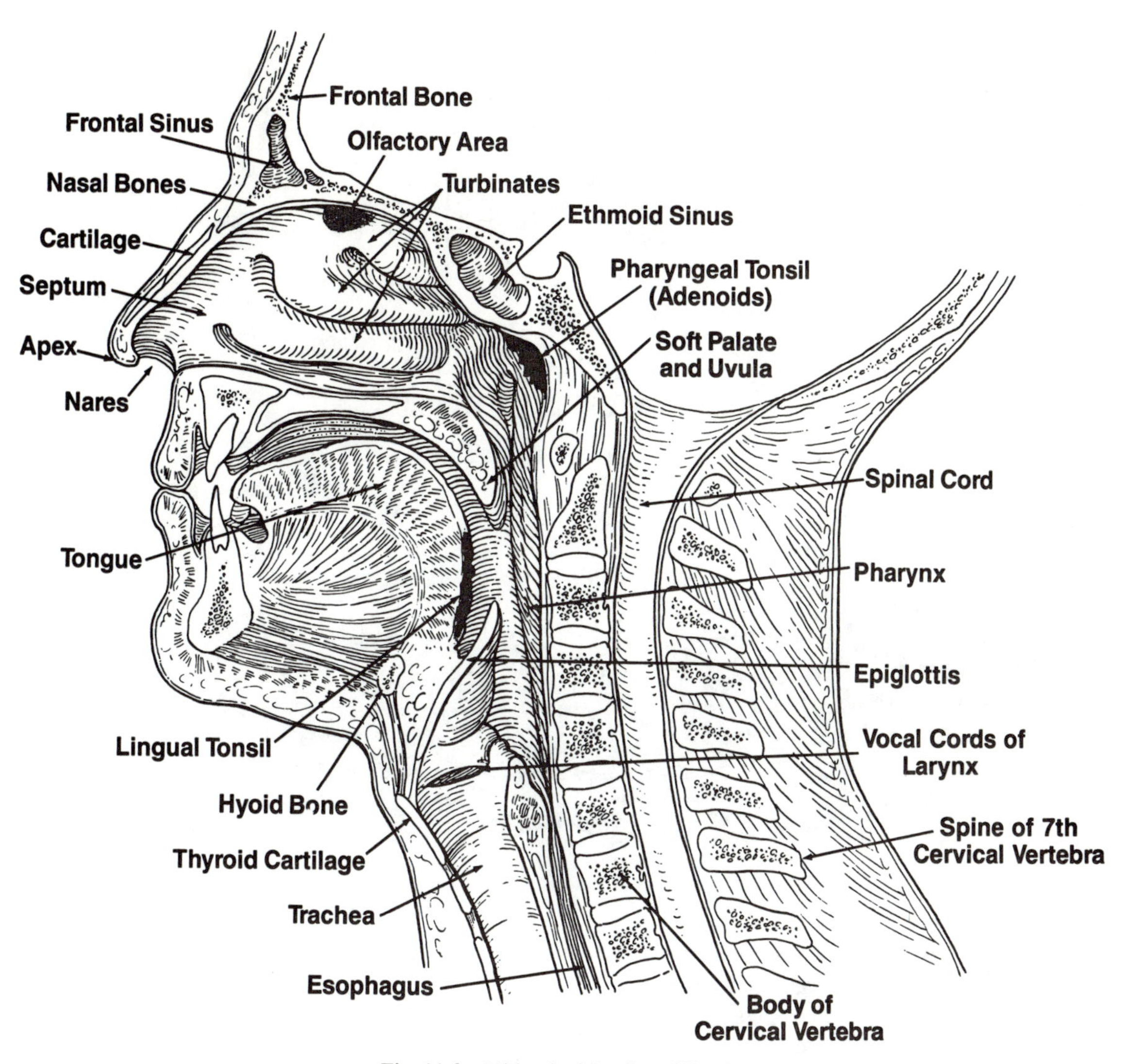

Fig. 11-2 Midsagittal Section of Head.

as in patients on mechanical ventilators, drying of the lungs can create discomfort and infection. Consequently, a vapor mist is often added to a ventilator.

6. Venous areas called *swell bodies* are located on the turbinates. About every half-hour, the swell bodies on one side of the nasal cavity engorge with blood, resulting in decreased air flow on that side, with good flow on the other side. Then it reverses. These periodic changes permit recovery from drying.
7. The structure of the nasal cavity above the vestibule permits the implantation of a wide variety of objects by children, such as peanuts, beads, beans and a diversity of insects, often leading to purulent infections.
8. The *septum* is richly supplied with sensory nerves and blood vessels. Most nosebleeds (epistaxis) originate here, the more common causes being trauma and nose-picking. Treatment involves the local application of neosynephrine nose drops, or epinephrine, for

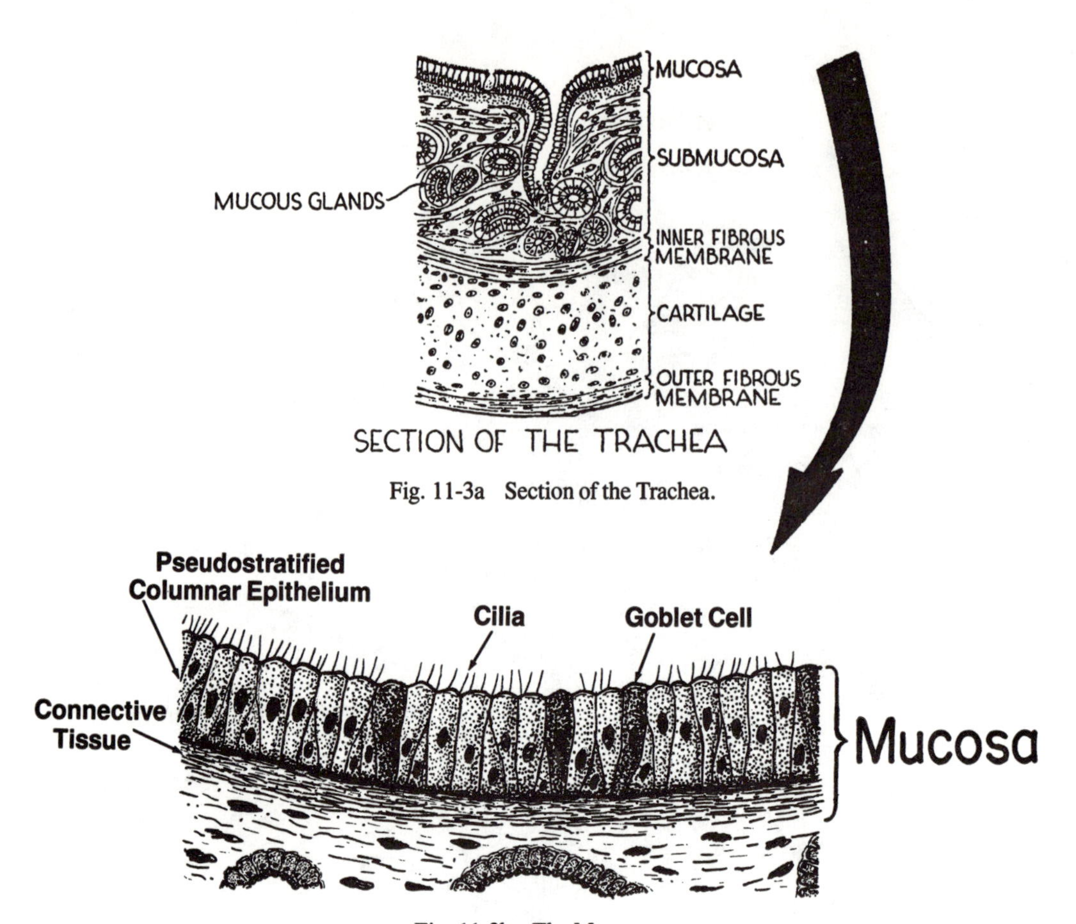

Fig. 11-3a Section of the Trachea.

Fig. 11-3b The Mucosa.

vasoconstriction. Sometimes cauterization (caut: burning) with a cotton applicator coated with silver nitrate is necessary. The accompanying burning and irritation of sensory nerve endings often triggers the *sneeze reflex* — the inhalation of air followed by an explosive exhalation, to clear the airway.

9. A "*deviated septum*" is a condition in which the septal cartilage is bent, usually from a blow, resulting in difficulty in breathing from one side of the nose. In earlier days, some boxers had the septal cartilage removed so that they could breathe easier, resulting in the pug-nosed appearance often seen in older fighters.

B. *Sinuses* are *spaces* in the frontal, ethmoid, sphenoid and maxillary *bones* that are lined with *mucosa* and communicate with the nasal cavity (Fig. 11-4). They make the skull light, thus making it easier for people to walk in the upright position. Since the sinus mucosa communicates with the nasal cavity, sinuses are prone to the same infections as the nasal cavity.

C. *The nasopharynx* is the continuation of the nasal cavity into the throat, or *pharynx* (the *pharynx* is discussed with the digestive system).

D. *The larynx*, or voice box, is the entrance to the trachea, and consists of *cartilages*, *ligaments*, *muscles* and the *vocal cords*. The cartilages provide a rigid structural framework for the larynx and the trachea below it, insuring air passage at all times.

1. The *thyroid cartilage* is the largest, and can be felt in front of the neck

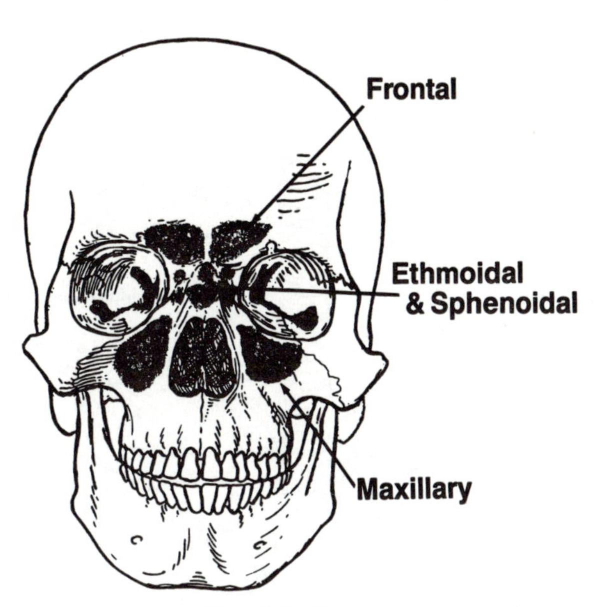

Fig. 11-4 Sinuses.

as the "Adam's apple". Directly below it is the *cricoid cartilage*, which is connected to the thyroid cartilage by a connective tissues membrane, the *cricothyroid membrane*. Inside the larynx, the *epiglottis* folds over the *vocal cords* during swallowing, allowing food to pass into the esophagus. The *glottis* is the opening between the two vocal cords (Figs. 11-5, 11-6, 11-2).

2. The *function of the larynx*, in addition to permitting air passage to and from the lungs, is to produce sound, or *phonation. The lips and the tongue create speech.* During expiration, the vocal cords vibrate to produce high or low sound, or *pitch*. In high pitch the glottis is more closed and taut, while in low pitch the glottis is more open and relaxed. *Laryngitis*, or clinical hoarseness, results from inflammation of the vocal cords by overuse, infection or tumor. Obstruction of the glottic area for any reason can rapidly be fatal. Common causes are blockage from food, and sometimes croup in children. Bacterial infection of the epiglottis (epiglottitis) in children is a life-threatening but fortunately rare cause.

E. *The trachea,* or windpipe, is a five-inch tube that begins at the glottis and ends at the junction of the two main bronchi (the *carina*) near the level of the *sternal angle*. The framework of the trachea consists of sixteen to twenty *cartilage rings*, with connective tissue between them (Fig. 11-5). Internally the trachea is lined with mucosa (Fig. 11-3). When a foreign particle enters the trachea, it is trapped by mucous and cilia, and the *cough reflex* is initiated: the glottis closes, the respiratory muscles contract, the glottis suddenly opens and the smooth muscle layer of the trachea contracts, making the tube narrower. This increases the velocity of expired air, and an explosive discharge takes place.

F. The *right and left main bronchi* are the continuation of the trachea, and have the same structural framework: *mucosa* (including more smooth muscle than in the trachea), *connective tissue*, and *cartilage rings*. Each main

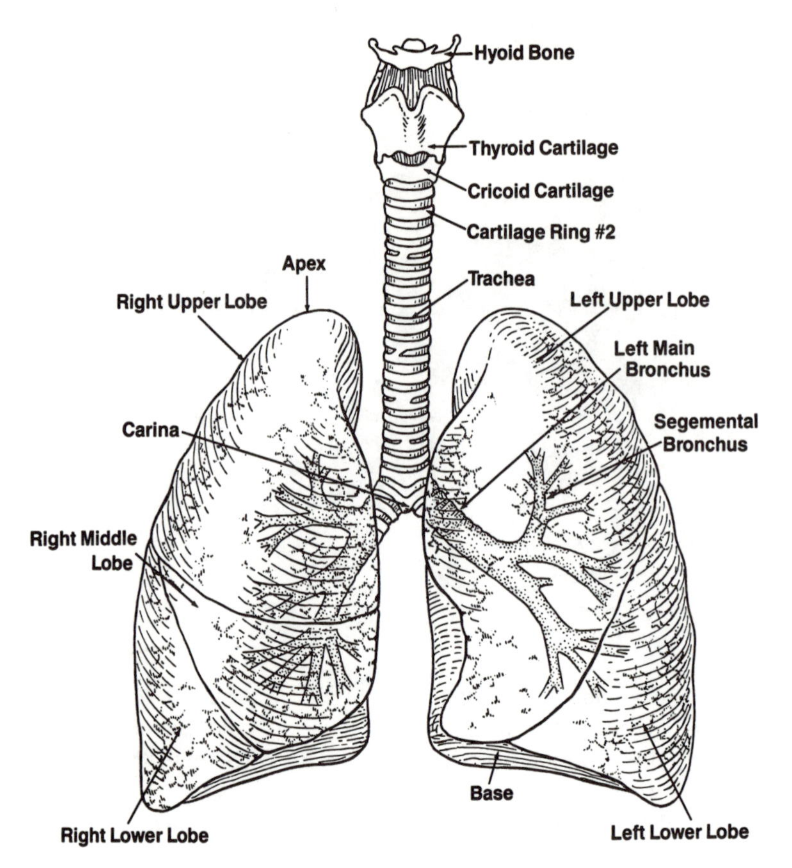

Fig. 11-5 The Respiratory Tree.

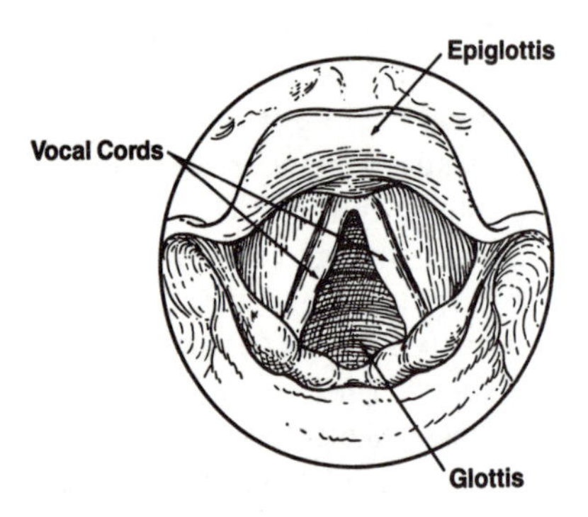

Fig. 11-6 Larynx (Viewed from Above).

bronchus divides into two (left lung) or three (right lung) *lobar bronchi*. Each lobar bronchus divides into ten segmental bronchi, which again further divide (Fig. 11-5). Gradually, cartilage is lost, and when the tubes become about *one millimeter in diameter*, they are *bronchioles*, which terminate in the air sacs, or *alveoli*. The right main bronchus is the natural continuation of the trachea; thus, the angle is less. If a child inhales (aspirates) a peanut, for example, it is more likely to enter the right main bronchus than the left.

G. *The lungs* are spongy, highly vascular organs that lie in the left and right pleural cavities. The *right lung* has *three lobes*: an *upper*, *middle*, and *lower*; the *left* an *upper* and *lower*. The upper part of each lung is the *apex*, located about an inch above the first rib. The lower curved part is the *base* and lies on the diaphragm (Figs. 11-5, 11-7). The medial portion, next to the heart, is the *root, or hilum*, and contains the *pulmonary artery, vein, main bronchus and lymph nodes*. The *alveoli* are surrounded by *capillaries*,

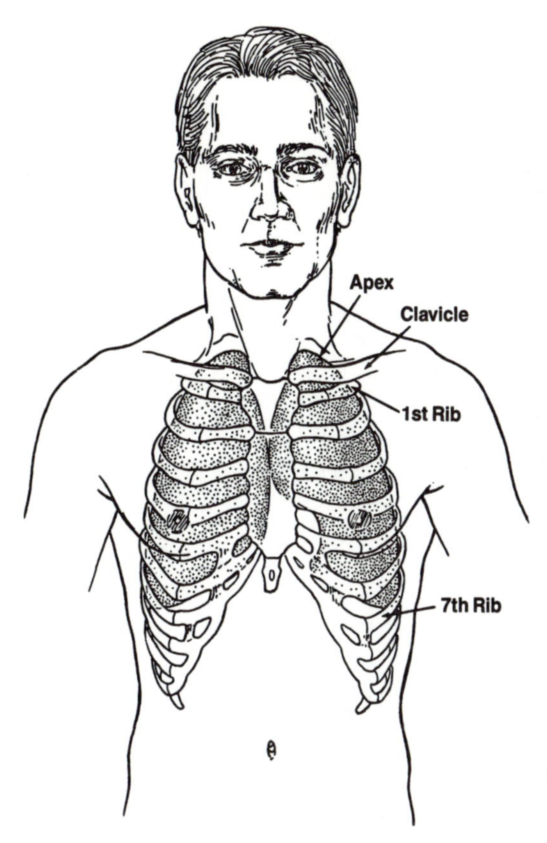

Fig. 11-7 Relation of Lungs and Thorax.

branches of the pulmonary arteries and veins. The *exchange of oxygen and carbon dioxide takes place here* (Figs. 11-8, 11-9).

H. *The thorax*, or chest cavity, is the area enclosed by the sternum, ribs and thoracic vertebrae. It includes the region posterior to the sternum containing the heart and great vessels (the *mediastinum*), as well as the *pleural cavities* containing the lungs (see p. 246).

I. *The pleura* is described in chapter one. The *pleural space*, or cavity, contains about 1/2 teaspoon of *fluid*, providing lubrication against friction. Actually the pleural cavity is not a cavity at all, since visceral and parietal pleurae are in contact with each other (see Fig. 1-5).

1. An increase of fluid in the pleural space (*pleural effusion*), seen in diseases such as lung cancer and pulmonary edema, can severely compromise breathing. A tube is inserted between the ribs into the pleural space and the fluid is drained off (thoracentesis).
2. Sometimes air enters the pleural space (*pneumothorax*) from a rupture of a part of a lung, as in emphysema, or from a penetrating injury by a knife or bullet. A chest tube (thoracostomy tube) is inserted between the ribs, connected to a pump, and the air evacuated. Blood in the pleural space (*hemothorax*) is drained in a similar manner.

J. *Nerves and vessels of the lungs*:

1. The *autonomic nervous system* supplies the bronchi and bronchioles. Stimulation of the *vagus nerve* (*parasympathetic*) contracts the smooth muscle and narrows the diameter of the tube (*bronchoconstriction*). Stimulation of *sympathetic nerves* causes smooth muscle relaxation and widening of the tube (*bronchodilation*).
2. The *pulmonary arteries and veins* participate in the *exchange of gases* between the capillaries and alveoli.
3. Branches of the aorta and upper intercostal arteries, the *bronchial arteries*, supply most of the *lung tissue*. *Venous drainage* is from the *azygous vein* on the right side of the thorax, and the first intercostal vein on the left.

II. *Respiratory Physiology*

A. *The mechanics of breathing*:

1. During *inspiration*, the *external intercostal* muscles between the ribs contract, lifting the lower ribs up and out. The *diaphragm*, the dome-shaped muscle separating the thorax from the abdomen, moves down, increasing the volume of the pleural cavities. Elastic fibers in the alveolar walls stretch, permitting expansion of the air sacs.

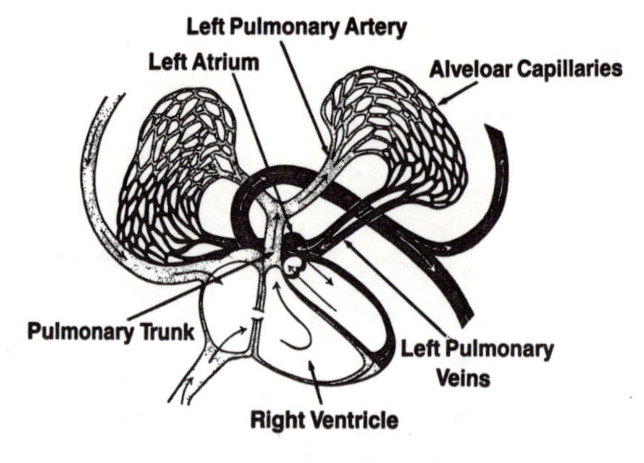

a. The Pulmonary Circulation

(Modified, with permission, from Circulation of Blood, General Biological, Inc., Chicago, 1947).

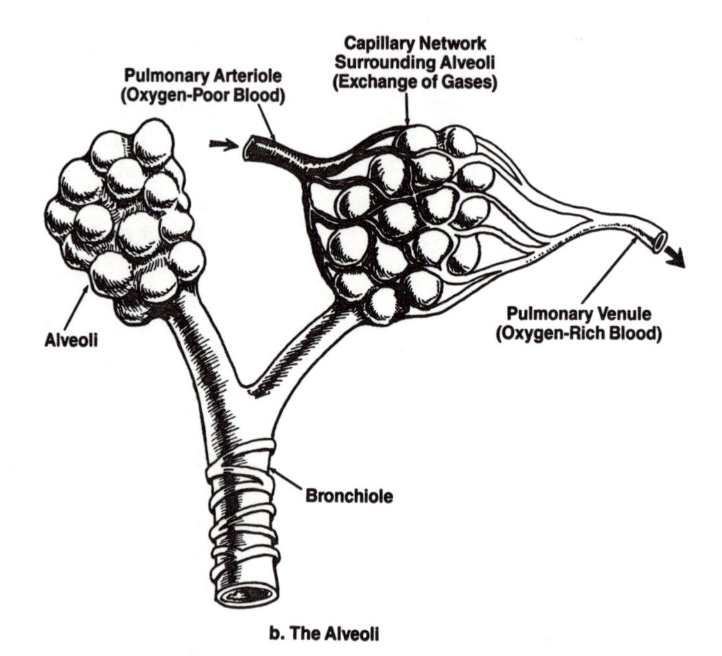

b. The Alveoli

Fig. 11-8 The Pulmonary Circulation and Alveoli.

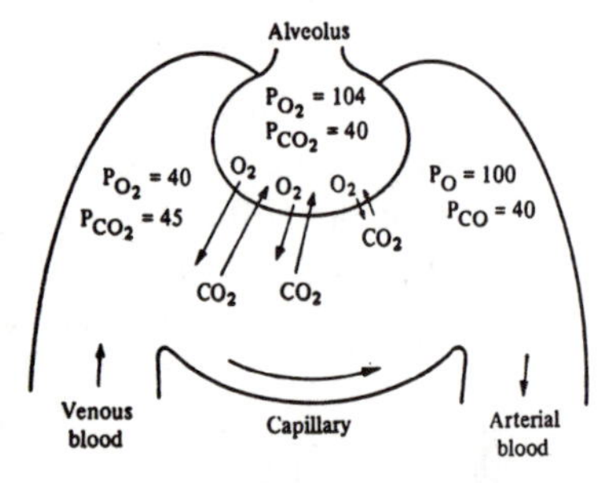

Fig. 11-9 Exchange of Gases at Alveolus. (Reproduced, with permission, from Thurber, Human Physiology, Wiley, 1969).

2. *Expiration* is passive; that is, relaxation of the external intercostals, diaphragm and alveolar walls (Fig. 11-10).
3. The *pressure in the pleural space* is below atmospheric pressure (760 mm Hg). This reduced pressure allows for expansion of the lungs during inspiration. However, the higher external pressure means that if a wound punctures the chest wall, air will enter the pleural space

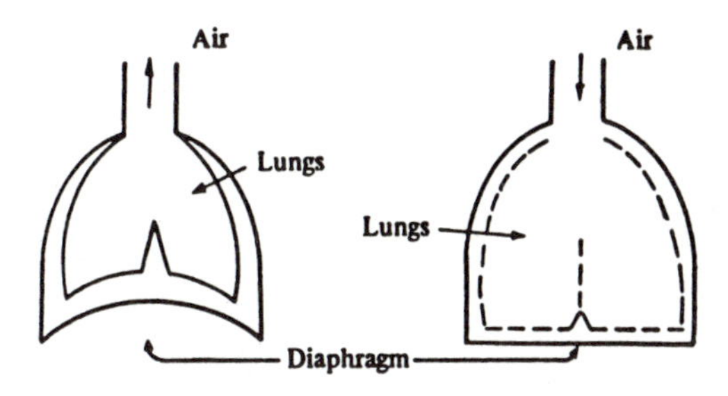

a. Expiration **b. Inspiration**

Fig. 11-10 Chest Movements in Normal Breathing. (See ref., Fig. 11-9).

(pneumothorax) and collapse the lung (atelectasis).

4. In *diseases* such as asthma, bronchitis and emphysema, accessory muscles of respiration are often used. Inspiration is aided by contraction of the sternocleidomastoid and other muscles of the neck; expiration is aided by the use of the internal intercostals and abdominal muscles.
5. The *diaphragm* receives its nervous supply from the neck region (*C3-5*) of the spinal cord (during fetal development, the diaphragm descends from the neck to the abdomen). A broken neck that injures the spinal cord below C-5 still allows the person to breathe, since the diaphragm does most of the breathing. Injury to both phrenic nerves, or a spinal cord injury above C3-5, severly compromises breathing.

B. *The lung volumes* (Fig. 11-11):

1. Breathing in and out creates volumes which are measured on a *spirometer*, a machine with a writing lever recording on graph paper. As one breathes in, the lever goes up; as one breathes out, it goes down. Fig. 11-11 shows various lung volumes. Three of these give a good evaluation of the patient's lung status: (1) the *tidal volume*, or the amount of air inspired in one breath in normal breathing, about 1/2 liter; (2) the *vital capacity* (VC), which measures how much air the lungs can actually hold, about 3.5 to 5.5 liters; and (3) the *forced expiratory volume* in 1 second (FEV_1), which measures the percentage of the vital capacity that the person can exhale in 1 second, about 70%.
2. In lung diseases such as *asthma* and *emphysema*, the VC and FEV_1 are abnormal. An asthmatic, for example, usually has a normal tidal volume and vital capacity but decreased FEV_1; while a person with emphysema may have a normal (but usually decreased) tidal volume and decreased VC and FEV_1.

C. *Transport of oxygen and carbon dioxide*:

1. The percentages of gases in room air are: nitrogen (N_2) — 79%, oxygen (O_2) — 20.96%; and carbon dioxide (CO_2) — 0.04%. A convenient way to measure the amounts of these gases is by partial pressures, in millimeters of mercury (mm Hg).
2. The *exchange of oxygen and carbon dioxide* takes place by *diffusion*. The pulmonary arteries bring oxygen-poor blood from the right ventricle to the lungs. Carbon dioxide diffuses from the bloodstream, through the capillary and alveolar membranes, and is blown off. Oxygen diffuses in the opposite direc-

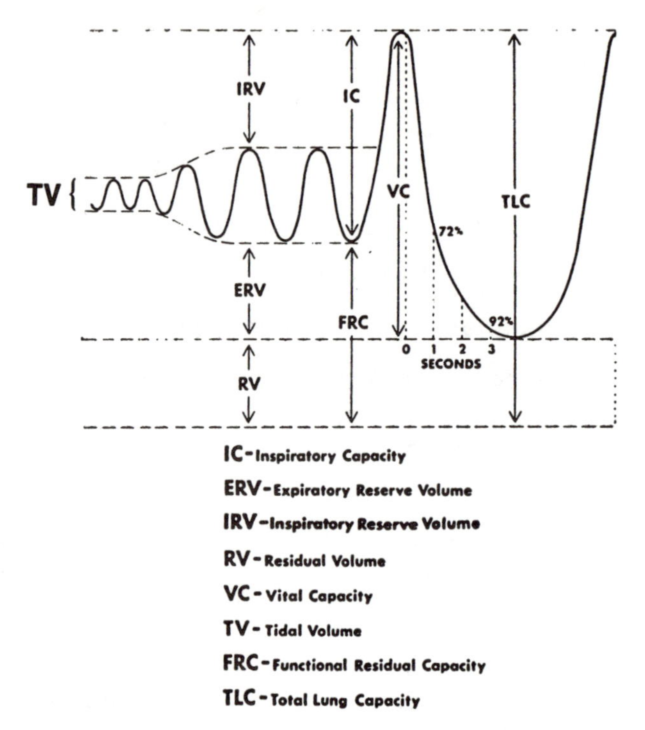

Fig. 11-11 Lung Volumes. (Reproduced and modified, with permission, from Ayers, A Guide to the Interpretation of Pulmonary Function Tests, Pfizer, 1974).

tion — from the alveoli through both membranes and into the bloodstream. The pulmonary veins return oxygen-rich blood to the left atrium (Fig. 11-8).

3. *Oxygen* is transported in the blood as *oxyhemoglobin*. Red cells move into the *capillaries*. At the arteriole end of the capillary, O_2 diffuses through the red cell, then through the capillary membrane *into the tissue fluid*. It then diffuses through the tissue cell membrane to be used as fuel for cellular metabolism (Fig. 11-12).
4. *Carbon dioxide* moves out of the tissue cell in the reverse direction (through the same membranes) into the red cell, where most is converted to *bicarbonate ion* (HCO_3^-). Bicarbonate is transported via the plasma to the lungs, where the process is reversed in the alveolus, and carbon dioxide is blown off.
5. In a person with *emphysema*, the respiratory center in the medulla is reset to a high carbon dioxide level, since he cannot blow it off. Instead of a high carbon dioxide being the stimulus to breathe, the main stimulus is now low oxygen (hypoxia). If a normal dose of oxygen (4-6 liters per minute) is given to a person with emphysema, it will take away his breathing stimulus and he may die. That is why low-dose oxygen (1-2 liters per minute) is given to emphysema patients.
6. A *SCUBA diver* normally breathes a combination of 80% nitrogen and 20% oxygen in his tank, sometimes with a small amount of helium. At sea level, about a liter of nitrogen is dissolved in the blood and other body fluids; at 100 feet down, about four liters. If he ascends too rapidly, the high concentration of dissolved nitrogen escapes from

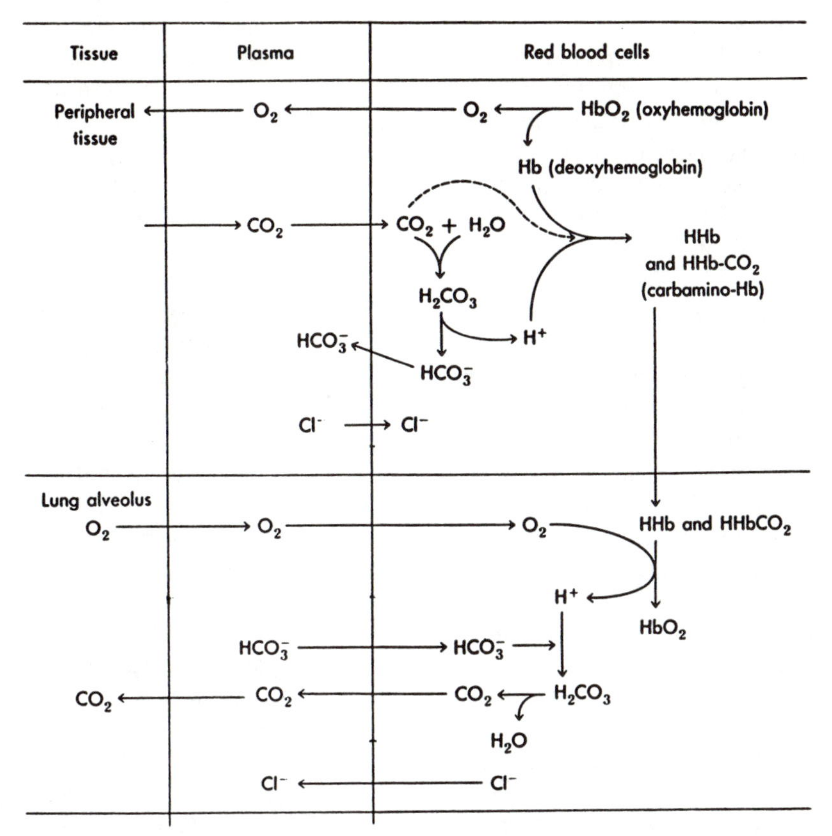

Fig. 11-12 Transport of CO_2 and O_2 in Body. (Reproduced and modified, with permission, from Rafelson, Basic Biochemistry, Macmillan Publishing Co., 1980).

solution. These gas bubbles, or *air emboli*, block capillaries in various parts of the body, particularly the joints and central nervous system. This is *decompression sickness (the bends)*. Treatment is to recompress the diver in a pressure chamber, with gradual decompression.

D. *Control of breathing*:

1. The *respiratory center* is a group of nerve cells in the medulla and pons. The center is affected by a variety of stimuli. Impulses from the cerebral cortex (voluntary control) modify respiration, as do changes in the carbon dioxide content and acidity of blood. An increase in carbon dioxide content (hypercapnia) causes an increase in acidity (acidosis) and the respiratory rate (hyperpnea); a decrease causes a decrease in the rate (bradypnea). The acidity (H^+ concentration) of blood affects the medulla by way of the CSF.
2. Certain cells in other areas of the body are also sensitive to acidity and oxygen concentrations. These are the *chemoreceptors*, and lie near the baroreceptors, discussed in chapter 9 (Fig. 9-21). Two are located near the arch of the aorta (*aortic bodies*), and one in each carotid artery (*carotid bodies*). A decrease in oxygen, or an increase in carbon dioxide or acidity, stimulates an increase in breathing rate; an increase in oxygen, or a decrease in carbon dioxide or acidity, causes a slight decrease in the res-

piratory rate.

E. *Vital sign #3* — the respiratory rate:

1. The *respiratory rate* in adults is about *12 to 16 breaths per minute*. In the newborn it is about 35, and decreases to adult values at about age 20. When a person is told that his respiratory rate is going to be checked, he often becomes self-conscious and begins to breathe in an odd fashion. Therefore, it is common to observe the breathing rate without mentioning it.
2. *Emotions* are a powerful stimulus for respiratory changes. Fear, grief and shock slow the rate; anger and sexual arousal increase it.
3. *Hyperpnea* (fast breathing) and *tachypnea* (rapid shallow breathing) are seen in exercise (increased oxygen requirement), obesity (increased vessel resistance), infections and fever (production of pyrogens and increased energy requirements), heart failure (decreased oxygen flow), pain (increased nervous stimulation), anemia (decreased oxygen), hyperthyroidism (increase in metabolic rate), emphysema and pneumothorax (blockage of oxygen).
4. *Bradypnea* (slow breathing) is seen in uremia, alcohol and other depressant-drug intoxications, because of the depressant action on the brain. It is also seen in increased intracranial pressure (from pressure on the respiratory center) and in diabetic coma (see metabolic acidosis).
5. Periods of hyperpnea alternating with periods of apnea (no breathing) are sometimes seen in the sleep of infants, particularly premature ones. They also appear in brain injury and in the terminally ill (*Cheyne-Stokes breathing*).
6. In some individuals, drowsy episodes accompanied by snoring and apneic spells occur (*obstructive sleep apnea* — OSA). Obesity is common in these people, as are large tonsils and adenoids, small jaw, large tongue and soft palate, and other subtle anatomic abnormalities. It appears that the tongue falls back during non-REM sleep and blocks the airway. *Sudden infant death syndrome (SIDS)* is probably a variation of this. Both central nervous system and obstructive problems may cause SIDS, but OSA seems to be an important component. Additional risk factors for SIDS include being male, low-birth-weight, decreased carotid body substance, decreased number of myelinated fibers in vagal nuclei of the brainstem, and an upper respiratory infection.

III. *Acid-Base Balance*.

A. *General*:

1. The acid-base status of the body has to do with buffers. A *buffer is a neutralizer*. Buffers are located in blood, tissue fluid and cells. The main buffers in blood and tissue fluid are hemoglobin (the most important), plasma proteins and bicarbonate. In the cell, the main buffers are proteins and phosphate. The kidneys contribute to buffering as a back-up system using bicarbonate and phosphate.
2. This perplexing and often confusing area of physiology and medicine can best be understood by looking at the *bicarbonate-buffer equation*:

$$\underset{\text{(acid)}}{H^+} + \underset{\text{(bicarbonate)}}{HCO_3^-} \leftrightarrow \underset{\text{(carbonic acid)}}{H_2CO_3} \leftrightarrow \underset{\text{(water)}}{H_2O} + \underset{\text{(carbon dioxide)}}{CO_2}$$

The above substances are in equilibrium with each other. The equation can move to the right or left, depending on the concentrations of each substance. If an acid (H^+) is added to the system, it will combine with HCO_3^- to form carbonic acid, which dissociates into water, and carbon dioxide is blown off. If a base (OH^-) is added, it combines with H^+ to form water, again neutralizing a toxic substance:

$$\underset{\text{(base)}}{OH^-} + \underset{\text{(acid)}}{H^+} \longrightarrow \underset{\text{(water)}}{H_2O}$$

3. The H^+ concentration in the bloodstream is expressed as the *pH, the negative logarithm of the H^+ concentration,* a simpler series of numbers with which to work. When the H^+ goes up, the pH goes down. A low pH means the condition is more acidic; a high pH, more basic.
4. For simplification, the bicarbonate buffer equation can be expressed in the following way:

$$H^+ + HCO_3^- \longleftrightarrow CO_2 + H_2O$$

If there is a high concentration of CO_2 (right side), the equation will shift to the left; a high concentration of either H^+ or HCO_3^{++} (left side) causes a shift to the right. The discussion of buffers is limited to the bicarbonate system, since it is the more important.

B. Blood for *gas analysis* is obtained by puncture of the radial, brachial or femoral artery, and is analyzed on a *blood gas analyzer*. This takes about five minutes—a truly emergency, or "STAT" procedure. Arterial blood gases (ABG's) have the following normal values (gases are expressed as partial pressures in mm Hg):
   1. pH—7.35 to 7.45
   2. pO_2 — 75 to 100
   3. pCO_2 — 35 to 45
   4. Other figures on a laboratory slip are *calculated values*. In many laboratories, the term "torr" has replaced "mm Hg" for partial pressure. *One torr* = one mm Hg.

C. *Respiratory acidosis*:
The modified formula:

$$H^+ + HCO_3^- \longleftarrow CO_2 \text{ (blocked)}$$

is what happens in respiratory acidosis. Ventilation is blocked. The person cannot blow off CO_2, which then accumulates; the equation shifts to the left, resulting in a high H^+ (low pH) and bicarbonate. This condition is seen commonly in *emphysema*, *bronchitis*, and in diseases where ventilation is poor, such as heart failure, pneumonia and respiratory failure from any cause. Labored breathing (dyspnea) is usually present. Primary treatment is to *increase ventilation* with a bronchodilator. In the severe case, a mechanical ventilator may be required.

D. *Respiratory alkalosis*:
The modified formula:

$$H^+ + HCO_3^- \longrightarrow CO_2 \text{ (blown off)}$$

is what happens in respiratory alkalosis. The person blows off CO_2 at an increased rate; the equation shifts to the right, resulting in a low H^+ (high pH) and bicarbonate. This condition is seen in *hyperventilation*, common in those who are anxious or are in pain, neurotic individuals (anxiety attacks), athletes doing sprints and in patients on respirators. Respiratory alkalosis causes a transient loss of ionized calcium, and patients frequently are in *tetany* (see chapters 10 and 12). A common scenario: the person is brought to the emergency room in carpo-pedal spasm, frightened, agitated, fingers tingling, "short of breath", sometimes with chest pain. Reassurance that nothing seriously wrong is necessary. Primary treatment consists of rebreathing expired air (high in CO_2) using a paper bag or a re-breathing mask. In the case of the patient on a respirator, the machine is turned down.

E. *Metabolic acidosis*:
The modified formula:

$$H^+ + HCO_3^- \longrightarrow CO_2 \text{ (blown off)}$$

is what happens in metabolic acidosis. In this case, the body produces an increased amount of H^+ and a shift to the right occurs, resulting in a high H^+ *(because of increased production)*, and low bicarbonate. There is an increase in the respiratory rate (a deep, regular breathing pattern called Kussmaul breathing) in order to blow off the accumulating CO_2. This condition is seen in *kidney disease* (accumulation of acidic substances), in *diabetes mellitus* (increased metabolism of fatty acids for energy) and in diarrhea (increased loss of HCO_3^- from the intestines). Several drugs, such as acetazolamide (Diamox) and aspirin, may also cause metabolic acidosis. Primary treatment: treating the underlying condition, and adding HCO_3^- if the pH is too low (usually below 7).

F. *Metabolic alkalosis*:
The modified formula:

$$H^+ + HCO_3^- \longleftarrow CO_2$$

is what happens in metabolic alkalosis. In this case, acid is lost, ventilation decreases, and the equation shifts to the left to restore the acid. There is thus a low H^+ and high HCO_3^-. This condition occurs in people who have been *vomiting* for a long time (depleting hydrochloric acid from the stomach) and in those using *diuretics* (losing H^+ from the kidney). Primary treatment is directed at relieving the vomiting and letting hydrochloric acid accumulate, or stopping the diuretic.

G. *Practical considerations regarding blood-gases and acid-base status*:

1. As a general rule, a patient with either an O_2 of less than or a CO_2 of greater than 50 torr should be considered for endotracheal intubation and mechanical ventilation. An exception is the patient with chronic obstructive pulmonary disease (emphysema or bronchitis). This person has adapted to a high CO_2 (possibly 70 torr).
2. A patient with a pH of less than 6.8 should probably be given bicarbonate regardless of the specific condition.

H. *Summary of acid-base disturbances*:

1. The body may compensate for changes in acid-base status (*compensated*), or it may not (*uncompensated*). Compensated adjustments take place each day in all of us. Buffers in the blood and in the kidney help achieve this. In the above conditions, however, compensated adjustments are not sufficient to control the situation.
2. In general, *disorders of acidosis are usually the life-threatening ones*, although occasionally alkalosis may cause death, as in intractable vomiting, prolonged use of diuretics, taking anti-ulcer alkaline drugs, or with mechanically assisted ventilation. The major effect of acidosis, regardless of the cause (emphysema, diabetes, uremia, diarrhea or heart failure), is a depression of the central nervous system with eventual coma and death.
3. Frequently, *two disorders take place at the same time*. In diabetes, for instance, *metabolic acidosis* creates an increased breathing rate which results in a superimposed *respiratory alkalosis* (hyperventilation). In cardiac arrest, there is both a *metabolic* and *respiratory acidosis* from the accumulation of acids in the blood as well as from the lack of breathing **(Table 11-1).**
4. An important point is that in *respiratory* problems the *increase or decrease of CO_2* creates the increase or decrease in H^+; in *metabolic* problems, the *increase or decrease in H^+* leads to the respiratory adjustment.

IV. *Common Disorders Of The Respiratory System.*

A. *Asthma*, like hay fever, is a common *allergic disorder*, affecting the small airways. It may be triggered by antigens, such as dog hair, house dust, etc. or an emotional upset. In either case, the *smooth muscle layer of the bronchi and bronchioles goes into spasm (constriction) and the glands of the bronchi*

Table 11-1: Summary of Acid-base Disturbances

pCO_2	HCO_3^- (meq) < 21	HCO_3^- (meq) 21 - 26	HCO_3^- (meq) >26
>45	METABOLIC ACIDOSIS + RESPIRATORY ACIDOSIS	RESPIRATORY ACIDOSIS	METABOLIC ALKALOSIS + RESPIRATORY ACIDOSIS
35 - 45	METABOLIC ACIDOSIS	NORMAL	METABOLIC ALKALOSIS
<35	METABOLIC ACIDOSIS+ RESPIRATORY ALKALOSIS	RESPIRATORY ALKALOSIS	METABOLIC ALKALOSIS+ RESPIRATORY ALKALOSIS

hypersecrete mucus (bronchioles have no glands). Air enters the alveoli but cannot leave because of the constriction. Dyspnea and wheezing (the sound of air flowing through a narrowed tube) occur as the person tries to force air out of the lungs. Arterial blood gases initially show a low CO_2 (respiratory alkalosis) because of the increase in breathing rate. A rise in CO_2 to normal and above may herald impending respiratory failure. Primary treatment: a beta-2 agent such as terbutaline, epinephrine (beta-1 and beta-2), a bronchodilator such as aminophyllin, fluid and an antibiotic if the trigger is an upper respiratory infection. Steroids may be required to reduce inflammation.

B. *Carbon monoxide poisoning* is the leading cause of gas deaths in the country. It is odorless, and *binds to hemoglobin 210 times more readily than oxygen*. The majority of deaths occur from smoke inhalation during fires. Some occur from automobile exhaust fumes, poorly ventilated gas heaters and charcoal stoves. Poisonings also occur in machine-shops where ventilation is poor. Symptoms are headache, dizziness, weakness and nausea, usually when the blood has about six to seven percent *carboxyhemoglobin*. Sadly, many city dwellers, particularly taxi drivers, auto mechanics and garage attendants, normally show more than six percent of carbon monoxide in the blood. Primary treatment: the use of 100% oxygen by mask to displace the carbon monoxide from hemoglobin.

C. In *choking*, often the person is drinking alcohol, the throat muscles are relaxed, he is talking while eating and inhales at the same time as he swallows. The piece of food, usually meat, impacts against the larynx. The term "cafe coronary" has been applied to

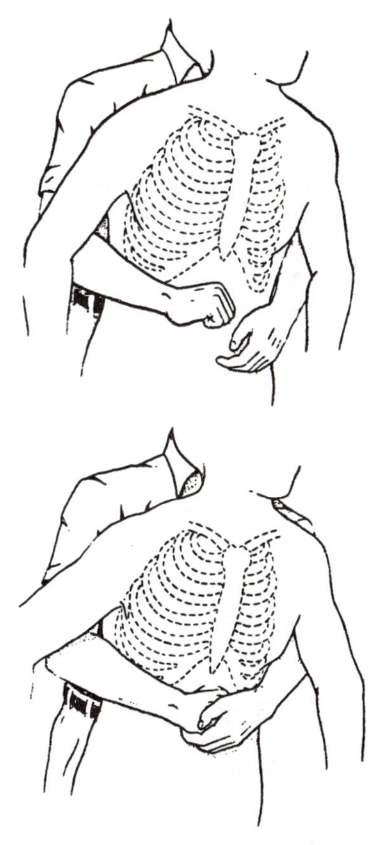

Fig. 11-13 Abdominal Thrust (Hemilich Maneuver). (Reproduced, with permission, from Suratt, Manual of Medical Procedures, Mosby, 1982).

such a situation because it superficially resembles a heart attack. The first thing to do is ask the person if he can speak. If he can, he is not obstructed; if he cannot, he is. Steps to take:

1. The food may be dislodged by one or more *abdominal thrusts (Heimlich maneuver)* (Fig. 11-13): the thumb side of the fist is placed against the victim's abdomen, slightly above the navel. Grasp the fist with the other hand, and press upward into the victim's abdomen quickly and forcefully. This produces a sudden intra-thoracic pressure, and will usually dislodge the food. If the person is on the floor, place him on his back and place yourself astride him. With one hand on top of the other, press quickly into his abdomen upwards towards his chest. Repeat the thrusts as often as necessary. This maneuver can also be done on oneself.
2. If the four minute point is approaching, and the above maneuvers have been unsuccessful, an emergency *cricothyrotomy* should be performed (Fig. 11-14). Find the *cricothyroid membrane* as an indentation between the thyroid and cricoid cartilages and thrust anything, such as a knife or ballpoint pen, with enough force so that the trachea is penetrated. The force must be substantial or the trachea will not be entered. Remove the inner pen piece and keep the barrel in place, or keep the space open with two keys, or a straw. Stay with the victim to make sure the wound is kept open until he reaches the emergency room.

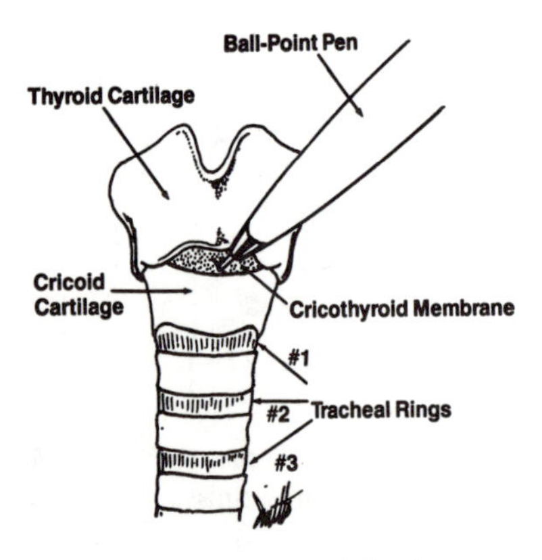

Fig. 11-14 Emergency Cricothyrotomy.

D. *Chronic obstructive pulmonary disease (COPD)*: Although the processes involved in the evolution of *emphysema* and *bronchitis* are different, the end result is irreversible respiratory insufficiency, sometimes called chronic obstructive pulmonary disease (COPD). *In emphysema the obstruction is in the alveoli; in bronchitis it is in the bronchi*. A component of each is seen in heavy smokers.
   1. *Acute bronchitis* is usually the result of an upper respiratory infection moving downward. *The bronchial mucosa is swollen and red, much mucus is secreted by the glands, and the sputum is often green or yellow*. Chronic bronchitis is often due to prolonged irritation by cigarette smoke. Wheezes may be heard. Primary treatment: a bronchodilator and an antibiotic.
   2. In *emphysema* (Gr: "to puff up"), resulting from long-term irritation of the bronchi and bronchioles, mucus and pus accumulate, and the *air in the alveoli becomes trapped*. When the pressure exceeds the elastic limit, *the alveoli become permanently ballooned-out*, producing the typically barrel-chested person who uses the internal intercostals as well as the abdominal and neck muscles to breathe. A chest x-ray shows flattening of the diaphragms. Primary treatment: increasing ventilation by physical therapy and bronchodilators. Emphysema is the most common cause of respiratory failure. It is often dramatically improved by stopping cigarette-smoking.

E. The *common cold* is caused by *one of 55 viruses*. Usually infecting the nasal mucosa, the viruses may spread to the sinuses, pharynx and down the respiratory tract. Many of the signs and symptoms of a cold are similar to those seen in hay fever, except that in addition there is frequently pharyngitis, with fever and cough. The fever is due to the infection, and the cough is caused by irritation of the pharyngeal mucosa. Primary treatment consists of aspirin, fluids and bed rest.

F. *Croup* is a *viral infection* of children ages six months to four years. The *larynx, epiglottis and trachea are red and swollen* and may block the glottis. A "seal bark" cough is usually present. Sometimes a high-pitched whistling inhalation is present (stridor), and the child must use his neck and abdominal muscles to breathe. Primary treatment: cold steam and oxygen. Occasionally a breathing tube must be inserted.

G. In *drowning*, the *victim both inhales and swallows water*. In 10% of cases, the larynx goes into spasm from the first small amount of fluid inhaled, and asphyxia (suffocation: no oxygen and an increase in carbon dioxide) takes place with no fluid in the lungs. Survival depends mostly on the presence of a pulse, and not so much on the time of immersion. The 3-4 minute rule about anoxia (no oxygen) causing permanent brain damage may not apply. This is illustrated by the survival with normal intelligence of a 5-year-old boy after a documented 40 minutes of submersion in *ice cold* fresh water. He regained consciousness in two days and recovered normally. Primary treatment in near-drowning consists of cardiopulmonary resuscitation, with 100% oxygen given at the hospital and bicarbonate for the acidosis.

H. *Hayfever* is an *allergic reaction* seen in people overly sensitive to pollen, house dust, chicken feathers, etc. *Nasal vessels are engorged with blood* (*congestion*), and fluid leaks from the capillaries into the tissue spaces (edema). From the tissue spaces, fluid drains into the nasal cavity (rhinitis, or runny nose). The congestion and edema cause the irritation, sneezing, redness and swelling seen in the weary sufferer in the spring and summer months. Primary treatment is with an anti-histamine and decongestant.

I. *Hiccups* (singultus) is a *sudden involuntary contraction of the inspiratory muscles, producing the sound of inspiration with the glottis closed*. Most cases involve food or alcohol, are short-lived, and resolve without therapy. This is a primitive reflex, like yawning, coughing, sneezing and vomiting. Premature infants hiccup more than usual, and hiccuping is documented in fetuses. The reflex is designed to dislodge a foreign object. Often, sedatives and home remedies work. Sometimes the use of a major tranquilizer is necessary.

J. *Influenza* (flu) is a *common viral infection of the entire body*, resulting in sore bones, tiredness, backache, weakness, cough and pleurisy. Primary treatment, as with most viral infections, is bed rest, aspirin and fluids.

K. *Pleurisy* (pleuritis) *is an inflammation of the pleura*, usually from a lung infection. A stabbing pain is felt on deep inspiration. The pain usually resolves when the infection resolves.

L. *Pneumonia* is an *inflammation of the lung, usually caused by infectious agents such as bacteria, viruses and fungi*. Chemicals may also cause pneumonia (*aspiration pneumonia*, seen in the elderly and debilitated, is when acid stomach fluid is inhaled into the lungs, resulting in a chemical erosion of lung tissue with subsequent infection). Common signs and symptoms are fever, chills, chest pain, coughing up yellow or green sputum, and rales (the sound of movement of air and fluid in the bronchial tree). A chest x-ray usually shows a white infiltrate. The white blood count is elevated. Primary treatment for bacterial pneumonia is an antibiotic, bed rest, aspirin and fluids. Sometimes oxygen must be given. Occasionally the person must be hospitalized and the antibiotic given intravenously.

M. In *pulmonary embolism*, a *clot (thrombus) from a deep leg vein (usually a calf vein) detaches (embolus) and travels to the right atrium, then the right ventricle, and to the pulmonary trunk*. If the clot is large, and lodges at the division of the pulmonary trunk to the right and left pulmonary arteries, death may ensue (saddle embolism). If the clot moves into one of the pulmonary arteries and impacts in one of its branches, destroying tissue, this is a pulmonary *infarction*. Common signs and symptoms are *chest pain, dyspnea and tachypnea*. Primary treatment: oxygen and an anticoagulant. Predisposing factors for clot formation are obesity, heart failure, surgery, immobility and a history of thrombophlebitis (see chapters 9, 10).

N. *Pulmonary edema* is the *accumulation of fluid in the lungs*. The most common cause is *heart failure*. When the left ventricle fails, blood backs up in the pulmonary circulation and fluid leaks into the lungs. Common signs and symptoms are dyspnea, cyanosis (bluish tinge to the skin), shortness of breath, rales and jugular venous distention (from accompanying right ventricular failure).The person often sits up to catch his breath. Primary treatment is the same as for heart failure: oxygen, a diuretic, a bronchodilator, a vasodilator and morphine.

O. A *sinus attack (sinusitis)* commonly *accompanies a nasal infection*. There is inflammation, congestion, edema, and pain because of irritation of the sensory nerve endings in the periosteum. Pain takes the form of a *headache*, particularly if the *frontal sinus* is involved. Drainage into the nasal cavity is blocked by congestion. The *maxillary sinus* lies over the upper teeth, and it is sometimes difficult to tell if one is having a sinus attack or a toothache; the pain often elicits a feeling of "long teeth". Primary treatment is the same as for a cold: an antibiotic and a decongestant.

P. *Sore throat (pharyngitis)*, perhaps the most common malady in office practice, is an *inflammation of the mucosa of the pharynx*. The cause is usually *viral*. If tonsils are present, the condition is *tonsillitis*. A throat culture (strep screen) will show if group A beta-hemolytic streptococcus is present (the organism that causes *rheumatic fever, rheumatic heart disease* or *glomerulonephritis* as complications in certain individuals). Common signs and symptoms are a red tender throat, occasionally with an exudate, enlarged cervical lymph nodes and fever. Primary treatment: rest, an analgesic, saline gargles and an antibiotic if the culture is positive for streptococcus.

Q. *Tracheotomy*: obstruction from many causes—infection, tumor, injury, etc., may make it necessary to insert a *breathing tube between two cartilage rings* of the trachea. A semi-permanent opening is a *tracheostomy*.

R. A *yawn* is an interesting physiological event with psychological overtones. One reason is said to be *underventilation of the alveoli; thus the sudden involuntary attempt to oxygenate them*.

-TWELVE-

ENDOCRINE GLANDS AND HORMONES

The *endocrine system* consists of ductless *glands* that *secrete hormones directly into the bloodstream*. In contrast, glands with ducts, or exocrine glands (e.g., salivary and sweat glands) discharge their products at the terminations of their ducts. *Hormones are amino acid derivatives (polypeptides, proteins) or steroids; like enzymes, they are not destroyed in the process of affecting cellular activity. They control the rate at which reactions occur.* Their presence in proper amounts is required for normal bodily functions. End products are excreted primarily by the kidneys. Some hormones act locally, but most act at sites distant to the site of origin.

I. *Actions of hormones*. Most hormones affect specific cells (e.g., ACTH stimulates only cells of the adrenal cortex) although some hormones stimulate almost all cells (e.g., thyroid hormone). The type of cell sensitive to the hormone is the *target cell*. Sensitivity is dependent on the *receptor site*, located either in the cell membrane or within the cell itself. Certain receptors are sensitive only to certain hormones. Hormones act by activating either the cyclic-AMP system, or genes and protein synthesis inside the cell.

A. *Activation of the cyclic-AMP system.* In this case, the *receptor* lies externally, on the *cell membrane* (Fig. 12-1). The *hormone acts as a "first messenger"* to stimulate and combine with the receptor. This binding activates the enzyme *adenyl cyclase*. Adenyl cyclase converts ATP to cyclic-AMP, which acts as a "*second messenger*", producing the physiological response in the cell. Most hormones act in this way.

B. *Gene activation and protein synthesis in the cell. Steroid hormones* act by inducing the synthesis of enzymes in target cells. They bind to a receptor protein in the cytoplasm.

C. *The negative feedback mechanism of hormone control.* The body seeks to maintain physiological balance, or *homeostasis*, insuring that there is not an excess or deficiency of a substance. Two examples of "negative feedback" are shown in Fig. 12-2:

1. *Parathyroid hormone*. Low blood calcium triggers calcium release from bone by parathyroid hor-

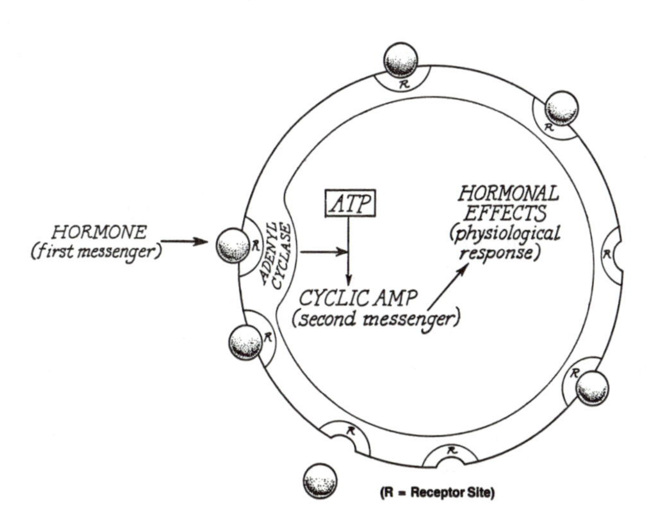

Fig. 12-1 Action of Hormone on Target Cell.

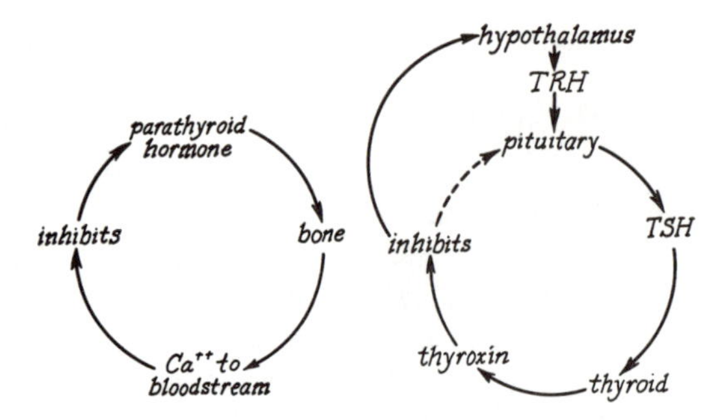

Fig. 12-2 Two Examples of Negative Feedback.

mone, restoring a normal blood calcium and shutting off further parathyroid hormone release.

2. *Thyroid hormone*. When the thyroxin level is low, the hypothalamus secretes thyrotropin-releasing hormone (TRH) which stimulates the pituitary to secrete thyroid stimulating hormone (TSH). TSH stimulates the thyroid gland to produce thyroxin which shuts off further hypothalamic and pituitary secretion.

II. Table 12-1 lists each endocrine gland, its major hormone(s), function(s), and the most common causes of excesses and deficiences.

III. *The pituitary* (Fig. 12-3):

A. *Anatomy*: The pituitary (hypophysis) is the inferior extension of the hypothalamus and lies in the sella turcica. It has an anterior portion (*anterior lobe*) and a posterior portion (*posterior lobe*).

B. *Functions*: The pituitary secretes hormones that regulate growth, fluid balance, lactation and childbirth, as well as other endocrine organs ("tropic" action). The pituitary is regulated by releasing and inhibiting hormones from the hypothalamus (an example is *luteinizing-hormone releasing hormone, LHRH,* recently emerging as a promising fertility and contraceptive agent. It has also been used to treat cryptorchidism and precocious puberty). The negative-feedback mechanism acts on the hypothalamus and not on the pituitary directly. The terms *primary* and *secondary* refer to *target organ problems* versus *problems in other organs* that affect the target gland. Commonly, the pituitary is the other organ, but the kidney may be involved. *Primary hyperthryroidism* means that the cause is in the thyroid gland. *Secondary hyperthyroidism* refers to a pituitary origin for the condition. The *anterior lobe* secretes six major hormones and the *posterior lobe* two.

C. *Simple tests of pituitary function*:

1. In *giantism and acromegaly,* the pituitary produces excessive growth hormone, and x-ray demonstrates an enlarged sella turcica (Figs. 12-4, 12-5). In *secondary Cushing's disease*, the pituitary produces excessive ACTH, resulting in an increased production of steroids by the adrenal gland (see Fig. 12-10). *In a midget*, the pituitary produces decreased growth hormone, and many tropic hormones are also low. In pituitary deficiency, there is a decrease in all pituitary hormones, resulting in the loss of target organ hormones (adrenal steroids, thyroxin and the gonadotrophins).
2. In *diabetes insipidus*, vasopressin release by the pituitary is decreased. The urine osmolality decreases and the urine output increases, sometimes to 20 liters per day.

D. *Primary treatment*:

1. *Treatment for gigantism, acromegaly and secondary Cushing's disease* is surgery and/or x-ray therapy. *Pituitary dwarfism* in the child responds to the administra-

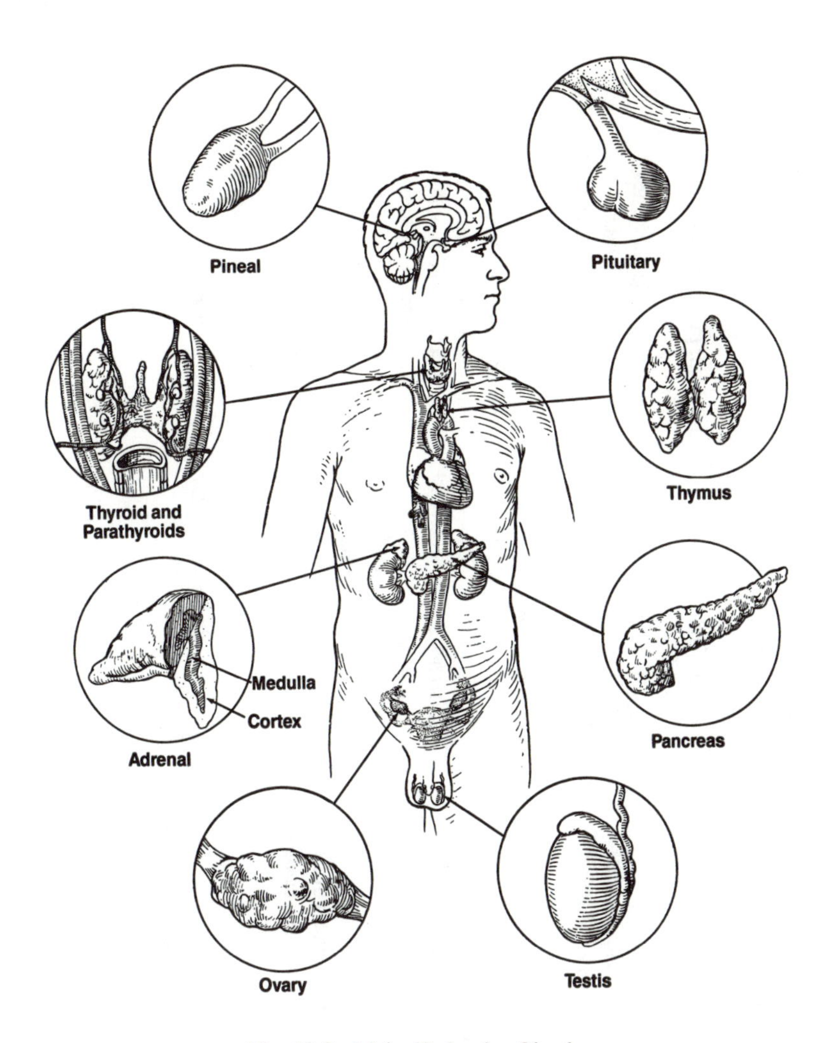

Fig. 12-3 Major Endocrine Glands.

tion of growth hormone. Target-organ hormones (thyroid, adrenal, sex steroids) are replaced. Post-partum necrosis sometimes resolves spontaneously. Otherwise, replacement therapy is begun.

2. In the case of *diabetes insipidus,* vasopressin is effective. An adequate fluid intake may control mild cases. If the cause of the diabetes insipidus is a tumor, x-ray therapy and/or surgery are employed.

IV. *The thyroid gland* (Fig. 12-3):

A. *Anatomy*: The thyroid gland lies on the trachea below the thyroid cartilage. It consists of a *right and left lobe* connected by a bridge (*isthmus*). The isthmus covers the second and third cartilage rings, and the lobes extend up to the mid-portion of the thyroid cartilage on each side. Many vessels supply the thyroid.

B. *Function*: The thyroid has one important function: it *regulates metabolism in all organs* by maintaining an adequate cellular level of oxygen consumption. The principal hormones are *thyroxine* and *triiodothyronine*.

Table 12-1: ENDOCRINE GLANDS
(*MCC* = MOST COMMON CAUSE)

I. *PITUITARY* A. *ANTERIOR LOBE HORMONES:*	*REGULATED BY*	*MAIN FUNCTION(S)*	*EXCESS*	*DEFICIENCY*
1. *Growth Hormone* (Somatotropin)	Hypothalamus: GHRH, GIH (Growth hormone-releasing hormone, growth hormone-inhibiting hormone)	Stimulates bone growth at epiphyseal plates	(applies mainly to growth hormone) MCC: tumor in child & adult a. *Child: Gigantism* 1) symmetrical growth of long bones 2) acromegaly 3) sometimes hyperglycemia b. *Adult: Acromegaly* 1) enlarged hands, feet and jaw 2) sometimes hyperglycemia c. In the above conditions *tropic hormones* may be decreased, increased or normal	a. *Child: pituitary dwarfism (midget)* MCC: genetic deficiency 1) symmetrical dwarf 2) normal intelligence 3) normal tropic hormones b. *Adult:* MCC: multiple thromboses in mother following childbirth. There is hypofunctioning of all tropic hormones resulting in atrophy of thyroid, adrenal and gonads with accompanying clinical symptoms
2. *ACTH* (Adrenocorticotropic hormone, Corticotropin)	CRH (Corticotropin-releasing hormone)	Stimulates release of hormones from adrenal cortex: aldosterone, cortisol		
3. *TSH* (Thyroid-stimulating hormone, Thyrotropin)	TRH (Thyrotropin-releasing hormone)	Stimulates release of thyroxin from thyroid gland		
4. *FSH* (Follicle-stimulating hormone)	FRH (Follicle stimulating hormone-releasing hormone)	*Female:* stimulates ovarian follicle growth and estrogen secretion *Male:* stimulates production of spermatozoa	[for excesses and deficiencies of adrenals, thyroid and gonads, see those sections]	
5. *LH* (Luteinizing hormone, formerly ICSH, or interstitial-cell-stimulating hormone, in male)	LHRH (Luteinizing hormone-releasing hormone)	*Female:* stimulates ovulation and secretion of progesterone by corpus luteum *Male:* stimulates secretion of testosterone by testes	rare	rare

Table 12-1: Cont'd

I. PIT (Cont'd)	*REGULATED BY*	*MAIN FUNCTION(S)*	*EXCESS*	*DEFICIENCY*
6. *Prolactin* (LTH, luteotropic hormone)	PRH, PIH (Prolactin-releasing hormone, prolactin-inhibiting hormone)	Stimulates production of milk by mammary gland	rare tumor	rare
B. *POSTERIOR LOBE HORMONES:*				
1. *ADH* (Anti-diuretic hormone, Vasopressin)	Hypothalamus: osmotic stimuli, Na^+	Reabsorption of water in collecting duct of kidney	*Inappropriate secretion*, MCC: lung cancer Hyponatremia	*Diabetes insipidus* *MCC;* tumor (breast cancer) 1) thirst (polydipsia) 2) increased urination(polyuria)
2. *Oxytocin*	Hypothalamus: via suckling reflex	Stimulates: 1) contraction of uterus 2) delivery of placenta 3) milk ejection	extremely rare	extremely rare
II. *THYROID GLAND:*				
1. *Thyroxin* (T_4) 2. Triiodothyronine (T_3)	TSH	Controls metabolic rate, growth	*Hyperthyroidism (Graves' Disease)* MCC: autoimmune 1) increased metabolism 2) goiter 3) exophthalmos	*Hypothyroidism;* a. *Child: Cretinism*, MCC congenital 1) dwarfism 2) enlarged tongue and abdomen 3) mental retardation b. *Adult: Myxedema*, MCC is treatment for hyperthyroidism 1) thick dry skin 2) weakness 3) fatigue 4) delayed tendon reflexes 5) slowed mentation

Note #1. Laboratory T_3 and T_4 show status of thyroid gland. TSH shows status of pituitary.

Note #2. "Endemic goiter," which may include some cases of cretinism, is a compensatory enlargement of the thyroid gland from a lack of *iodine*.

Table 12-1: Cont.

	REGULATED BY	MAIN FUNCTION(S)	EXCESS	DEFICIENCY
III. *PARATHYROID GLANDS:*				
Parathormone (PTH)	Blood Ca^{++} level	Stimulates release of calcium from bone	*Hyperparathyroidism* (hypercalcemia) 1) MCC: calcium loss from kidney disease (thus secondary) a) fractures & cysts of bone b) calcium deposits in tissues 2) Another common cause is lung cancer	*Hypoparathyroidism* (hypocalcemia) MCC: removal of thyroid gland Tetany: flexion of wrist and joints (carpopedal spasm)
IV. *PANCREAS:*				
Insulin (from beta cells in islets of Langerhans)	Blood glucose level	Permits glucose to enter cell	Hyperinsulinism. MCC: diabetic with too much insulin injected and/or too little food	*Diabetes mellitus* MCC: hereditary 1) hyperglycemia 2) glycosuria 3) ketoacidosis
V. *ADRENAL GLANDS:*				
A. *MEDULLA:*				
Epinephrine	Sympathetic nervous system	Sympathomimetic emergency ("fight-or-flight") reaction; elevation of blood glucose	Episodic hypertension MCC: tumor	very rare
B. *CORTEX:*				
1. Mineralocorticoids: *Aldosterone*	ACTH, Renin/ Angiotensin	Na^+ reabsorption/ K^+ loss by kidney	*Conn's Syndrome.* MCC: tumor 1) hypertension 2) hypokalemia 3) metabolic alkalosis	*Addison's Disease* (chronic adrenal cortical unsufficiency) MCC: autoimmune, resulting in mineralocorticoid and glucocorticoid deficiencies 1) hypotension 2) skin pigmentation 3) loss of NaCl 4) muscle weakness 5) unable to resist stress
2. Glucocorticoids: *Cortisol, Corticosterone*	ACTH	1) converts some amino acids to glucose 2) maintains body's ability to counteract stress	*Cushing's Disease.* MCC: lung cancer, causing ectopic foci of ACTH-secreting tumors: 1) central obesity 2) moon face & buffalo hump 3) hypertension and hypokalemia 4) hyperglycemia	

Table 12-1: Cont'd

V. ADRENAL GLANDS (Cont'd)	*REGULATED BY*	*MAIN FUNCTION(S)*	*EXCESS*	*DEFICIENCY*
3. Sex steroids	ACTH	Secretion of androgens may be responsible for initial secondary sex characteristics in males, and early body hair growth in young females	*Adreno-genital Syndrome.* MCC: Rare genetic enzyme deficiency resulting in increased secretion of androgens. *Prepuberty:* Female: pseudo-hermaphroditism Male: precocious puberty *Post-puberty:* Female: change toward secondary male sex characteristics	(Addison's, as above)
Note: serum cortisol level evaluates adrenal cortex. ACTH evaluates pituitary. Common synthetic steroids (hydrocortisone, prednisone) are used to suppress inflammatory response.				
VI. *TESTES:*				
Testosterone (from intersitial cells of Leydig)	LH	Male puberty: secondary sex characteristics	Precocious puberty. MCC: tumor both pre- and post-puberty	*Eunuchoidism.* MCC: 1) pre-puberal: *hypopituitarism* 2) post-puberal: *Klinefelter's Syndrome:* a) XXY chromosome b) sterility c) enlarged breasts d) lack of many secondary male sex characteristics
VII. *OVARIES:*				
Estradiol (from theca cells of follicle)	FSH	Female puberty: secondary sex characteristics	Precocious puberty. MCC: tumor	*Amenorrhea.* MCC: (in order) 1) menopause 2) castration 3) Turner's Syndrome: a) XO chromosome b) sterility c) dwarfism d) webbed neck e) lack of many female secondary sex characteristics
Progesterone (from corpus luteum)	LH	Maintains uterine endometrium, suppresses ovulation	very rare tumor	(as above)

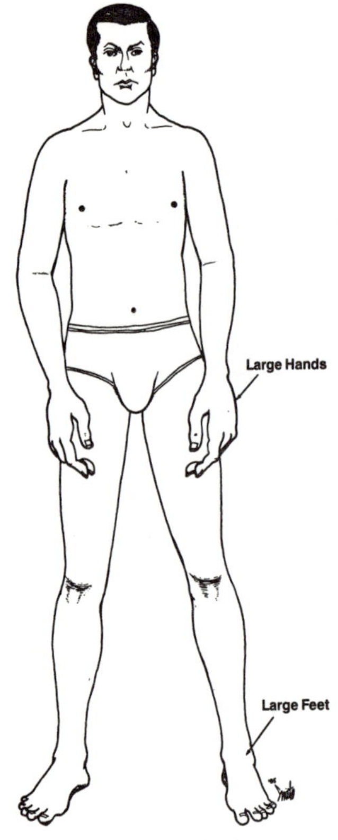

Fig. 12-4 Giantism.

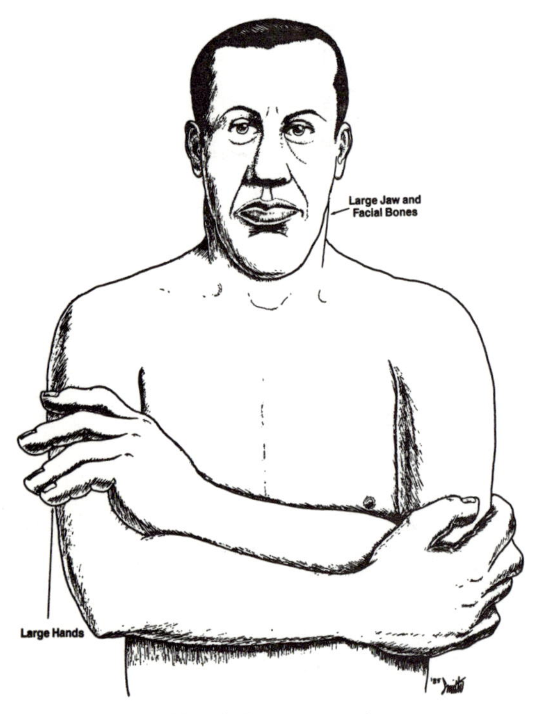

Fig. 12-5 Acromegaly.

Iodine is necessary for the synthesis of both hormones.

C. *Hyperthyroidism* (thyrotoxicosis) is the *second most common endocrine disorder (behind diabetes mellitus)*, affecting mostly females. The most common cause is *autoimmune*.
   1. Several immunoglobulin autoantibodies, such as long acting thyroid stimulator (LATS) and thyroid-stimulating immunoglobulin (TSI), apparently bind to TSH receptor areas on the thyroid cell membrane, causing hypersecretion of thyroxin. Symptoms include sweating, weight-loss (even with 4 to 5 large meals per day, plus snacks), nervousness, loose stools, tachycardia, warm moist velvety skin, hand tremor and hyperactivity. Hyperthyroidism mimics manic-depressive psychosis, and is diagnosed with certainty only by thyroid tests. If thyrotoxicosis is accompanied by *goiter* and *exophthalmos*, it is *Grave's disease* (Fig. 12-6). Rarely is the pituitary involved (secondary hyperthyroidism). Hyperthyroidism is almost always accompanied by a goiter.
   2. The cause of *bulging eyes (exophthalmos, proptosis)* is *an antigen-antibody reaction* in the external eye muscles. An inflammatory infiltrate develops, and the orbital muscles enlarge. The reaction of TSI on orbital tissue causes excessive water binding. This edema, plus the inflammatory reaction of the ocular muscles, pushes the eyes forward.

D. *Hypothyroidism.* The most common cause of hypothyroidism is treatment of hyperthyroidism by radioactive iodine or subtotal thyroidectomy. The second most common cause is autoimmune. This is interesting, since excluding treatment for hyperthyroidism the most common cause is the same as for hyperthyroidism. Symp-

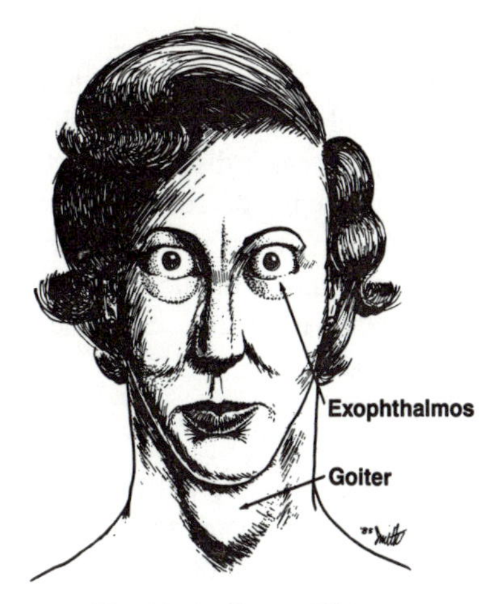

Fig. 12-6 Graves Disease.

toms include weakness, fatigue, constipation, hoarseness, bradycardia, obesity, sluggishness and slowed mentation (sometimes with psychotic behavior). A goiter is often present. Severe hypothyroidism accompanied by thick leathery skin, puffy face and coarse hair is *myxedema.* (Fig. 12-7). Untreated hypothyroidism in an infant results in dwarfism and mental retardation (*cretinism*) from decreased metabolism of bone epiphyses and cerebral cortical neurons (Fig. 12-8). Again, the cause is rarely secondary (pituitary).

E. *Simple tests of thyroid function*:
The easiest test is a *T-4* (thyroxine) and

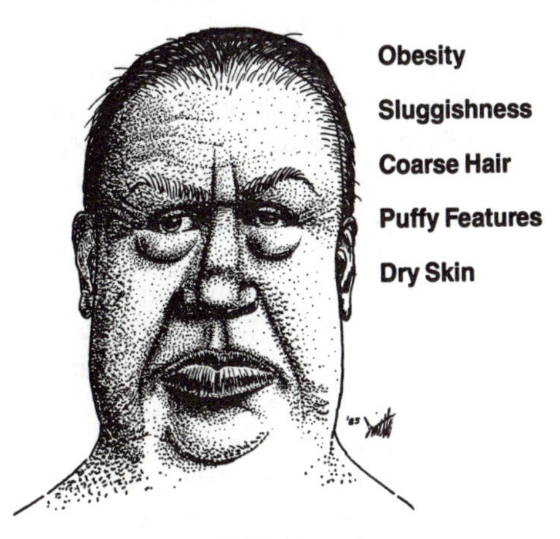

Fig. 12-7 Myxedema.

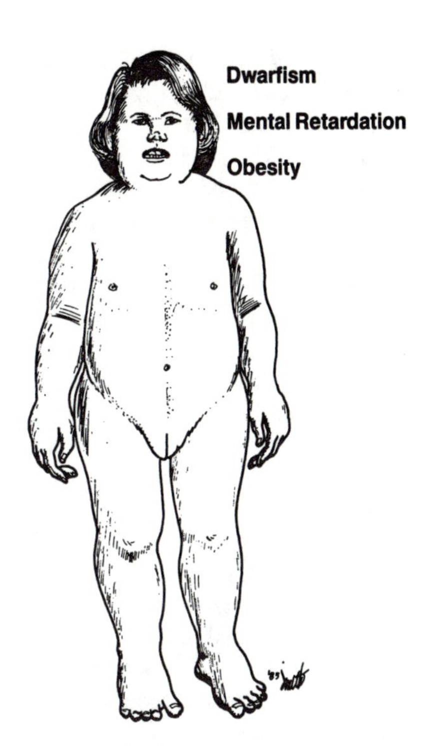

Fig. 12-8 Cretinism.

a *TSH*. In *primary hyperthyroidism* the T-4 is high and the TSH low; in secondary hyperthyroidism, the TSH and T-4 are high. In *primary hypothyroidism,* the T-4 is low, but the TSH is high (no negative feedback from thyroxine); in secondary hypothyroidism, both the TSH and T-4 are low.

F. *Primary treatment*:
   1. Depending on the severity of the *hyperthyroidism*, several methods of treatment are employed: *thyroidectomy*, the use of anti-thyroid drugs such as *propylthiouracil* (which blocks iodine from being incorporated into thyroxin), or the use of *radioactive iodine (I-131)* to destroy the overfunctioning thyroid.
   2. *Hypothyroidism* responds to oral thyroid medication. In myxedema, a high replacement dose of thyroxin may precipitate a heart attack. Therefore, in patients with chronic hypothyroidism, a *low*

dose of thyroxin with a gradual increase is indicated.

V. *The parathyroid glands*:

A. *Anatomy*: The parathyroid glands are four small round, pea-sized bodies located on the posterior portion of the thyroid lobes.

B. *Function*: *The parathyroids control the level of calcium in the bloodstream* (see Fig. 12-2). An excess of parathormone causes excessive removal of calcium from bone, resulting in weak bones. A deficiency causes hypocalcemic tetany (see chapter 10). Vitamin D is necessary for the parathyroid resorption of bone.

C. *Simple tests of parathyroid function*: In *primary hyperparathyroidism*, calcium is increased in blood and urine; in *secondary hyperparathyroidism* from kidney disease, blood calcium is decreased and urine calcium is increased. In *hypoparathyroidism*, both blood and urine calcium are low, resulting in spasms of skeletal muscles (Fig. 12-9).

D. *Primary treatment*: Since the most common cause of hyperparathyroidism is kidney disease, treatment is directed at correcting the kidney problem. Emergency treatment of hypoparathyroidism is intravenous calcium chloride. For maintenance therapy, either calcium or vitamin D tablets are used.

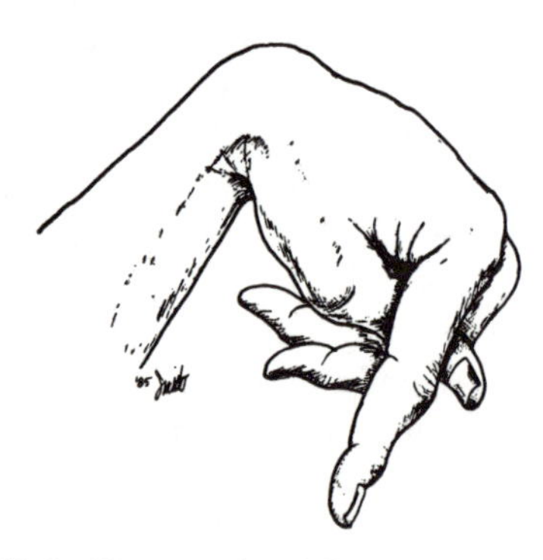

Fig. 12-9 Hypoparathyroidism (Carpal Spasm).

VI. *The pancreas*:

A. *Anatomy*: The pancreas is both an *exocrine* and *endocrine gland* (the exocrine pancreas is discussed in chapter 13). Islands of cells (*islets of Langerhans*) that produce the hormones *insulin* and *glucagon* are interspersed within the exocrine parenchyma.

B. *Functions*:

1. *Beta cells* of the islets of Langerhans secrete the hormone *insulin* in response to an increase in blood glucose, permitting glucose to enter cells to be used for energy. Insulin receptors are present on most cell membranes. Binding is necessary for glucose transport across the cell membrane. Insulin also permits *potassium to enter the cell*, thus lowering the potassium in blood.
2. *Alpha cells* of the islets secrete the hormone *glucagon*, which increases blood glucose when needed (the opposite of insulin).

C. A *high insulin level* is occasionally caused by a benign tumor. More commonly, it is seen in a diabetic who takes insulin without eating ("*insulin reaction*"). The body is flooded with insulin. Glucose enters the cells at an increased rate, and the blood glucose falls (*hypoglycemia*). Since the brain is deprived of glucose, confusion and weakness are present.

D. *Hypofunction*: *Diabetes mellitus* is the most common cause of low insulin. Glucose is elevated in the bloodstream (*hyperglycemia*), floods the kidneys and appears in the urine (*glycosuria*). Glucose is a potent osmotic diuretic; when glucose is lost in the urine, water accompanies it. Thus, *dehydration*, thirst (*polydypsia*), increased urination (*polyuria*) and an increased appetite (*polyphagia*) result. Since the body is unable to utilize glucose, fats are used for energy. The breakdown of *fats* results in the formation of *keto-*

acids as byproducts, increasing body acidity (*keto-acidosis*). The combination of dehydration and acidosis may depress the cerebral cortex to the point of *coma*. This *metabolic acidosis* stimulates the respiratory center to increase the breathing rate (*Kussmaul breathing)*. The two types of diabetes are:

1. *Type I*, or insulin-dependent diabetes, is usually severe and occurs at a young age. Keto-acidosis may be the first manifestation.
2. *Type II*, or non-insulin-dependent diabetes, is usually milder and is seen in an older age group. Common signs and symptoms are polyuria, polydypsia, blurred vision and fatigue.
3. *Complications* of diabetes are vascular disease and infection. Diabetes increases the development of *arteriosclerosis*, and glucose is a good medium for *bacterial growth*. Other complications include *kidney disease*, *heart attacks* (twice average), *eye problems* (diabetic retinopathy) and *impotency*. Gangrene of the feet is twenty times that in the general population. Diabetes accounts for more amputations of the feet than any other condition, including trauma.

E. *Simple tests for endocrine pancreatic function*:
   1. The test for *hypoglycemia* is a *blood glucose*.
   2. The test for *diabetes* is a *fasting blood glucose*, and, if necessary, a *glucose tolerance test*. *Urinalysis* reveals *glycosuria* and sometimes *ketonuria*. Blood gases from a patient in *keto-acidosis* shows a *low pH and bicarbonate*.

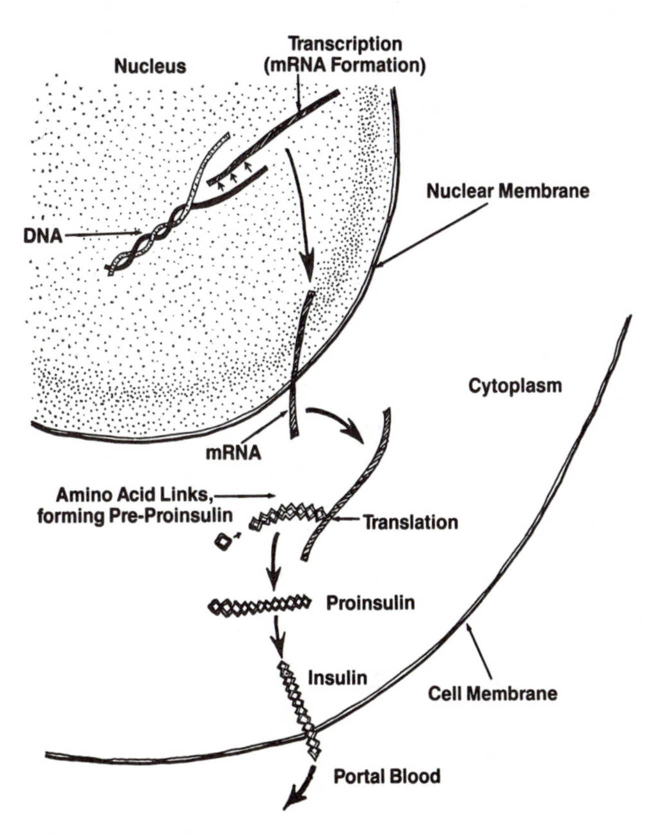

Fig. 12-10 Synthesis of insulin in the beta cell of the pancreas (see p. 12)

F. *Primary treatment*:

1. Treatment for *hypoglycemia* is *intravenous glucose*. This condition sometimes resembles alcoholic intoxication, and has been the source of some emergency mismanagements. A diagnosis of diabetic-coma versus insulin-reaction may not be possible if a diabetic is seen when unconscious. Glucose should be given immediately. This will overcome the insulin reaction and will not harm a patient in diabetic coma, since the coma results from the low pH due to fatty acid breakdown and not the increased glucose.
2. *Treatment for diabetes*: In *type I*, *insulin* is begun immediately. In *type II*, *weight-loss* in obese patients is the first step. The obese person has a reduced number of insulin receptors per cell. When weight is lost, the number of insulin receptors increases. In mild cases of diabetes, *oral hypoglycemic agents*, such as chlorpropamide (Diabinese) or tolbutamide (Orinase) may be used. The American Diabetes Association (ADA) has developed a diet that is a fundamental part of therapy. *Ketoacidosis* is treated with intravenous saline, bicarbonate, insulin and potassium. Treatment for the complications of diabetes include *meticulous attention to the hygiene of the feet*, and an exercise program for weight loss and fitness.

VII. *The adrenal glands*:

A. *Anatomy*: The adrenal glands lie on the kidneys. Each gland consists of an *outer cortex* and *inner medulla*.

B. *Functions*:

1. The *adrenal medulla* secretes the catecholamines, *epinephrine* and *norepinephrine*, which cause the sympathetic "*fight-or-flight*" *response*.
2. The *adrenal cortex* secretes several major steroid hormones (steroid: derived from cholesterol):
   a. *Cortisol (hydrocortisone)* is the main "glucocorticoid" (glucose-producing steroid). Important effects of the hormone are conversion of certain amino acids into glucose (gluconeogenesis) causing a rise in blood sugar, mobilization of free fatty acids for energy, and a *resistance to stress* ("stress" is physiological or emotional trauma: ACTH is secreted, causing a rise in cortisol). Without a functioning adrenal cortex, an animal succumbs to stress and dies.
   b. *Aldosterone* is the main "mineralocorticoid" (sodium and potassium-regulating steroid). It regulates the *reabsorption of sodium and water* (in exchanges for potassium and hydrogen) *in the kidney tubules*, and is necessary for survival.
   c. *Sex steroids* similar to those of the testes (androgens) and ovaries (estrogens) are secreted in small amounts, with androgens predominating.

C. *Simple tests of adrenal function*:

1. A common test for pheochromocytoma (a tumor causing hypersecretion of catecholamines) is an increase in a 24-hour urine sample of urinary catecholamines.
2. In *Cushing's disease*, plasma cortisol and urinary 17-hydroxycorticosteroids (excretory products of cortisol) are *elevated* (Fig. 12-11). If the cause is *primary*, *ACTH is low;* if secondary (pituitary), ACTH is high. In *Conn's syndrome* (primary hyperaldosteronism) aldosterone is elevated in the plasma and urine, potassium is decreased, and

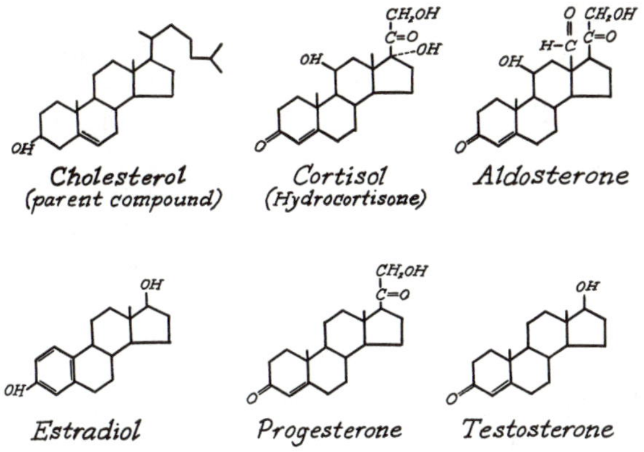

Cholesterol (parent compound) — Cortisol (Hydrocortisone) — Aldosterone

Estradiol — Progesterone — Testosterone

a. Naturally occurring steroids

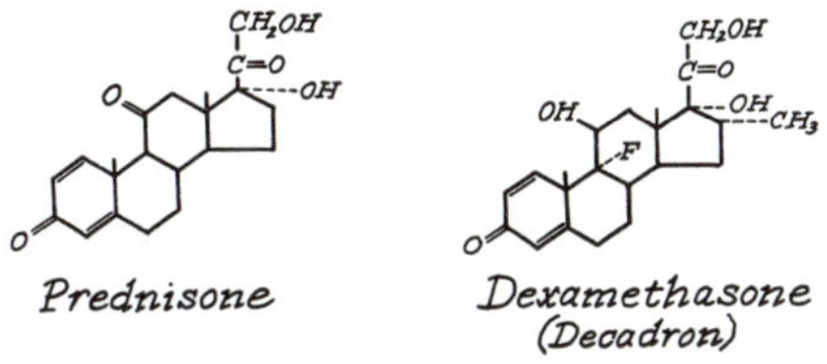

Prednisone — Dexamethasone (Decadron)

b. Synthetic steroids

Fig. 12-11 Steroid Hormones

sodium is elevated in the plasma. *Addison's disease* shows a *low plasma cortisol*, a low sodium and high potassium. Urinary 17-hydroxycorticosteroids are low and the blood glucose is low. *Primary Addison's disease shows a high ACTH* (no cortisol to oppose ACTH); secondary Addison's shows a low ACTH (nonstimulation of the adrenal gland by the pituitary).

D. *Primary treatment*:
   1. *A pheochromocytoma* is removed *surgically*.
   2. In *Cushing's disease*, the *tumor is removed surgically*. If the cause is *secondary* (pituitary), *x-ray therapy* to the pituitary is the treatment of choice. *Conn's disease* is treated by *surgical removal* of the tumor. In *Addison's disease*, the treatment is with hydrocortisone or fludrocortisone (*synthetic steroids* with both glucocorticoid and mineralocorticoid activity).

E. *Synthetic adrenocorticosteroids (corticosteroids, steroids)* (Fig. 12-11):
   1. *Main actions*: synthetic steroids are used to *decrease the effects of inflammation*. They do this by reducing capillary dilatation and permeability. They also prevent the release of vasoactive substances such as histamine and kinins.
   2. *Common diseases treated with steroids*: Allergic disorders such as asthma, bee stings, contact dermatitis, drug reactions, hay fever and hives are treated with steroids. They are also used in arthritis, bursitis, and autoimmune disorders

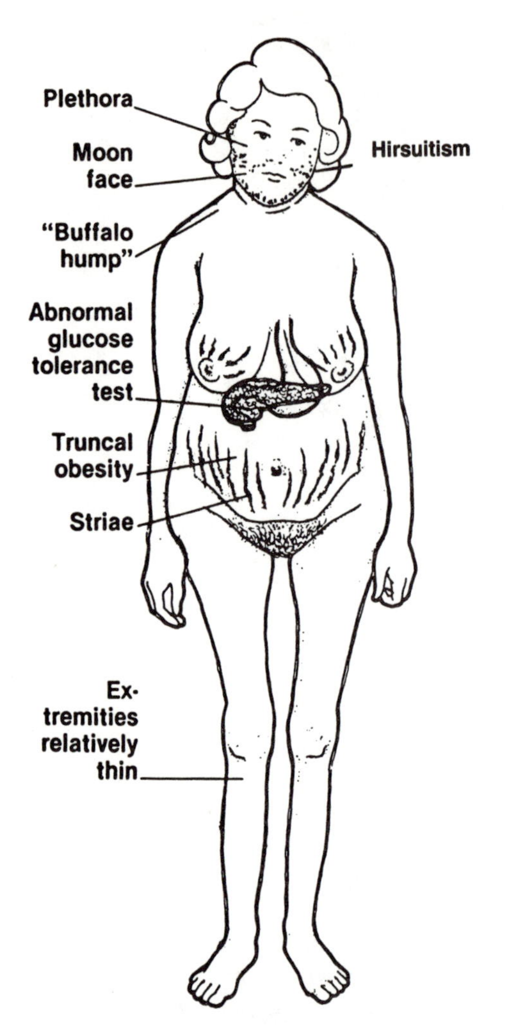

Fig. 12-12 Cushing's Syndrome. (Reproduced and modified, with permission, from Apostolopoulos, "Female Hirsuitism: Evaluation and Current Methods of Treatment", Hospital Physician, Feb. 1984).

such as lupus erythematosus and rheumatoid arthritis. They are useful in leukemia, thrombocytopenia purpura, multiple myeloma, Crohn's disease, ulcerative colitis, kidney failure, infections, and skin disorders. It is dangerous to use steroids for a long time without re-evaluating and perhaps decreasing the dosage periodically. *They suppress the pituitary and ACTH by negative feed-back*. Since ACTH is suppressed, the adrenal glands are not functioning. When steroid treatment is stopped, sometimes the adrenals do not rebound, and the person may lapse into Addison's disease. It is usual to wean the patient from large doses of steroids. Another problem with oral steroid therapy is *GI bleeding*. Protection of the stomach lining with cimetidine (Tagamet) is often necessary (see chapter 13).

VIII. *Testes and ovaries* (see chapter 15, and Table 12-1).

IX. *Gastrointestinal hormones*, the first hormones discovered, are secreted by cells along the GI tract. The presence of food in the GI tract stimulates their secretion (see chapter 13).
   A. *Gastrin* is produced by the stomach and duodenum; it stimulates the *release of HCl and pepsin* from the stomach.
   B. *Cholecystokinin-pancreozymin (CCK-PZ)* is produced by the proximal small intestine. It is responsible for *contraction of the gall-bladder* and the release of bile for fat emulsification, as well as the *secretion of enzymes from the pancreas*.
   C. *Secretin*, also produced by the proximal small intestine, stimulates the *release of pancreatic digestive enzymes*.

X. *Prostaglandins* are a group of about 14 unsaturated fatty acid hormones that perform a myriad of functions, the physiology of which is poorly understood. Although a high concentration is found in seminal fluid (from which the name "prostaglandin" was derived), they are found and synthesized in all tissues of the body.
   A. *Actions in the cell*. In the cell, they seem to act as modified second messengers by interacting with cAMP. They increase cAMP in platelets, thyroid, corpus luteum, pituitary and lung, but decrease it in fat cells.
   B. *The role of non-steroidal anti-inflammatory agents and prostaglandins*.

1. During inflammation, in addition to histamine, a prostaglandin is released which contributes to vasodilation and pain. *Aspirin and other anti-inflammatory agents act as analgesics by inhibiting the synthesis of this prostaglandin.*
2. Aspirin acts as an anti-coagulant by preventing the release of the prostaglandin that causes platelet clumping.

C. *Uses*. At present, several prostaglandins are under trial as a second trimester abortifacient and as a morning-after contraceptive.

XI. *Hormonal effects of certain cancers*:
Some cancers produce hormone-like substances that cause endocrine syndromes. *Bronchogenic carcinoma of the lung* may produce an *ADH-type substance*, resulting in an "*inappropriate secretion of ADH*", or water intoxication. Lung cancer may produce an *ACTH-type substance, causing Cushing's syndrome*. Both lung and breast cancer may produce a *parathyroid-type substance*, which results in *hypercalcemia*. Sometimes, the above syndromes may be the first manifestation of lung or breast cancer.

-THIRTEEN-

DIGESTION

The digestive tract consists of the oral cavity, pharynx, esophagus, stomach and the small and large intestines. Accessory structures include the salivary glands, pancreas, liver and gall-bladder (Fig. 13-1). The stomach and 24 feet of intestines lie in the abdominal cavity mostly suspended by a double layer of peritoneum. The function of the digestive system is to break down foods to be assimilated by the body. Small products of digestion diffuse across the walls of the small intestine, enter the bloodstream, and are transported to the liver. Products from the liver move through the bloodstream to the tissues, were they are used in the metabolism of cells.

I. *Digestion is the conversion of large food-stuffs into smaller units*. Digestion begins in the mouth and ends in the small intestine. It is accomplished by *hydrolytic enzymes (protein catalysts) that split large substances into small ones.*

II. *Main food groups*.

A. *Proteins* are large high-molecular-

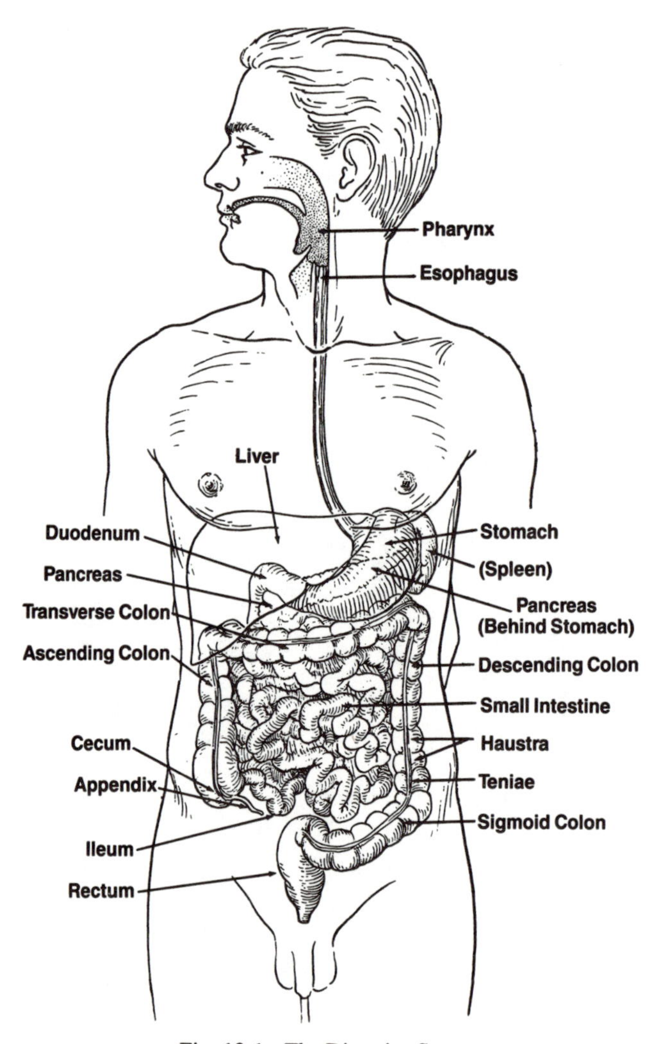

Fig. 13-1 The Digestive System.

weight substances containing carbon, hydrogen, oxygen and nitrogen, as well as smaller amounts of other elements. Proteins are hydrolysed to *polypeptides*, then *di-peptides* (two amino acids) and finally *amino acids*, which are then absorbed. The bond between two amino acids (between the amine group of one and the carboxyl group of the other) is the peptide bond (Fig. 13-2). The body uses 24 amino-acids for its metabolic requirements. Most can be manufactured in the liver from others, but 8 can not, and are the *essential amino acids* (phenylalanine, valine, threonine, leucine, isoleucine, methionine, tryptophan and lysine). In addition, histidine and arginine are required for growth and development. The liver makes proteins (e.g., plasma proteins) from amino acids, and converts some amino acids to glucose for energy (glucogenic amino acids). Proteins are the chief structural components of the body. Enzymes, some hormones, muscle tissue and a substantial portion of chromosomes are proteins. Proteins are essential components of the cell membrane. Important compounds such as epinephrine and acetylcholine are derived from amino acids.

B. *Carbohydrates (CHO)* have, as basic components, carbon, hydrogen and oxygen in definite proportions (Fig. 13-3). Long chains of glucose molecules form a *starch*, found in rice, vegetables, pastes, etc. *Animal starch*, or *glycogen*, is the storage form of glucose in liver and muscle. Starch is hydrolyzed in the oral cavity and small intestine to *monosaccharides* and *disaccharides*. Two linked monosaccharides consititute a disaccharide.

1. Important disaccharides are:
   a. *Maltose*: 2 glucose units
   b. *Sucrose* (plant sugar): glucose and fructose
   c. *Lactose* (milk sugar): glucose and galactose
2. Important *simple sugars*, or monosaccharides, are:
   a. The hexoses (6-carbon sugars):
      1) Glucose
      2) Galactose
      3) Fructose
   b. The pentoses (5-carbon sugars):
      1) Ribose — found in RNA

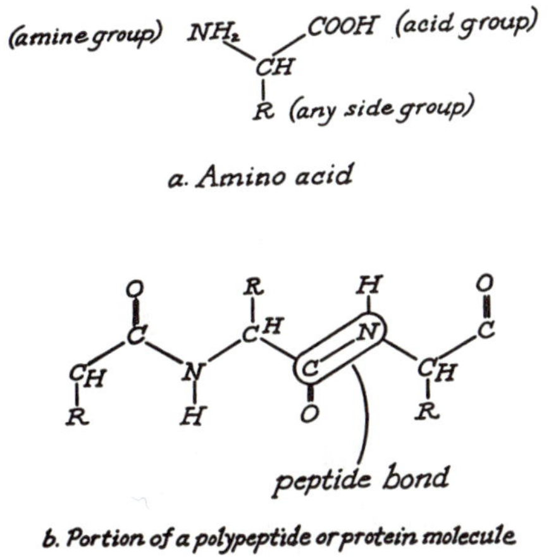

(amine group) NH₂ COOH (acid group) CH R (any side group)

a. Amino acid

peptide bond

b. Portion of a polypeptide or protein molecule

Fig. 13-2 Components of Proteins.

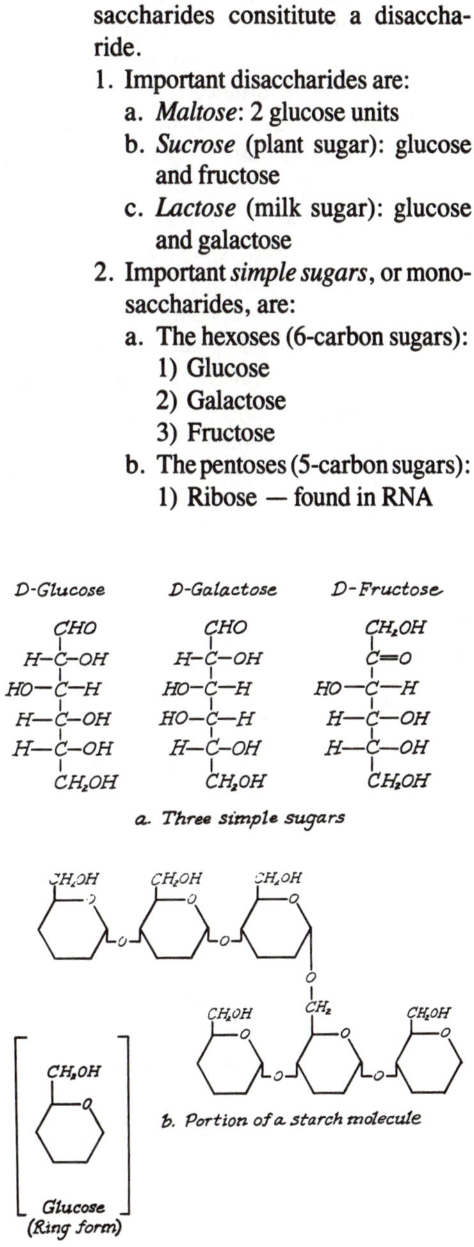

D-Glucose D-Galactose D-Fructose

a. Three simple sugars

b. Portion of a starch molecule

Glucose (Ring form)

Fig. 13-3 Components of Carbohydrates.

2) De oxyribose — found in DNA

3. *Glucose is the main fuel for the* manufacture of *ATP* in the cell. The hexoses, fructose and galactose, are converted to glucose in the liver. In some infants, the enzyme that converts galactose to glucose is missing. Galactose from milk accumulates in tissues (*galactosemia*). This condition may result in brain damage. Treatment is with a milk substitute.

C. The common *fats, or triglycerides*, consist of three fatty acids bound to glycerol, a 3-carbon sugar (Fig. 13-4). Fats are hydrolysed to fatty acids and glycerol. A fatty acid is a molecule consisting of a chain of carbons, with *no double-bonds (saturated*) or *several double-bonds (unsaturated)*. The unsaturated fats most closely resemble body fat; therefore, they are more easily assimilated and utilized. Saturation makes fat more solid. One fatty-acid (*linoleic*, shown above) is essential to human nutrition. In addition to serving as a reservoir of stored energy, fats are essential components of the cell membrane and myelin sheath of the nerve fiber.

D. *Vitamins and minerals*.

1. *Vitamins are growth factors needed in small amounts for daily body metabolism*. They are classified as "fat soluble" or "water soluble". Many vitamins act as enzyme activators (*co-enzymes*). Deficiencies are rare in this country, except for occasional insufficiences in the elderly. More common are excessive intakes (*hypervitaminoses*), particularly by health faddists. The fat soluble vitamins are the more toxic, as the water soluble ones are absorbed and excesses excreted more easily. Hypervitaminosis A causes headache, as well as nausea, vomiting, vertigo and bone pain. Hypervitaminosis D is sometimes

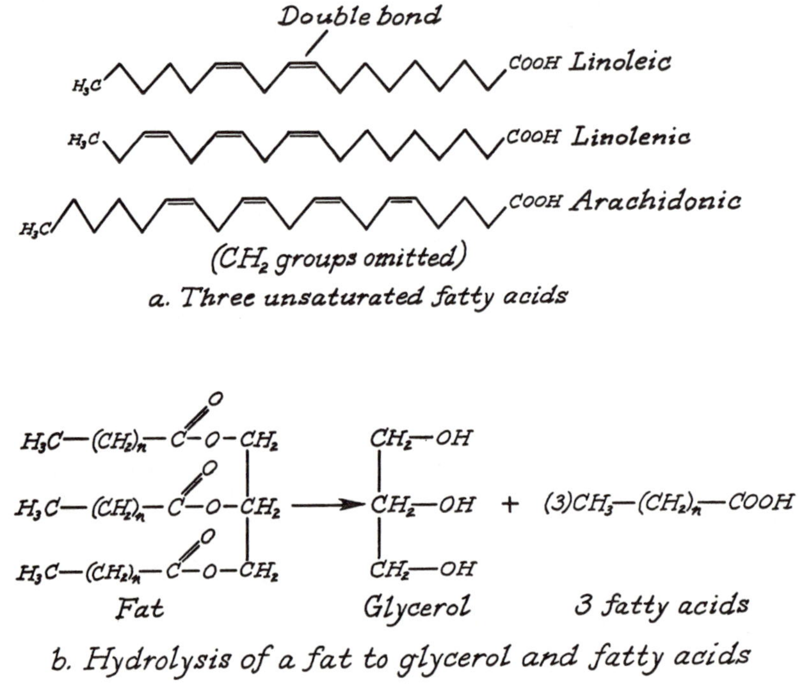

Double bond

H_3C — COOH Linoleic

H_3C — COOH Linolenic

H_3C — COOH Arachidonic

(CH_2 groups omitted)

a. Three unsaturated fatty acids

$$\begin{array}{l} H_3C-(CH_2)_n-\overset{O}{\overset{\|}{C}}-O-CH_2 \\ H_3C-(CH_2)_n-\overset{O}{\overset{\|}{C}}-O-CH_2 \\ H_3C-(CH_2)_n-\overset{O}{\overset{\|}{C}}-O-CH_2 \end{array} \longrightarrow \begin{array}{l} CH_2-OH \\ CH_2-OH \\ CH_2-OH \end{array} + (3)CH_3-(CH_2)_n-COOH$$

Fat Glycerol 3 fatty acids

b. Hydrolysis of a fat to glycerol and fatty acids

Fig. 13-4 Components of Fats.

seen in people who drink large quantities of milk. The increased vitamin D causes calcium to precipitate in the soft tissues of the body. Kidney stones may result. Table 13-1 gives the vitamin, source, function and deficiency.

2. The more important *minerals* are listed in Table 13-2.

III. *Absorption.*

A. *Amino acids, simple sugars and small fatty-acids pass through the intestinal villi into the bloodstream.* The larger fatty acids are reconstituted to fats in the intestinal wall and pass into the lymphatic system as *chylomicrons.* The chylomicrons enter the the thoracic duct, which drains into the left

Table 13-1: Vitamins

	Source	*Function*
1. *Fat-soluble:*		
A	carrots	needed to prevent *night blindness*
D	milk *skin	needed for absorption of calcium from GI tract. deficiency: *rickets* (child) and *osteomalacia* (adults)
E	milk meat	an antioxidant; deficiency syndrome rare, observed in premature infants characterized by edema, anemia, thrombocytosis and skin lesions.
K	leafy green vegetables	needed for synthesis of prothrombin. deficiency: *hemorrhage*
2. *Water-soluble:*		
C (Ascorbic acid)	citrus fruits	lack causes bleeding gums, skin disorders: *scurvey*

B-complex:
Includes *thiamine* (B-1), *riboflavin* (B-2), *niacin, pyridoxine* (B-6) *pantothenic acid, biotin, folic-acid* and *cobalamin* (B-12). All are found in necessary quantities in meat, particularly liver. Deficiency syndromes include dermatitis of some sort, often with nervous system abnormalities. Two of the more well-known syndromes are *pellagra* from niacin deficiency exhibiting diarrhea, dementia and dermatits (the 3 "D"s), and *beriberi* from B-1 deficiency exhibiting neuritis. *Vitamin B-12* is involved with the formation of the red cell. It is absorbed in the presence of *"intrinsic factor"* secreted by the stomach. *Folic acid* has a similar function, and lack of these vitamins or intrinsic factor causes faulty red cell development and *pernicious anemia.* Vitamin B-12 and folic acid have little in common functionally with the B-complex, except being water-soluble.

*Provitamin D (7-dehydrocholesterol), present in the epidermis of the skin, is converted to vitamin D_3 by ultraviolet radiation from the sun. Thus, those living in sunny climates are rarely deficient.

Note #1: The letters A, D, E and K (rather than the names of the vitamins) are commonly used with fat-soluble vitamins, since several compounds in each group have similar biological activity and each has a separate name (K_1—phylloquinone, K_2—menaquinone, D_2—ergocalciferol, D_3—cholecalciferol, etc.).[2]

Note #2: The most common cause of *thiamine deficiency* in this country is *alcoholism.* Alcoholics have low vitamin stores, and some are acutely sensitive to a thiamine deficit. Thiamine functions as a co-enzyme in carbohydrate reactions, such as the oxidation of pyruvate to acetyl CoA in the breakdown of glucose. Giving glucose (an IV of D_5W) depletes what little thiamine is present for glucose metabolism, precipitating disturbances of gaze, ataxia and mental confusion known as *Wernicke's encephalopathy.* This situation is prevented by adding 100 mg. of thiamine to the glucose solution.

Table 13-2: Common Minerals

Calcium	— necessary for development and function of bone, blood coagulation, nerve activity and muscle contraction
Phosphorous	— high energy bonding
Iron	— hemoglobin structure and function
Iodine	— proper functioning of thyroid gland
Sodium, Chloride and *Potassium*	— for proper functioning of the cell (nerve, muscle, etc.)

subclavian vein.

B. Capillaries of the intestinal villi become venules, then veins, and, finally, the large *portal vein* carries absorbed foodstuffs to the *liver* (Fig. 13-5). The liver converts these substances into others, and synthesizes compounds required for bodily functions.

C. *Vitamins* and *minerals* are absorbed in the small intestine. *Calcium* absorption is enhanced by the presence of *vitamin D*. "*Intrinsic factor*" is necessary for the absorption of *vitamin B-12*.

IV. *Organs of the digestive system.*

The gastrointestinal tract from the oral cavity to the anus is lined with *mucosa*, described in chapter 11. It contains glands which secrete mucous and hydrolytic enzymes. Products of digestion are propelled along the tract from the esophagus to the anus by the rhythmical contraction of smooth muscle (*peristalsis*).

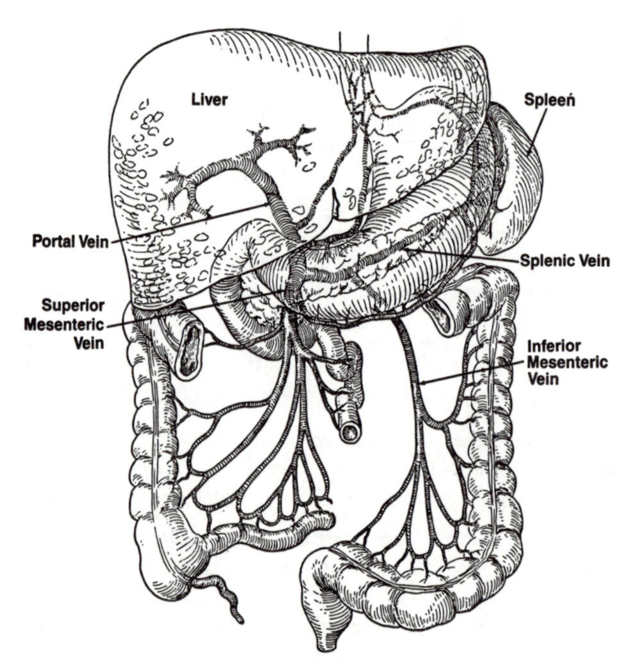

Fig. 13-5 The Portal System.

A. The *oral cavity* includes the lips, cheeks, gums, hard and soft palate, teeth, tongue, and ducts from the salivary glands. The anterior portion between lips, cheeks, teeth and gums is the *vestibule* (Fig. 13-6).
   1. *Structure:*
      a. Superior:
         1) Hard palate, anterior
         2) Soft palate, posterior, terminating in the fleshy *uvula*.
      b. Inferior: tongue and mandible
   2. The *tongue* is composed of *skeletal muscle*, contains taste buds, and receives motor innervation from cranial nerve 12. Cranial nerves 7 and 9 innervate the taste buds. *Function*: the tongue is necessary for speech (*phonation*) as well as swallowing (*deglutition*).
   3. The *teeth* are imbedded in the gums (gingiva).
      a. *Baby (deciduous) teeth* are 20 in number, in by 2 years of age. Dental formula per quadrant: 2 incisors, 1 canine, 2 molars ($\times$ 4) = *20*.
      b. *Permanent teeth*: 32 in number, all but 3rd molar in by 12 years of age. Dental formula per quadrant: 2 incisors, 1 canine, 2 premolars, and 3 molars (3rd molar is "wisdom tooth", in by age 25) $\times$ 4 = *32*.
   4. *Salivary glands*.
      a. *Anatomy*.
         1) The *parotid gland* (right and left) is situated anterior to the ear. The duct enters the oral cavity near the upper second molar.
         2) The *sublingual gland* (right and left) lies under the tongue. Several ducts enter the oral cavity near the *frenulum* of the tongue (Fig. 13-7).
         3) The *submandibular (submaxillary) gland* (right and left) lies on the floor of the mouth near the angle of the mandible. The duct of the gland terminates at the *frenulum*.
      b. *Functions of saliva*.
         1) Lubrication of food *(main function)*.
         2) Secretion of salivary amylase (ptyalin) that breaks down carbohydrates.
         3) Smell, sight, taste and thought of food stimulate parasympathetic fibers to increase the secretion of saliva.
      c. *Importance*. Mumps is a viral infection of the parotid gland. Salivary glands may swell if the

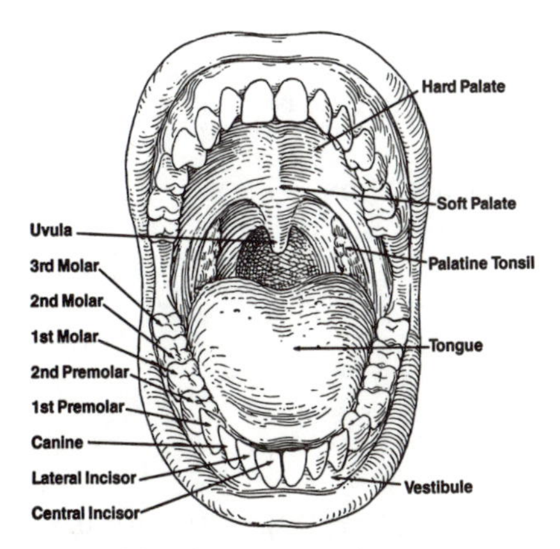

Fig. 13-6 The Oral Cavity.

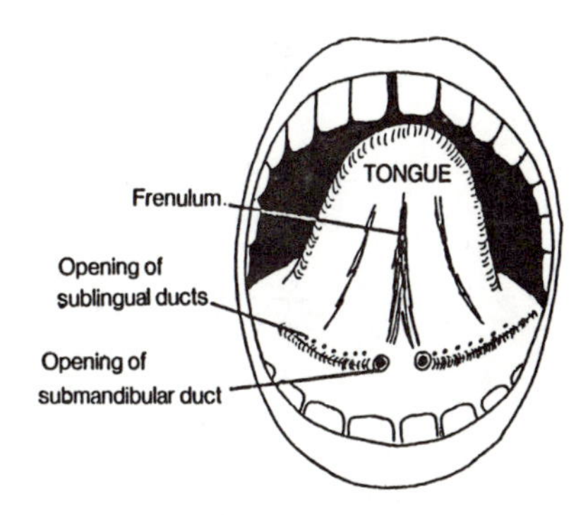

Fig. 13-7 Underside of the Tongue. (Reproduced, with permission, from Goldberg, Clinical Anatomy Made Ridiculously Simple, MedMaster, 1984).

ducts are blocked.

B. The *pharynx, or throat*, is the five-inch long continuation of the nasal and oral cavities (naso-pharynx and oro-pharynx). The *Eustachian tube* from the middle ear opens into the pharynx on each side (Fig. 8-11). The *tonsils* form a ring of lymphatic tissue around the pharynx. Three sets are present (see also Fig. 11-2, 13-6).
   1. Several *lingual tonsils* lie on the back part of the tongue.
   2. One large *pharyngeal tonsil (adenoids)* is located at the nasopharynx.
   3. Two *palatine tonsils* are on the right and left side of the throat.
   4. *Importance*: since the adenoids sit near the nasal cavity, enlargement from infection may impair breathing in children ("pursed-lip" breathing). Multiple bouts of tonsilitis may necessitate their removal.

C. The *esophagus* is a 9-inch long tube running from the pharynx to the stomach. It enters the abdominal cavity through an opening in the diaphragm (esophageal hiatus). The *function* of the esophagus is to transport food to the stomach (Fig. 13-1).

D. The *abdomen*, or *abdominal cavity*, contains the major organs of digestion. The cavity is lined with a mucous membrane, the *peritoneum*, whose function is to prevent friction (Fig. 13-8). The portion of the peritoneum lying on the body wall is the *parietal peritoneum*; the portion surrounding each organ is the *visceral peritoneum.* The intestines are suspended by a double layer of peritoneum, the *mesentery* (small intestine) and the *mesocolon* (transverse colon). Another double layer is the *greater omentum* (from the stomach to the transverse colon). A storehouse for fat, it accounts for much of the girth in obesity. The *lesser omentum* runs from the stomach and duodenum to the liver. Organs lying in the abdominal cavity are *intra-abdominal (intraperitoneal):* the distal portion of the esophagus, the stomach, the small and large intestine, the spleen, most of the liver, the gall-bladder and the appendix. Organs lying on the posterior body wall behind the parietal peritoneum are *retro-peritoneal:* the kidneys, adrenal glands, ureters, pancreas, most of the duodenum, inferior vena cava,

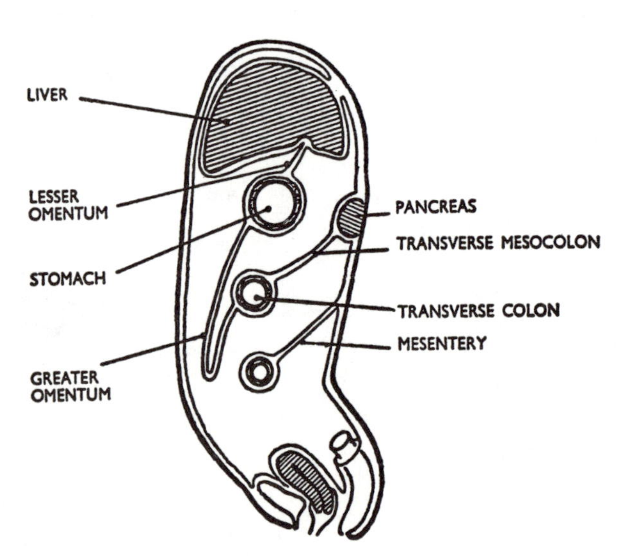

Fig. 13-8 Sagittal View of the Abdomen. (Modified, with permission, from Ellis, Clinical Anatomy, Blackwell Science Pub. Ltd., 1971).

abdominal aorta, thoracic duct and the sympathetic trunk (Fig. 13-9) The abdominal muscles are listed in Table 5-4, chapter 5. The lower abdomen, or *pelvis*, contains the reproductive organs and bladder. The abdomen is divided into surface regions, or *quadrants*. These are discussed in more detail in chapter 16.

1. The *right upper quadrant* (RUQ) contains the liver and gall-bladder.
2. The *left upper quadrant* (LUQ) contains the stomach and spleen.
3. The *right lower quadrant* (RLQ) contains the appendix, and in the female the right ovary and fallopian tube.
4. The *left lower quadrant* (LLQ) contains the sigmoid colon, and in the female the left ovary and fallopian tube.

E. *Stomach* (Fig. 13-10).

1. *Anatomy*. The stomach lies in the left upper quadrant of the abdomen. The pancreas lies posteriorly (Fig. 13-11). Important parts are the *lesser* and *greater curvatures, fundus, body, antrum, pylorus, cardiac orifice* and *pyloric sphincter*. Several specialized cell types are present (parietal, chief and mucous cells are in the body and fundus; gastrin cells are in the antrum and pylorus).
2. *Vessels*. The arterial supply is from the celiac trunk, the first main branch of the abdominal aorta (Figs. 13-11, 13-12). Veins drain to the *portal vein* (Fig. 13-5).
3. *Nerves*. *Parasympathetic stimulation* from the vagus nerve causes *increased peristalsis and secretion of gastric juice;* sympathetic stimulation has the opposite effect.
4. *Functions*:
   a. *Stores* and churns food
   b. Chief cells secrete *pepsinogen*, the inactive form of the enzyme, *pepsin*. Pepsin cleaves the peptide bonds of proteins and polypeptides.
   c. Parietal cells secrete *hydrochloric acid* (pH of 1, thus very acidic). HCl converts the inactive pepsinogen to the active pepsin. Parietal cells also secrete "*intrinsic factor*", responsible for the absorption of

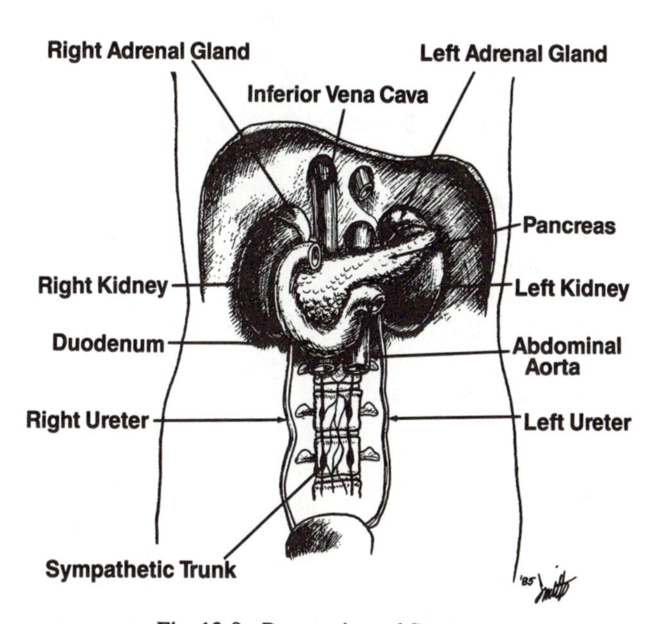

Fig. 13-9 Retroperitoneal Structures.

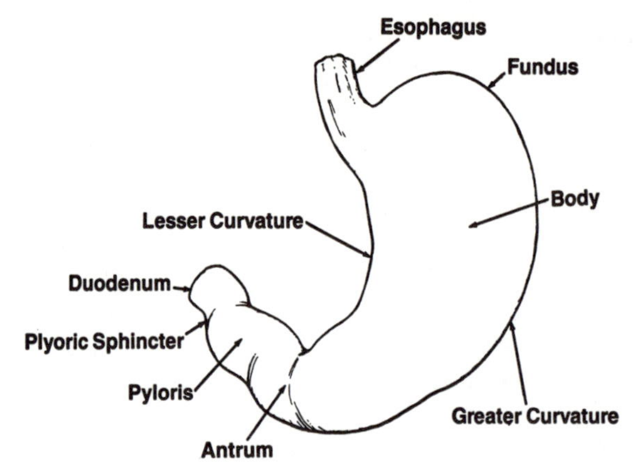

Fig. 13-10 The Stomach.

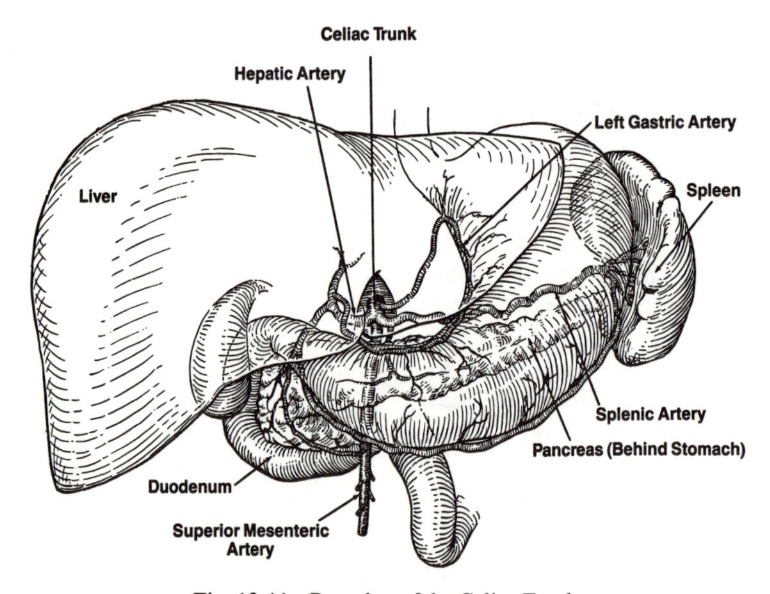

Fig. 13-11 Branches of the Celiac Trunk.

vitamin B-12 from the small intestine.

d. Mucous cells secrete *mucus* that protects the stomach wall against HCl and self-digestion by pepsin.
e. Absorbs water, glucose and alcohol.
f. Gastrin cells secrete the hormone gastrin, that stimulates the release of HCl and pepsinogen.
g. Some gastric cells have *histamine* receptors. Irritation of the stomach appears to liberate histamine, a potent stimulator of gastric acid secretion. Cimetidine (Tagamet) competes with histamine for receptor sites, thus blocking the secretion of acid. It is therefore effective in treating *ulcers*.

F. *Small intestine* (Fig. 13-1).

1. *Anatomy*. The small intestine is the largest organ in the abdomen. The first part is the 10-inch *duodenum* (lying retroperitoneally) into which open the pancreatic and

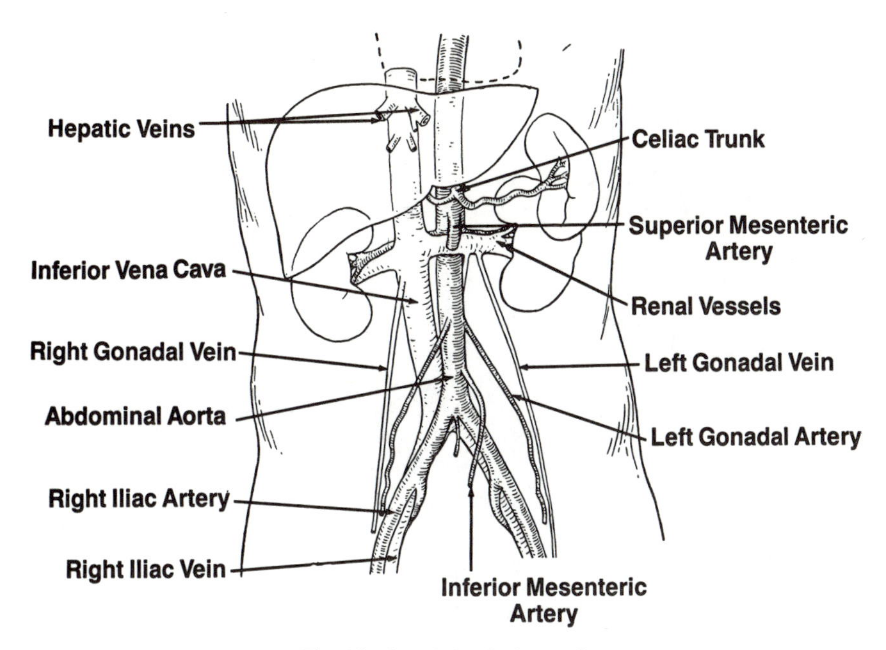

Fig. 13-12 Abdominal Vessels.

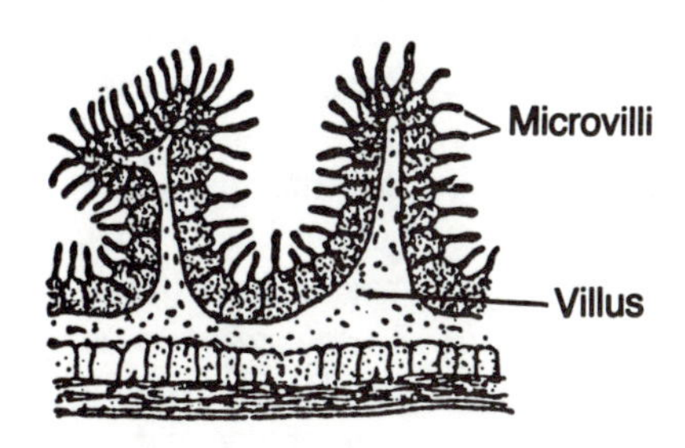

Fig. 13-13 Section of Small Intestine. (Mucosal Surface Showing Villi)

common bile ducts. The second part is the *jejunum*, about 8 feet long. The distal segment is the *ileum*, about 12 feet long. Numerous glands secreting mucus and enzymes are present. The mucosal surface is folded into fingerlike projections, the *villi*, with numerous microscopic projections, *microvilli* (Fig. 13-13). Villi and microvilli substantially *increase the absorbing surface of the small intestine.*

2. *Vessels*. The arterial supply is the *superior mesenteric artery*, a major branch of the abdominal aorta (Figs. 13-11, 13-12). The mesenteric veins drain to the *portal vein*. The vessels lie between layers of mesentery.
3. *Nerves. Parasympathetic stimulation* from the vagus nerve causes *increased peristalsis* and secretion of mucus, which protects the intestinal wall; sympathetic stimulation has the opposite effect.
4. *Functions*:
   a. *Absorption* of foodstuffs takes place mostly in the jejunum, but also in the ileum.
   b. *Bile* from the gall-bladder, as well as pancreatic *enzymes*, enter the duodenum at a common site, the *ampulla of Vater* (hepatopancreatic ampulla) (Fig. 13-14).
   c. The presence of the products of digestion from the stomach (chyme) stimulates the secretion of intestinal and pancreatic enzymes.
   d. The gastrointestinal hormones, gastrin, cholecystokinin-pan-

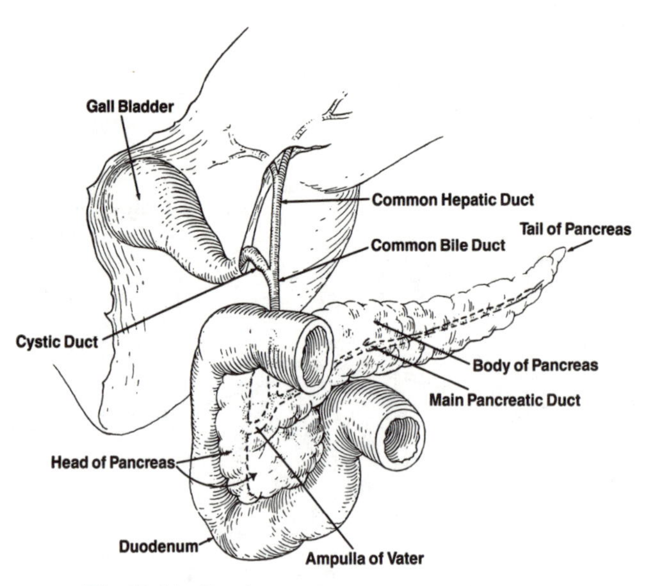

Fig. 13-14 Duodenum, Pancreas and Gall Bladder.

creozymin (CCK-PZ) and secretin are produced by the small intestine in response to food products in the GI tract.

e. Enzymes that split disaccharides into monosaccharides are secreted by the wall of the small intestine.

G. *Large intestine (colon)* (Fig. 13-1).

1. *Anatomy*. The colon is the wide 4-foot-long continuation of the small intestine, and begins in the right lower quadrant of the abdomen as a pouch, or *cecum*. The *appendix* is a fingerlike lymph organ attached to the cecum.

a. The *ascending colon* lies retroperitoneally and ascends to the liver (hepatic flexure).

b. The *transverse colon* crosses from the right upper quadrant to the left upper quadrant (splenic flexure) and is suspended by *mesocolon*.

c. The *descending colon* descends to the left lower quadrant behind the peritoneum, and emerges in the pelvic area as the "S-shaped":

d. *The sigmoid colon* is suspended by mesocolon. The colon terminates in the:

e. *Rectum* (15cm) and *anus* (last 3 cm).

f. Important parts of the colon are outpouchings (haustra) and muscle bands (teniae).

g. A *colostomy* is an artificial opening between the colon and skin of the abdomen for the evacuation of feces. Usually done to relieve tumor obstruction, it is occasionally performed temporarily after inflammation or trauma.

2. *Vessels*. The arterial supply to the right side of the colon is from the *superior mesenteric artery*; to the left side from the *inferior mesenteric artery* (Fig. 13-12). Veins drain to the *portal vein* (Fig. 13-5).

3. *Nerves*. *Parasympathetic* stimulation from the vagus and pelvic nerves (from the sacral part of the spinal cord) causes *increased motility*; sympathetic stimulation has the opposite effect.

4. *Functions*:
    a. The *main function* of the colon is the *absorption of* water and sodium.
    b. Undigested matter passes as *feces*. About 3 days transit time is required for food to pass from the oral cavity to the anus. The brown color of stool is caused by the breakdown products of bile pigments.
    c. Large numbers of *bacteria* are found in the colon. They may have a role in the production of *vitamins*.
    d. Medicines are often given as rectal suppositories, since the colon has great absorptive capacity.

H. The *pancreas* (Fig. 13-14).
1. *Anatomy*. The pancreas lies *retroperitoneally*. The *head* of the pancreas lies in the curve of the duodenum, the *body* behind the stomach, and the *tail* touches the spleen. The pancreas is both an *exocrine* and *endocrine* gland (the endocrine portion is discussed in chapter 12). The *main pancreatic duct* (Wirsung) combines with the *common bile duct* to form the *ampulla of Vater* at the duodenum.
2. *Vessels*. The arterial supply is from branches of the splenic artery. Venous drainage is to branches of the superior mesenteric and portal veins.
3. *Nerves*. *Parasympathetic* stimulation from the vagus nerve causes *secretion of digestive enzymes*, insulin and glucagon; sympathetic stimulation has the opposite effect.
4. *Functions*:
    a. The *main function* of the exocrine pancreas is the secretion of the *major enzymes of digestion*.
        1) *Trypsin* breaks down proteins to peptides and amino acids. It is secreted in the inactive form, trypsinogen, and is activated by enterokinase, an enzyme from the small intestine.
        2) *Pancreatic amylase* breaks down large carbohydrates to disaccharides. Hydrolysis of disaccharides to glucose, galactose and fructose is accomplished by enzymes in the small intestine.
        3) *Lipase* splits fats into glycerol and fatty acids.
    b. The hormones secretin and CCK-PZ from the small intestine, trigger the release of pancreatic enzymes. Vagus nerve stimulation has a less pronounced but similar effect.

I. *The liver* (Fig. 13-15).
1. *Anatomy*. The liver, the largest gland of the body (about 3½ pounds), is located in the *right upper quadrant* of the abdomen, lying against the diaphragm and extending down to the *right costal margin* (RCM) (Fig. 13-1). Parts: a large *right lobe*, a smaller *left lobe*, and small *quadrate* and *caudate lobes*. The portal vein, hepatic artery and common bile duct are situated on the posterior-inferior surface. The gall-bladder lies between the right and quadrate lobe. The inferior vena cava lies between the caudate and right lobe.
2. *Vessels*. The arterial supply is from the *hepatic artery* (Fig. 13-11). Foodstuffs absorbed from the stomach and intestines flow into the liver from the portal vein. Three *hepatic veins* drain into the inferior vena cava (Fig. 13-12).
3. *Nerves*. *Parasympathetic* stimulation from the vagus nerve causes a slight increase in the synthesis of glycogen (*glycogenesis*); sympathetic stimulation causes the

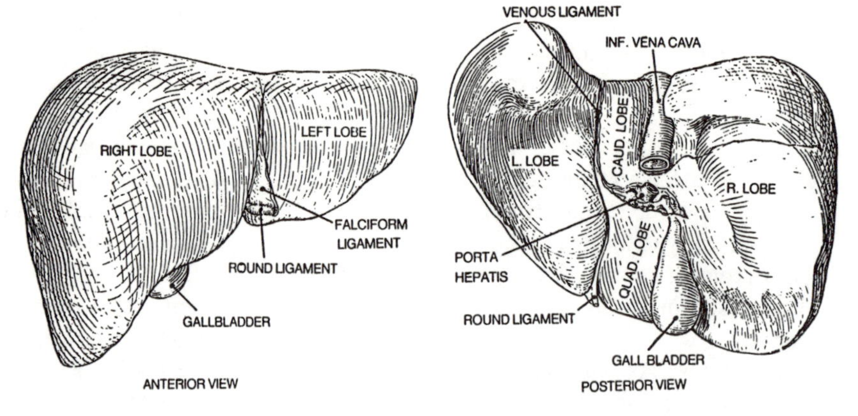

Fig. 13-15 The Liver. (See ref., Fig. 13-7).

release of glucose from the liver (*glycogenolysis*).

4. *Functions*:
   a. *The formation of bile*. About a liter of bile is formed per day and some is stored in the gallbladder. It is composed mostly of water, and contains the pigment, *bilirubin* (the breakdown product of hemoglobin), as well as *bile salts* which emulsify fats. *Emulsification*, the formation of small particles from larger ones, takes place in the small intestine. In the case of fats, bile salts lower surface tension, permitting large globules to break into smaller units. This facilitates the action of *lipase*.
   b. *Carbohydrate metabolism*. An important function of the liver is maintaining a normal glucose concentration in the bloodstream. It does this in 3 ways:
      1) Glucose is released from glycogen stores.
      2) Galactose and fructose are converted to glucose.
      3) Some amino acids are converted to glucose (gluconeogenesis).
   c. *Fat metabolism*.
      1) Fatty acids are broken down for energy if glucose is not available.
      2) Cholesterol is formed for steroid hormone production.
      3) Some carbohydrates and proteins are converted to fats.
   d. *Protein metabolism*.
      1) Urea is formed.
      2) Plasma proteins, including the clotting factors, are manufactured.
   e. Iron is stored as ferritin.
   f. The removal, excretion, detoxification and inactivation of drugs, hormones and toxins take place in the liver.

J. *The gallbladder* (Fig. 13-14).
   1. *Anatomy*. The gallbladder lies on the right lobe of the liver. It has a fundus, body and neck. The neck becomes the *cystic duct* which unites with the *hepatic duct* to become the *common bile duct*.
   2. *Vessels*. The cystic artery to the gallbladder is a branch of the hepatic artery. Venous drainage is to the portal vein.
   3. *Nerves*. Parasympathetic stimulation from the vagus nerve causes a weak contraction of the gallbladder; sympathetic effect is relaxation.
   4. *Functions*:
      a. The *main function* is *storage of bile*.
      b. Fats entering the small intestine

trigger the release of CCK-PZ from the intestinal wall, causing contraction of the gallbladder and the release of bile into the common bile duct.

IV. *Common disorders of the digestive system.*

A. *Appendicitis. Obstruction of the lumen, usually by feces, results in inflammation, infection and infarction of the wall (appendicitis).* Occasionally, perforation occurs, resulting in infection of the peritoneum (peritonitis). If peritonitis has occurred, pressing in gently on the abdomen and releasing the hand quickly causes pain (rebound tenderness). The usual sequence of events is:

1. Loss of appetite (anorexia).
2. Pain.
3. Vomiting.
4. Tenderness in the right lower quadrant of the abdomen.
5. Fever and leukocytosis.

Primary treatment is removal of the appendix.

B. *Cirrhosis is the infiltration of connective tissue into the liver parenchyma, resulting in a hard, small, barely functional organ.* The most common cause is alcoholism. The liver first enlarges (alcoholic hepatitis), then decreases in size. The loss of function illustrates several important antatomical and physiological concepts:

1. *Portal hypertension* is the back-up of blood in the portal vein. Injury and scarring causes obstruction of blood flow through the liver.
   a. *Enlargement of the spleen* (splenomegaly) takes place because blood backs up in the splenic vein.
   b. The venous circulation connecting to the portal system also engorges with blood. *Esophageal varices* and *hemorrhoids* result.
   c. Seepage of plasma back through the portal area, (particularly from the liver) causes an accumulation of fluid in the abdominal cavity (*ascites*).
2. In males, small amounts of female hormone are normally produced by the testes and adrenal cortex. These estrogens are inactivated by the liver. In liver disease, failure to inactivate these estrogens results in breast enlargement (*gynecomastia*), testicular atrophy and body hair loss (alopecia).
3. When bile excretion is blocked, bilirubin accumulates in the bloodstream, causing *jaundice*.
4. As the damaged liver can not produce plasma proteins, *edema* develops, particularly in the ankles.
5. Blood-clotting factor production is inadequate, resulting in a *bleeding tendency*. A poor prognostic sign is an elevated prothrombin time, signifying extensive hepato-cellular damage.
6. Since the liver is not able to detoxify substances such as *ammonia to urea*, they accumulate in the bloodstream and then in the brain, resulting in the altered mental state, *hepatic encephalopathy*.
7. Alcohol also affects other organs as well as the liver. Damage to the cerebral cortex causes memory and recall problems (*Wernicke-Korsakoff syndrome*), and damage to the *cerebellum* causes tremor and balance difficulties. The toxic effect of alcohol on the bone marrow may cause *anemia*. *Primary treatment* is abstinence from alcohol.

C. *Colon cancer* is the most common cancer overall, affecting males and females about equally. Tumors of the ascending colon (right-sided) usually cause rectal bleeding; those in the descending colon constipation and obstructive symptoms. *Polyps* and ulcerative colitis are important risk factors. Screening for occult blood in the stool, as well as sigmoidoscopy, detects most lesions—70% are located in the sigmoid colon and rectum. Surgical resection of the bowel is often curative.

D. *Cystic fibrosis is a genetic disease involving exocrine gland dysfunction.* Secretions from the pancreas, mucus glands of the respiratory tract and sweat glands are defective. The sodium and chloride content of sweat is two to three times normal. Trypsin, amylase and lipase are decreased. Growth is poor. Stools are foul-smelling and float. Thick mucus is present in the respiratory tract, and pulmonary problems are common. Rarely does a child survive beyond his teens; most die from pulmonary infections. *Primary treatment* is a high protein, high caloric diet, accompanied by the replacement of pancreatic enzymes (e.g., Pancreatin). Antibiotics, inhalation and physical-therapy are useful. Continuous home-pulmonary care is often necessary.

E. *Diverticular disease. Diverticula are small outpouchings of the intestinal wall*, usually about 1 cm, found in weak areas of the colon near the teniae, where vessels are located. Most occur in the sigmoid colon. *When multiple diverticula are present, the condition is diverticulosis. If they become inflammed and infected, it is diverticultis.* Signs and symptoms of diverticulitis are:
   1. Pain in the left lower quadrant of the abdomen.
   2. Constipation is common, but diarrhea may occur.
   3. Occasionally, bleeding may be present.
   4. Perforation of a diverticular sac may cause peritonitis. *Primary treatment* consists of a high fiber diet (bran, etc.), a bulk-forming laxative such as psyllium (Metamucil) and an antibiotic. In severe cases, hospitalization may be required.

F. *Gallbladder disease is almost always the result of a bile salt and/or cholesterol stone formed in the gallbladder and lodged in the cystic duct.* Stone formation is cholelithiasis. *The pain of a stone in the cystic duct is biliary colic. Persistent obstruction leading to inflammation and infection of the gallbladder is cholecystitis.* A fatty meal often percipitates an attack, since CCK-PZ stimulates contraction of the gallbladder. Signs and symptoms are:
   1. Pain in the right upper quadrant of the abdomen, often radiating to the right scapula.
   2. Nausea and vomiting.
   3. Fever.

   Mild cases are treated with an analgesic. Fatty foods are eliminated from the diet. Severe cases require intravenous fluids, nothing by mouth (NPO) and nasogastric suctioning. If infection is present, antibiotics are used. Definitive treatment is removal of the gallbladder (cholecystectomy).

G. *Gastroenteritis is a general term for inflammation and/or infection of the GI tract.* If the stomach is involved, the condition is *gastritis*. If the intestine is affected, it is *enteritis*. Usually both are involved, so the term gastroenteritis is used. The most common cause is a virus. If the cause is a bacterial toxin, the condition is *food poisoning*. Occasionally, it is caused by a bacterial infection. Rarely is it due to a protozoal infection (dysentery). *Diarrhea* and generalized *cramping abdominal pain* are hallmarks of the condition. Primary treatment is rehydration and relaxing the hyperactive bowel. Fluids for 24 hours relax the intestine, since food stimulates gastrointestinal hormone release and peristalsis. Several compounds slow the bowel: bismuth subsalicylate (Pepto-Bismol), diphenoxylate with atropine (Lomotil) and anticholinergic antispasmodics such as dicyclomine (Bentyl). If a stool culture shows a bacterial or protozoal infection, an antibiotic or an antiprotozoal agent is given.

H. *Hemorrhoids are dilated, varicosed areas of the rectal venous plexus.* External hemorrhoids lie distal to the ano-rectal margin; internal hemorrhoids lie proximal. Occasionally, a thrombus or clot forms, resulting in a painful, bluish mass. Primary treatment: sitz-baths, suppositories and stool softeners, such as docusate sodium (Colace). The clot may be surgically removed under local anesthesia.

I. *Inflammatory bowel disease.* Two inflammatory diseases of the GI tract affect young men and women, ages 20 to 40, causing ulcerative lesions and thickening of the intestinal wall. The cause for both is unknown, but an autoimmune etiology has been postulated. *Ulcerative colitis* affects primarily the sigmoid colon, with symptoms of lower abdominal pain and bloody diarrhea. *Regional enteritis* (Crohn's disease) affects primarily the ileum, with symptoms of right lower quadrant (RLQ) pain and intermittent diarrhea. Primary treatment for both is an adequate diet, antibiotic therapy, and steroids. Occasionally surgical resection is necessary.

J. *Irritable bowel syndrome.* Also known as *spastic, or irritable colon,* this disease is a common GI problem. Symptoms include abdominal pain, alternating constipation and diarrhea, and frequent emotional problems. Primary treatment includes a diet high in bran, restriction of alcohol and tobacco, a psyllium laxative such as Metamucil, and perhaps an anticholinergic tranquilizer such as Librax.

K. *Pancreatitis.* Most cases of acute pancreatitis involve *alcoholism* and *gallstones.* The onset is often precipitated by alcohol excess and/or a stone at the ampulla of Vater. *Lipase, amylase and trypsin back up in the pancreas and are released into the surrounding tissue.* This causes auto-digestion of the pancreas and necrosis of tissues, including the peritoneum. Hemorrhage and shock may develop. Fats are hydrolysed to glycerol and fatty acids, which form soaps with calcium. Massive destruction of tissue accompanied by fluid and blood loss may lead to shock and death. *Signs and symptoms* include:
   1. Epigastric abdominal pain radiating to the back.
   2. Nausea and vomiting.
   3. The abdomen is distended and tender. The person feels better sitting rather than lying.
   4. Amylase and lipase levels are elevated in the bloodstream.

   Primary treatment: morphine or meperidine (Demerol) for pain, intravenous feedings and naso-gastric suctioning to quiet the GI tract. Calcium replacement may be needed.

L. *Peptic ulcer disease (PUD) is an epithelial crater, found mainly in the duodenum but also in the lesser curvature or anterior part of the stomach.* The crater is usually less than 2 cm in diameter. The term "peptic" means that pepsin in involved in the pathogenesis of the disease. Risk factors include being male, smoking, heredity, alcohol and stress. An increased secretion of HCl and pepsin, as well as decreased tissue resistance, contribute to the process. The normal protective mechanisms of the duodenal and gastric mucosa against HCl and pepsin are blocked. Excessive vagal stimulation is present. The ulcer causes pain and may erode into a vessel and cause bleeding. It may perforate through the wall, causing peritonitis and shock.
   1. *Common signs and symptoms are*:
      a. Heartburn or epigastric pain following a meal, relieved by

antacids.

b. Vomiting of coffee-ground material (hematemesis) or the passage of dark stools (melena), indicating the presence of blood.

c. Tenderness on palpation of the epigastric region of the abdomen.

2. *Primary treatment:*

Intravenous fluids and a nasogastric tube are usually required. Antacids such as a magnesium-aluminum hydroxide mixture (Maalox, Mylanta) are useful. Cimetidine (Tagamet) or ranifidine (Zantac) inhibit gastric acid secretion. Common surgical procedures are:

a. *Vagotomy denervates vagal nerve fibers* to the body and fundus of the stomach. The *stimulus for HCl and pepsin secretion as well as peristalsis is thus reduced.*

b. A *partial gastrectomy* usually involves removal of the antrum (antrectomy), thus *decreasing gastrin secretion.*

M. *Reflux esophagitis is the regurgitation of gastric acid up through an incompetent esophageal sphincter, causing heartburn.* A *hiatal hernia* (a small portion of the stomach protruding up through the esophageal opening of the diaphragm) is often present (Fig. 13-16). Heartburn is aggravated when lying flat and relieved when sitting upright. It is frequently seen in obese people. Primary treatment is weight-loss to help the hiatal hernia, and an antacid.

N. *Viral hepatitis is a viral infection of the liver.* It is divided into two major types: Type A ("infectious") is commonly transmitted by people living under poor sanitary conditions (fecal-oral route). Type B ("serum") is usually transmitted by infected blood or contaminated needles. Medical personnel and intravenous drug abusers are at high risk. Signs and symptoms include jaundice, malaise, and other "flu-like" manifestations. Blood levels of liver enzymes are elevated. Primary treatment is symptomatic: bed rest and fluids. The disease usually lingers from one to two months.

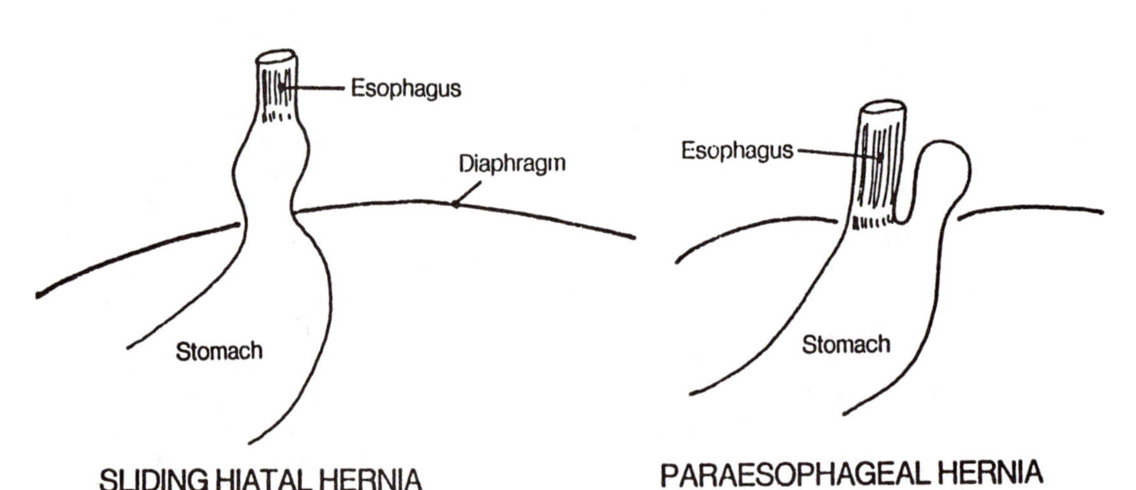

Fig. 13-16 Hiatal Hernias. (See ref., Fig. 13-7).

-FOURTEEN-

EXCRETION

The *kidneys filter waste products* from blood. They also *filter nutrients*, which are reabsorbed into the bloodstream. The *filtrate* passes down through the two *ureters* to the *bladder*. When a certain volume is reached, the urge to void is present. Urine is expelled from the bladder through the *urethra*. Several hormones affect the kidneys, and the kidneys, themselves, produce hormone-like substances. They also contribute to acid-base balance (Fig. 14-1).

I. *Organs of the excretory system.*

A. *Kidneys.*

1. *Gross anatomy*. The kidneys lie retroperitoneally against the back body wall musculature, imbedded in fat, at about the spinal level of T-11 to L-3 (Fig. 14-2). The right kidney, positioned behind the liver, is slightly lower than the left. The average kidney is bean-shaped, about 4.5 inches in length, and one inch thick (Fig. 14-3). Important parts are the *hilum* (containing the *renal artery, vein and pelvis*), *capsule*, an outer *cortex* and an inner *medulla*. The filtrate from the medulla is collected in *calyces* (singular "calyx") which form the

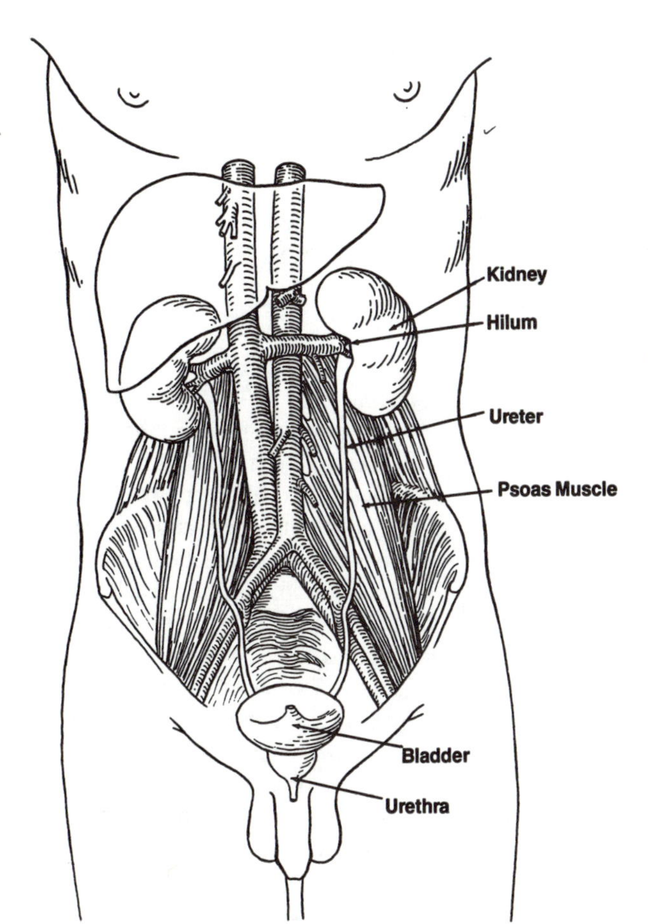

Fig. 14-1 The Excretory System.

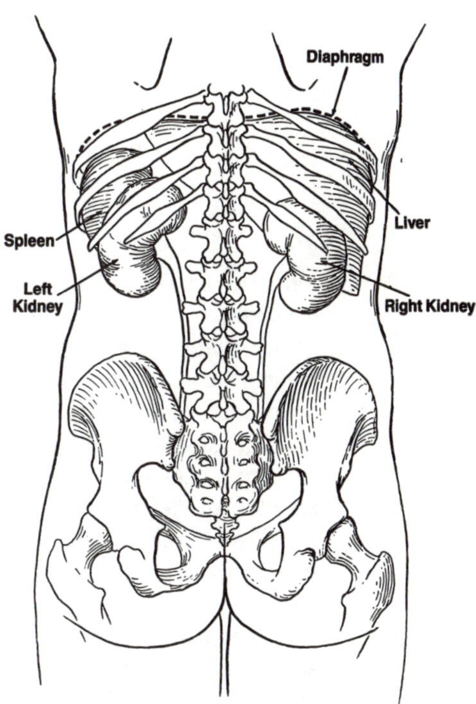

Fig. 14-2 Posterior View of the Kidneys.

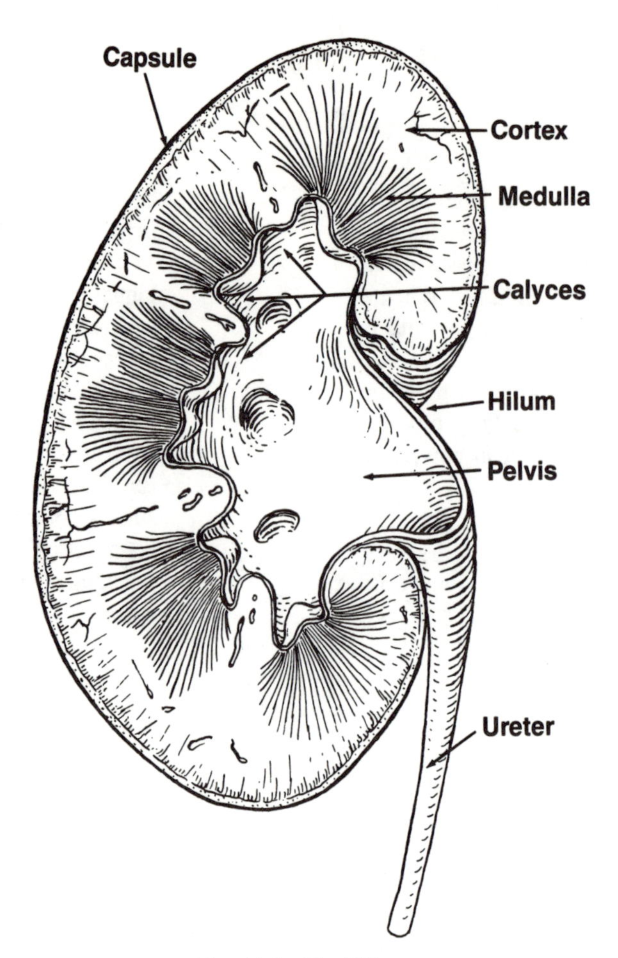

Fig. 14-3 The Kidney.

renal pelvis. The pelvis becomes the *ureter*.

2. *Microscopic anatomy*. After entering into the hilum, the renal artery divides into smaller units. At the arteriole stage, an *afferent arteriole* leads to a capillary tuft, the *glomerulus*; an *efferent arteriole* leads away. The glomerulus filters blood substances into a small tubule. The glomerulus and tubule is a *nephron*. Important parts of the nephron are *Bowman's capsule*, a *proximal tubule*, *loop of Henle* and a *distal tubule*. The *cortex* is composed of glomeruli, Bowman's capsules, proximal and distal tubules. Several distal tubules feed into a *collecting duct* (Fig. 14-4). Many collecting ducts make up the *renal pyramids*, the major part of the *medulla*. The filtrate passes down the collecting ducts and empties into the calyces. The medulla contains the loops of Henle. Each kidney contains about a million nephrons.
3. *Vessels*. Two *renal arteries* arise from the abdominal aorta. Two *renal veins* empty into the inferior vena cava (Fig. 13-11).
4. *Nerves*. Although both sympathetic and parasympathetic fibers are present in the kidney, the *important component is sympathetic, causing vascoconstriction*

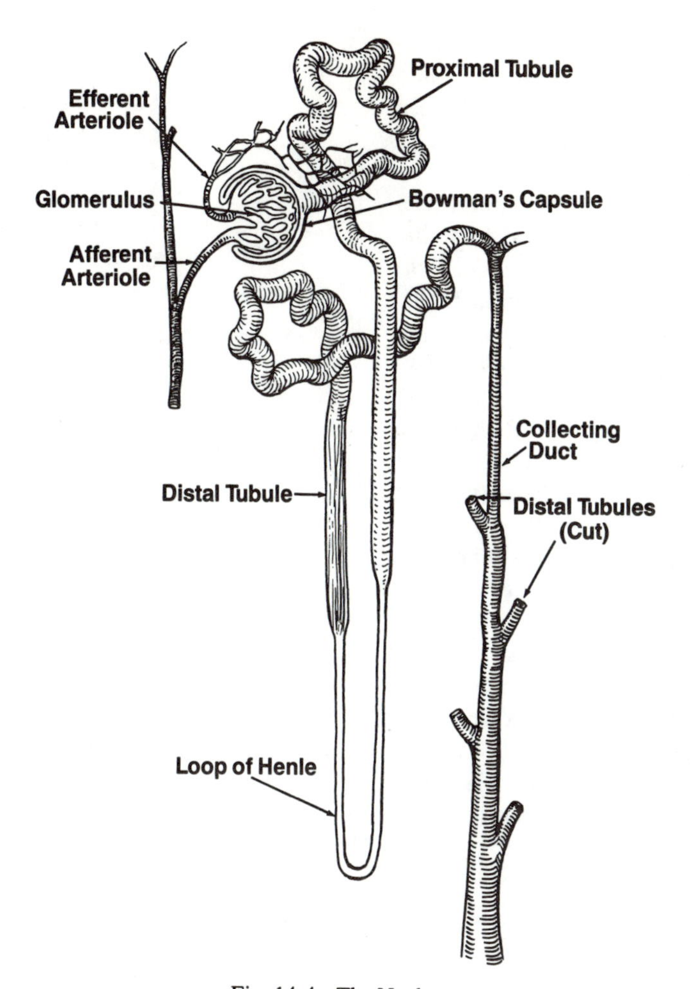

Fig. 14-4 The Nephron.

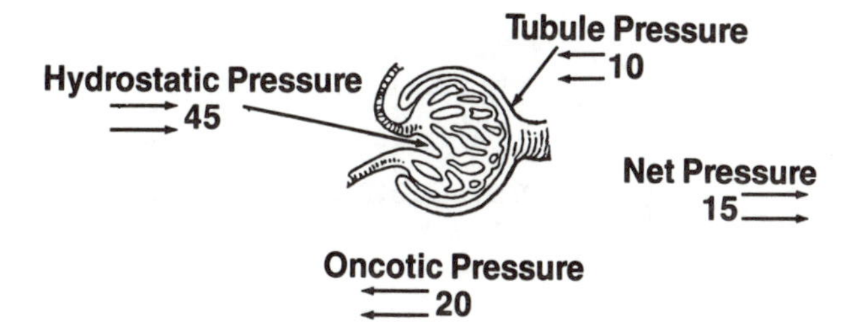

Fig. 14-5 Glomerular Filtration.

and the release of renin.

5. *Functions*:
   a. *Glomerular filtration* (Fig. 14-5). Plasma is filtered by *ultrafiltration*. The hydrostatic pressure in the glomerulus is high (45 mm Hg), since the capillary tuft is interposed between two arterioles. This is opposed by the tubule pressure (about 10 mm Hg) and the *oncotic pressure* of plasma proteins (about

20 mm Hg). The *net pressure* across the glomerular membrane is thus about 15 mm Hg. In the average person, about 100 liters of blood are filtered per day; 99 liters of filtrate are reabsorbed, leaving about one liter of urine. The *glomerular filtration rate (GFR)* measures the status of the glomeruli.

b. *Tubular reabsorption and secretion*. Many substances are *reabsorbed* from the glomerular filtrate into capillaries surrounding the tubules (peritubular capillaries) by *diffusion* or *active transport*. Other substances are *secreted* from the capillaries into the filtrate. If there is less of a substance on one side of a membrane, it is necessary to have a mechanism whereby it can still always be transported. This is *active transport, and permits movement against a concentration gradient, using ATP for energy*. Substances actively transported are sodium, potassium, chloride, calcium, glucose and amino acids. If the concentration of a substance in the filtrate exceeds the transport rate, it will appear in the urine. This is the *renal threshold*. For example, in the normal person, all of the glucose in the filtrate is reabsorbed. In *diabetes mellitus*, the increase in glucose in the filtrate overwhelms the transport mechanism, and glucose appears in the urine. Substances *secreted* from the capillaries into the tubular filtrate include hydrogen, potassium, ammonia and penicillin. Hydrogen and potassium are secreted actively.

c. *Anatomical areas of tubular exchange* (Fig. 14-6).

1) *Proximal tubule. All glucose*

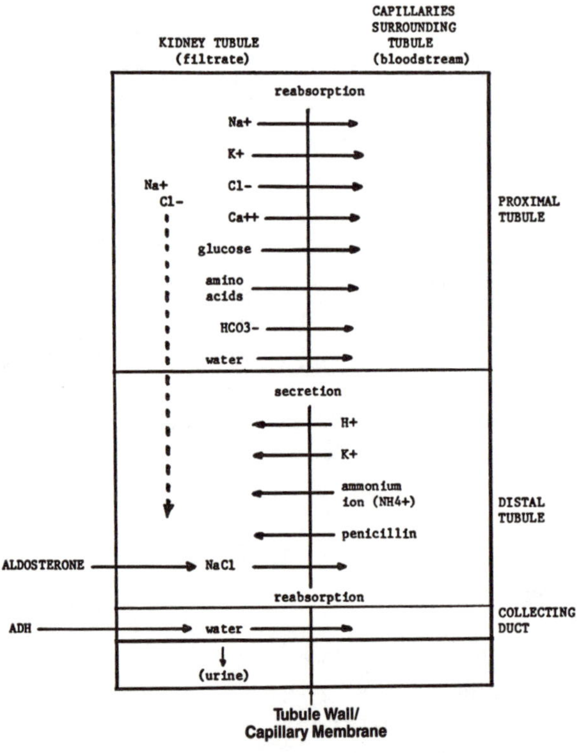

Fig. 14-6 Tubule Reabsorption and Secretion.

and amino acids are reabsorbed, as well as 80% of water. Most of the sodium chloride and potassium is reabsorbed here. Hydrogen and ammonia are secreted into the tubules.

2) *Loop of Henle.* In the first part of the loop, increased osmotic pressure moves water into the interstitial tissue, concentrating the urine. In the last part of the loop, NaCl may pass from the tubule into the tissues.
3) *Distal tubule. More sodium, chloride and water are reabsorbed; potassium and hydrogen are secreted.* This secretion is actually an exchange of sodium for potassium and hydrogen. Ammonia is also secreted. If NaCl is decreased in the bloodstream, *aldosterone* secretion is stimulated and more NaCl is reabsorbed.
4) *Collecting duct.* A high osmotic pressure of blood and tissue fluid stimulates secretion of the posterior pituitary hormone *ADH*, and *more water is reabsorbed.* This happens when NaCl in the bloodstream is increased, as in dehydration and in the hypovolemia of hemorrhage.

d. *The juxtaglomerular apparatus (JGA)* is a group of cells at the junction of the afferent arteriole and glomerulus (Fig. 14-7). It secretes the hormone, *renin*, and probably *renal erythropoietic factor* (REF).

1) *Renin* converts the plasma protein, angiotensinogen, to *angiotensin* (an important vasoconstrictor), stimulating the adrenal cortex to secrete aldosterone. Aldosterone causes the reabsorption of sodium from the distal tubule, increasing blood pressure. Physiologically, renin is secreted in response to low blood pressure, but there is evidence that it plays a role in "*essential*" (viz. 'cause unknown') *hypertension.*

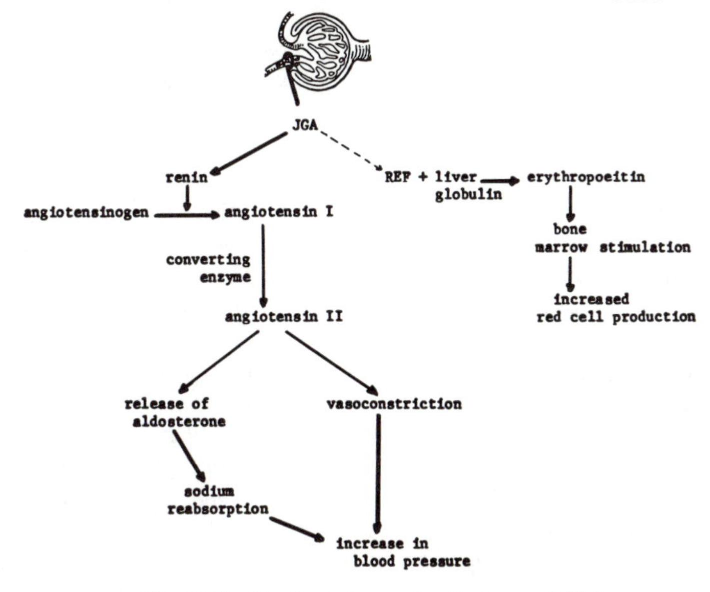

Fig. 14-7 The Juxtaglomerular Apparatus (JGA).

2) *Erythropoetin* is a hormone that stimulates red blood cell production in the bone marrow. Low oxygen stimulates it; a high oxygen shuts it off. Erythropoetin is formed from a liver globulin and REF. If the JGA is damaged, as in uremia, anemia develops.

e. *The kidney in acid-base balance.* The kidney is able to filter or reabsorb HCO_3^-. It also secretes H^+, using several buffers. The kidney has two buffers besides bicarbonate: *phosphate* and *ammonia*. Bicarbonate acts primarily in the proximal tubule, phosphate acts principally in the distal tubule and collecting duct, and ammonia is found in all three areas. The following reactions take place:

$$H^+ + HCO_3^- \longrightarrow H_2CO_3 \longrightarrow CO_2 + H_2O$$
$$H^+ + HPO_4^= \longrightarrow H_2PO_4^-$$
$$H^+ + NH_3 \longrightarrow NH_4^+ \text{ (ammonium ion)}$$

Exchange rules:

1) For each H^+ secreted, a Na^+ and HCO_3^- are reabsorbed.
2) Na^+ is exchanged for either H^+ or K^+, or both.
3) Whenever K^+ is lost, some H^+ is lost.

B. *Ureters.* The two ureters lie on the psoas muscles (Fig. 14-1). Each is a tube about 12 inches in length and 1/8 to 1/4 inch in diameter, abundantly supplied with nerves. Peristalsis moves urine down into the bladder.

C. *The bladder* is a muscle receptacle (the *detrusor muscle*) for urine, lined with mucosa, lying in the suprapubic region of the abdomen (Fig. 14-1). A muscular *sphincter* surrounds the urethra. The bladder is supplied with sensory, motor and autonomic nerve fibers. The bladder can hold more than a liter of urine, but the urge to void begins when about 200 ml accumulates.

1. *Micturition* (voiding, urination) is a parasympathetic response, modified by voluntary control. It is initiated when afferent impulses from stretch receptors in the bladder stimulate the sacral portion of the spinal cord. *The detrusor contracts* and *the sphincter relaxes*.
2. *Obstruction.* Urethral obstruction, occuring commonly in older males with prostate problems, may result in 2 or more liters of urine in the bladder. Sodium and urea accumulate. Rapid evacuation may create an osmotic diuresis resulting in shock. This can be prevented by clamping the catheter for 15 minutes after draining 400 ml, and so on. Sodium losses are replaced with intravenous saline.

D. *The urethra* is a narrow tube leading from the bladder to the outside. The male urethra is about 8 inches long and runs through the prostate gland and penis (Fig. 15-1). The female urethra lies posterior to the clitoris and is about 1½ inches in length (Fig. 15-5). The close proximity of the female urethra to the anus allows anal bacteria to migrate up the urethra to the bladder, ureters and kidneys (*ascending urinary tract infection*).

II. *Renal function tests.*

A. *Urinalysis (UA)* is a simple, inexpensive and informative series of laboratory tests. Color, appearance, and a microscopic examination of urine is done. A chemically impregnated color strip, or "dipstick" (e.g., Multistix), is dipped into a urine sample and the results (protein, blood, glucose, ketones, bilirubin) are read in seconds.

1. *Color.* Normal urine is a yellow or amber color. Many drugs cause a change in color. *Bilirubin* and *blood* cause a dark color.
2. *Appearance.* Cloudy urine often indicates the presence of bacteria, thus *infection*.
3. The *urine sediment* (after centrifu-

gation) under the microscope may show abnormal constituents, such as *bacteria*, *white and red blood cells*.

4. *Protein* (proteinuria, albuminuria) is an important indicator of *kidney disease*. A normal small amount may be detected after exercise.
5. *Blood* (hematuria) is commonly present in a bladder infection (*cystitis*), and with a stone in the ureter (*ureteral calculus*). Hematuria is present in glomerulonephritis and cancer.
6. *Glucose* (glycosuria) usually indicates *diabetes mellitus*.
7. *Ketones (acetone)* (ketonuria) are seen in *fasting*, because the body uses fats for energy. They are also seen in *diabetes mellitus*, because the body is unable to use glucose and relies on fatty acid metabolism for energy.
8. *Biliburin* is present in *hepatitis*.

B. *The blood urea nitrogen (BUN)* reflects kidney function (Fig. 14-8). The breakdown of proteins causes the formation of ammonia, derived from the amine groups of amino acids. Normally, ammonia is converted to urea in the liver, and urea is excreted by the kidneys. If the excretion of urea is blocked, as in kidney disease, urea accumulates in the bloodstream. The amount of urea in blood is calculated by measuring the amount of nitrogen in the body (since 90% of the nitrogen is in the form of urea). The most important cause of a greatly elevated BUN is *kidney disease* or *failure*, although other conditions may cause an elevation (e.g., dehydration, GI bleeding).

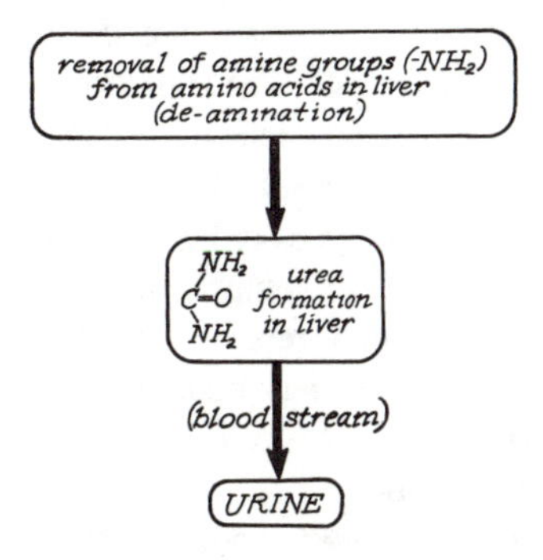

Fig. 14-8 Urea Formation.

C. *The glomerular filtration rate (GFR) is the amount of filtrate cleared from the glomeruli of both kidneys in one minute*. In order to determine this, a substance is used that is neither reabsorbed nor secreted by the kidney tubules (urea or creatinine). Commonly, creatinine is employed. Urine samples are taken at 1, 2, 4, 6 and 24 hour intervals. clearance is then calculated as follows:

$$\text{Clearance} = \frac{U \times V}{P}$$

U = concentration of creatinine in urine
P = concentration of creatinine in plasma
V = urine volume converted to ml/min

The average *creatinine clearance* is between 85 to 125 ml/min. Low values occur in glomerulonephritis and renal insufficiency.

D. *An intravenous pyelogram (IVP)* is an *x-ray of the kidneys, ureters and bladder*. An iodine-containing radio-opaque dye is injected into a peripheral vein. After a few minutes it appears in the filtrate. The kidneys, ureters and bladder are outlined. The test is frequently used to determine if a kidney stone is lodged in a ureter (*ureteral calculus*). Contrast material does not appear beyond the obstruction. An IVP is also useful in determining kidney size, shape, position and structure.

IV. *Diuretics*. Diuresis means an increase in urine production. Most diuretics *block the reabsorption of sodium from the proximal tubule, resulting in the excretion of sodium, and with it, water*. An important side-effect is the loss of potassium.

A. *Mechanism of potassium-loss with diuretics* (Fig. 14-9). Most diuretics

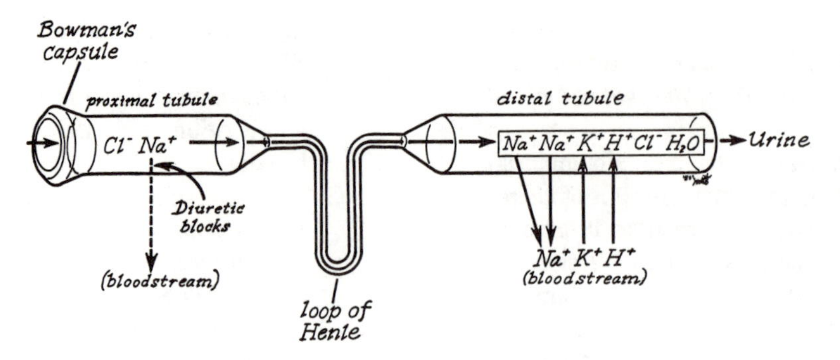

Fig. 14-9 Action of Most Diuretics.

block sodium reabsorption at the proximal tubule, so more sodium appears at the distal tubule. Sodium reabsorption is thus increased at the distal tubule, and *potassium and hydrogen are exchanged for it and lost.* However, the increased load of sodium at the distal tubule results in the loss of much sodium, and water. Chronic potassium loss may eventually result in *hypokalemia* and a life-threatening cardiac arrhythmia. The continuous accompanying loss of hydrogen may cause *metabolic alkalosis* (digitalis and insulin also cause potassium loss: insulin permits potassium to enter the cells, thus lowering it in the extracellular fluid; digitalis increases intracellular sodium, favoring potassium loss).

B. *Common diuretics:*
   1. Potassium-losing diuretics are the *thiazides* (e.g., hydrochlorothiazide) and the potent, fast-acting *furosemide* (Lasix).
   2. *Caffeine* blocks the reabsorption of sodium from the proximal tubule.
   3. *Alcohol* blocks ADH secretion from the posterior pituitary.

C. *Uses of diuretics.* Diuretics are used primarily in the treatment of *hypertension* and *congestive heart failure.*

Common disorders of the excretory system.

A. *Bladder infection (cystitis)* is a common infection, mostly affecting females. The reason is because the female urethra is only a few centimeters in length, and bacteria from the perineal region have easier accessibility to the bladder than in the male. Symptoms include pain in the lower abdomen (suprapubic, or hypogastric tenderness), pain on urination (dysuria), frequent urination (frequency) and a continuous urge to void (urgency). The UA shows bacteria, white cells and often blood. Antibiotics such as nitrofurantoin (Macrodantin), sulfa containing agents such as trimethoprim/sulfamethoxazole (Bactrim, Septra) and synthetic penicillins (ampicillin or amoxicillin) are effective.

B. *Glomerulonephritis* is a disease involving *antigen-antibody reactions affecting the glomeruli.* The antigen may be an external one, such as B-hemolytic streptococci, or it may involve an autoimmune reaction. The most common type is post-streptococcal, in which antibodies are formed that react with streptococci. Immune complexes are deposited in the glomeruli. The condition may follow a "strep" infection such as pharyngitis, tonsillitis or impetigo. Various degrees of hematuria, cells in the urine, proteinuria, edema, hypertension and renal insufficiency are present. Primary treatment: an antibiotic if due to strep, bed rest and salt restriction.

C. *Kidney stones (calculi). Most stones consist of calcium salts and precipi-*

tate in the renal calyces and pelvis. Causes of stone formation include an impaired tubular reabsorption of calcium, resulting in increased excretion (renal hypercalciuria), and increased calcium absorption from the small intestine (absorptive hypercalciuria). Other factors contributing to stone formation are dehydration, urinary tract infection, a positive family history and immobilization. Kidney stones are usually undiscovered until one passes into a ureter, causing the sudden, severe, excruciating, flank pain of *renal colic*. A urinalysis shows blood in the urine. Primary treatment includes forcing fluids to pass the stone. An IVP is often done, and frequently helps pass the stone. Occasionally, surgical removal may be necessary. Morphine or meperidine (Demerol) may be necessary to relieve the pain. Prevention involves a high fluid intake and the dietary restriction of calcium and vitamin D. As a substitute for surgery, a new device (lithotripter — Gr. "stone crush") focuses an ultrasonic beam on the stones and pulverizes them.

D. *Pyelonephritis is an infection of the kidney*. Bacteria may reach the kidney from the bladder, or by spreading through the bloodstream from another infected site, such as the tonsils, middle ear, sinuses, prostate, etc. (most cases *ascend* from the bladder). Common manifestations are flank and back pain, usually on one side, abdominal pain and, sometimes, fever. The UA shows bacteria and white cells in the urine. Primary treatment: an appropriate antibiotic.

E. *Uremia (kidney failure) is the terminal stage of renal insufficiency from any cause*, the most common being chronic pyelonephritis and chronic glomerulonephritis. The glomerular filtration rate, tubular absorption and secretion are decreased.

1. *Laboratory tests* show the following:
   a. *Anemia*. Erythropoietin secretion is decreased.
   b. *Acidosis*. The ability of the kidney to clear normal metabolic acids from the bloodstream is reduced. Impaired reabsorption leads to sodium loss (hyponatremia). Potassium and hydrogen are retained (hyperkalemia), adding to the acidosis.
   c. *Hypocalcemia*. Calcium reabsorption from the tubule is decreased.
   d. *Nitrogen retention*. The BUN, creatinine and uric acid are increased.
   e. *Proteinuria and hematuria*.

2. *Signs and symptoms:*
   a. Weakness and fatigability because of sodium, potassium, and calcium abnormalities, as well as from anemia and acidosis.
   b. Polyuria and nocturia are present because of the excretion of an increased solute load. Proteins and sodium are lost, resulting in an osmotic diuresis.
   c. Hypertension, because of injury to the JGA and the release of renin.
   d. Pruritis, from the accumulation of waste products in skin vessels.
   e. Dehydration may occur from the water loss.
   f. Generalized edema (anasarca) due to the loss of proteins.

3. *Primary treatment*:
   a. Cautious administration of amino acids, adequate calories, sodium, and calcium, as well as anti-hypertensive medication.
   b. Blood transfusions may be necessary.
   c. *Hemodialysis* is the clearing of wastes from blood, using a

series of tubes and filters (the *artificial kidney*). An in-place arteriovenous fistula in the arm makes possible the filtration of blood once or twice per week.

d. *Kidney transplants* are among the most successful of grafts. Careful matching of similar blood and genetic types, as well as up-to-date immunosuppressive drug therapy, may result in long-term survival rates.

-FIFTEEN-

REPRODUCTION

I. *The male reproductive system* consists of the two gonads (testes), the two vasa deferentia, the penis, and accessory structures.

A. *Gross anatomy* (Fig. 15-1).

1. Each *testis*, or testicle, is an almond-shaped gland, about 2 inches long and one inch thick, lying in a sac, the *scrotum*. The main artery to each testis is the testicular artery from the abdominal aorta. A network of veins in the scrotum is called the *pampiniform plexus*. The testis is covered by a peritoneal-like membrane, the *tunica vaginalis*. The *cremaster* muscle raises the testis. The *epididymis*, the continuation of the tubule system, lies posterior to the testis.
2. From the epididymis, an 18 inch duct, the *vas deferens*, passes up the inguinal canal lateral and posterior to the bladder, runs posteriorly, and enters the *prostate gland*.
3. Two glands, the *seminal vesicles*, join the vas deferens as it enters the *prostate*. The vas deferens is now the *ejaculatory duct*. It traverses the prostate and joins the urethra (*prostatic urethra*). At the root of the penis, ducts from two small *bulbourethral glands* (Cowper's glands) enter the urethra.
4. The *penis* is composed of three meshworks of erectile tissue; two large *corpora cavernosa*, and a smaller *corpus spongiosum* con-

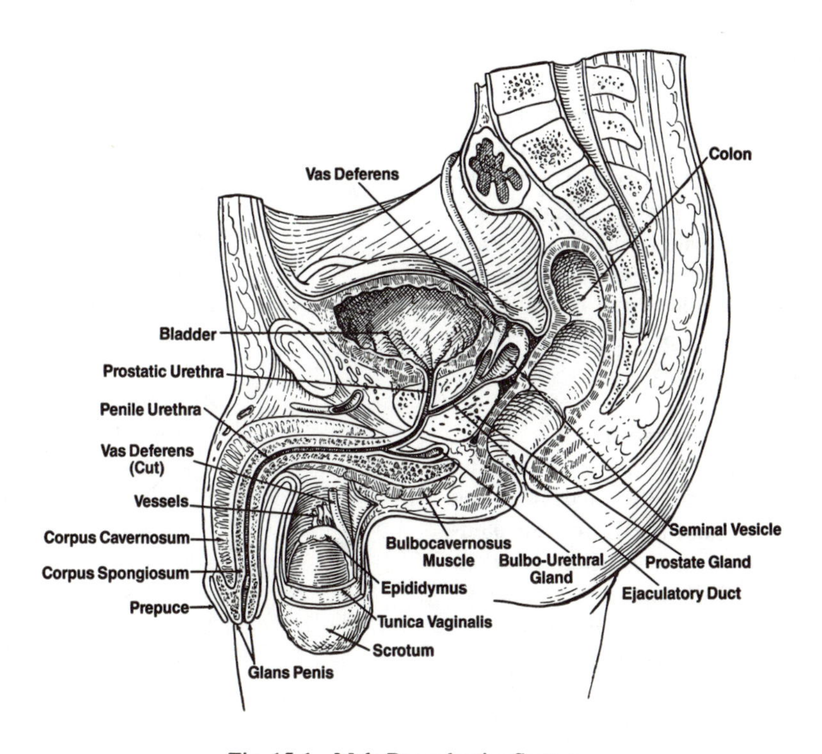

Fig. 15-1 Male Reproductive System.

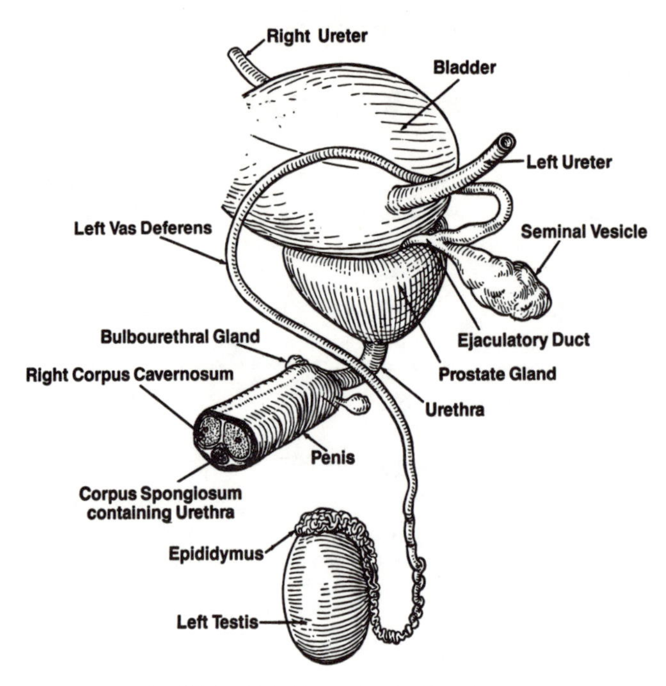

Fig. 15-2 Male Reproductive Structures.

taining the *penile urethra* (Fig. 15-2). The large tip of the penis is the *glans*. The *foreskin*, or prepuce, covers the glans.

B. *Microscopic anatomy*.
Each testis is composed of about 800 small tubules, the *seminiferous tubules*, surrounded by loose connective tissue that contains groups of *interstitial cells* (of Leydig). These cells secrete the male hormone, *testosterone*. *Sperm* (spermatozoa) are formed along the walls of the seminiferous tubules by the process of spermatogenesis (Fig. 15-3).

C. *Physiology*.

1. *Hormonal control*.

   a. *General*. Follicle stimulating hormone (FSH) and luteinizing hormone (LH), from the pituitary control testicular function. *FSH stimulates spermatogenesis, and LH stimulates the secretion of testosterone from interstitial cells. Gonadotropin-releasing hormone* from the hypothalamus stimulates production of FSH and LH. Testosterone has a negative feedback control on LH secretion.

   b. *Puberty*. Males under the age of ten produce little testosterone, because no releasing-hormone is secreted by the hypothalamus. During puberty, or adolescence, the *hypothalamus matures* and gonadotropin-releasing hormone stimulates the production of FSH and LH. LH increases the number of interstitial cells, and *testosterone* production is accelerated. Testosterone increases the synthesis of protein in cells (*anabolic effect*). Male secondary sex characteristics appear: body growth accelerates, muscle and bone mass increase, the penis and scrotum enlarge, the larynx develops (voice deepens) and hair appears on the face, chest, axillae, abdomen and pubis. The

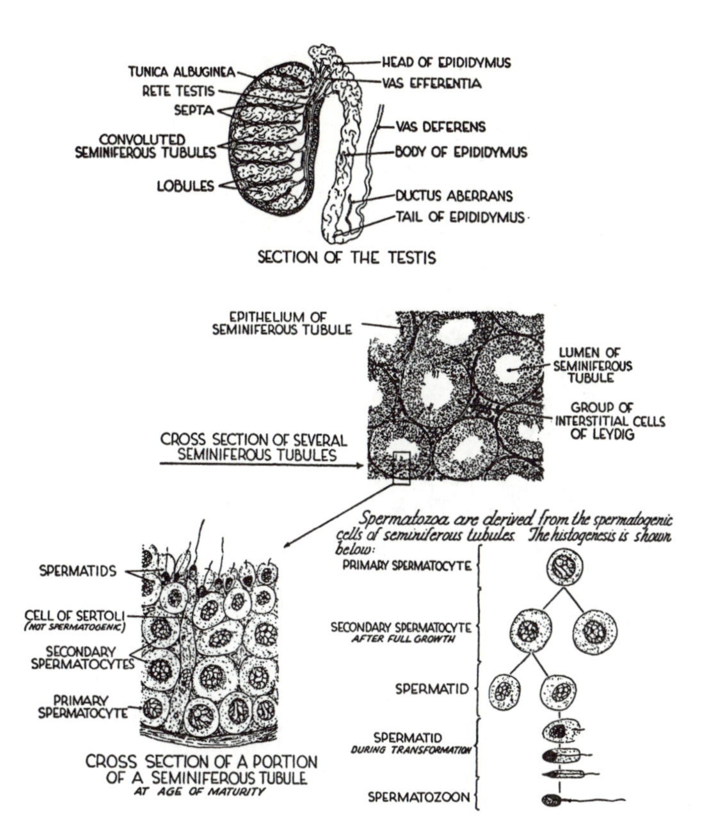

Fig. 15-3 Histology of the Testes.

sebaceous glands of the skin are stimulated (contributing to acne). Testosterone causes the male sexual drive (*libido*). Boys begin to exhibit more aggresive sexual and social behavior. The production of sperm (*spermatogenesis*) *is accelerated*. Testosterone and sperm are produced throughout life, gradually diminishing after age 40.

2. *Spermatogenesis is the formation and development of spermatozoa* in the walls of the seminiferous tubules (Fig. 15-4). *Spermatogonia* mature at puberty to large *primary spermatocytes*. Spermatogonia and primary spermatocytes contain the diploid number of chromosomes (*46*). Reduction (meiosis) of this number to *23* (haploid number) takes place as primary spermatocytes become *secondary spermatocytes*, which then become *spermatids*. The small spermatids undergo transformation to *spermatozoa*, each with a head, mid piece and tail. The head contains the nucleus, with 23 chromosomes. 50% of sperm have 22 plus an X chromosome; 50% have 22 plus a Y. Spermatogenesis takes place at a temperature lower than body temperature; hence, the location of the testes in the scrotal sac, where the temperature is cooler. During cold weather, the cremaster muscle contracts and elevates the testes closer to the body.
3. *Storage of sperm*. Sperm migrate from the seminiferous tubules to the epididymis, where they become motile. They are stored in

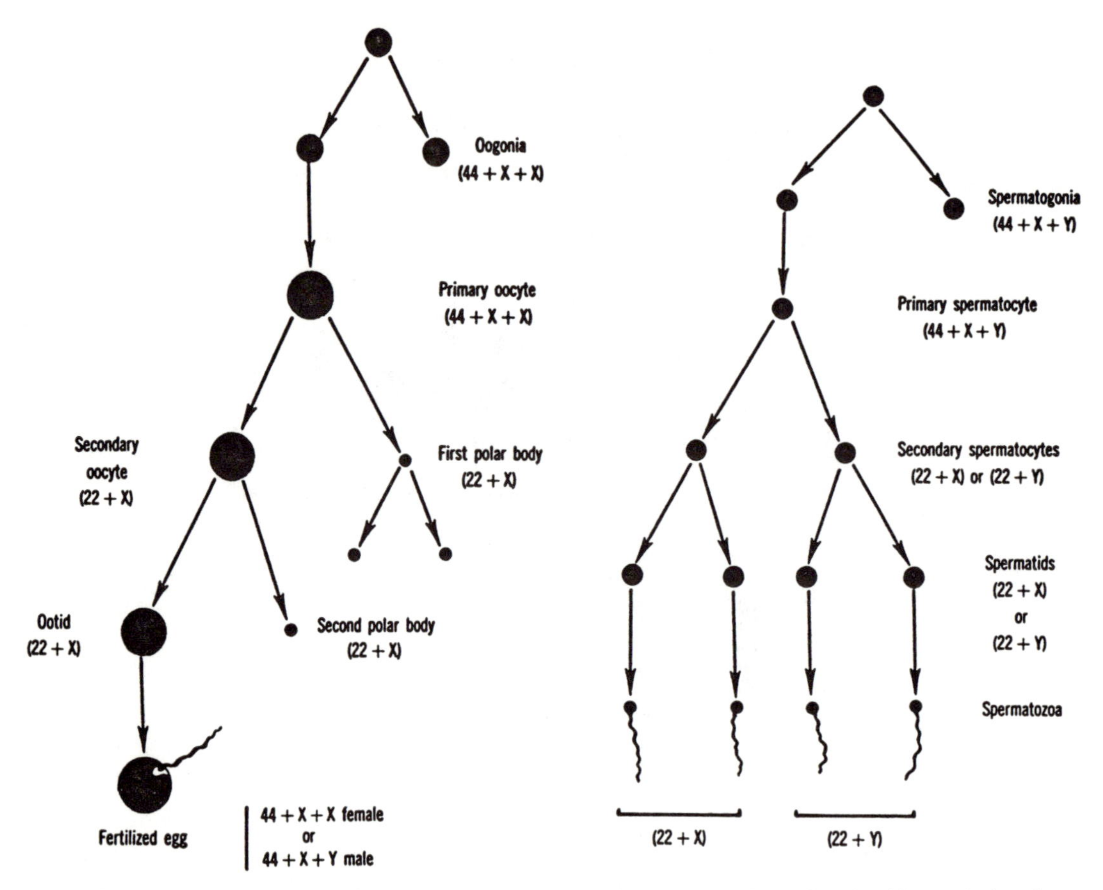

Fig. 15-4 Formation of Egg and Sperm. (Oogenesis and Spermatogenesis) (Reproduced, with permission, from Davies, Human Developmental Anatomy, Ronald Press, 1963).

the vas deferens.

4. *Erection* is a parasympathetic response in which arteries of the penis dilate and veins constrict; blood flows into the erectile tissue and venous outflow is blocked. A variety of stimuli cause erection.
5. *Ejaculation* is a 2-stage spinal cord reflex. The first stage, emission, is sympathetic and involves stimulation of L-1 and L-2 spinal cord levels. The second stage, ejaculation, is parasympathetic and involves the sacral portion of the spinal cord. Semen, an alkaline fluid containing fructose and prostaglandins, is secreted by two seminal vesicles, the prostate and two bulbourethral glands. Semen contains sperm. About *300 million sperm* are present in a five-day abstinent ejaculate. The *life of a sperm is about two days*.
   a. *Emission* involves the contraction of the epididymis, the vas deferens, the prostate and the seminal vesicles. Semen moves into the urethra.
   b. *Ejaculation* consists of contraction of the muscles at the base of the penis (bulbocavernosus, ischiocavernosus). Approximately *3 ml of semen* is propelled at high pressure through the penile urethra.

D. *Male contraceptive methods*.
Several are available: abstinence, the condom, the condom with spermici-

dal jelly or foam, withdrawal, and vasectomy. The condom alone and withdrawal are not very reliable. The condom with a spermicide is a fairly reliable method. Additionally, the condom protects against venereal disease. Vasectomy is a procedure done in the doctor's office under local anesthesia: a 2 cm piece of vas is removed, and the ends are tied. This should be considered a permanent method of sterilization, even though it is possible to re-join the cut ends. In most cases, after re-joining, fertility is less than 50%.

II. *The female reproductive system* (Fig. 15-5). *External structures*, collectively called the *vulva, or pudendum*, are the mons pubis, the labia, the vaginal opening, the clitoris and glands. The vulva lies in the outer skin-and-muscle-region called the perineum, through which also pass the urethra and anus (an *episiotomy* is a cutting of the perineum and vaginal mucosa to facilitate delivery of the baby). *Internal organs* are the vagina, the uterus, two fallopian tubes, two gonads (ovaries) and the breasts. The reproductive organs lie in the pelvis (see chapter 16).

A. *Anatomy* (Figs. 15-6, 15-7).

1. The *mons pubis* is the anterior, rounded portion of the vulva. It contains fat and is covered with hair at puberty.
2. Two large skin folds, the *labia majora*, lie posterior to the mons.
3. Medial to the labia majora are two smaller skin folds, the *labia minora*, which extend posterior to the clitoris and form the *vaginal opening* (orifice).
4. The *clitoris* is an erectile structure, homologous to the penis in the male, variably supplied with sensory nerves (a homologue is a structure similar in basic structure but not in function). The *urethra* lies posterior to the clitoris.
5. The *hymen* is a thin mucous membrane often present at the vaginal opening in the virgin female. It may persist after coitus, or it may be entirely absent. Its presence or absence does not signify virginity or non-virginity.
6. *Bartholin's glands*, homologues of the bulbourethral glands in the male, are two small round bodies situated in the labia majora. Each

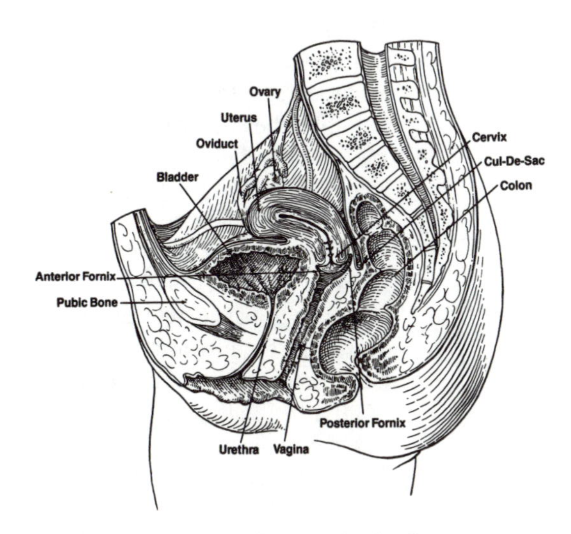

Fig. 15-5 Female Reproductive System.

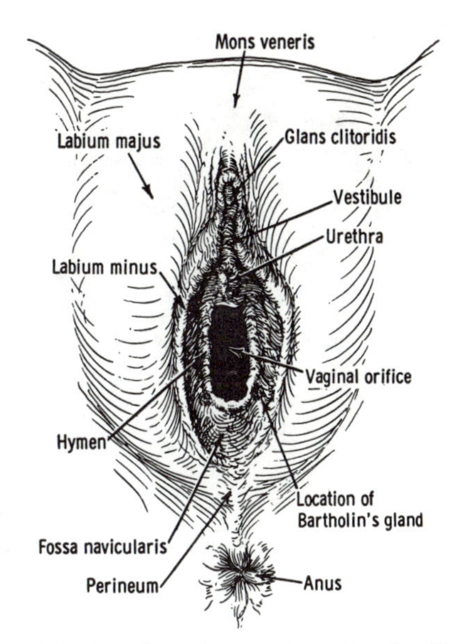

Fig. 15-6 Female Perineum. (Reproduced, with permission, from Benson, Handbook of Obstetrics and Gynecology, Lange, 1974).

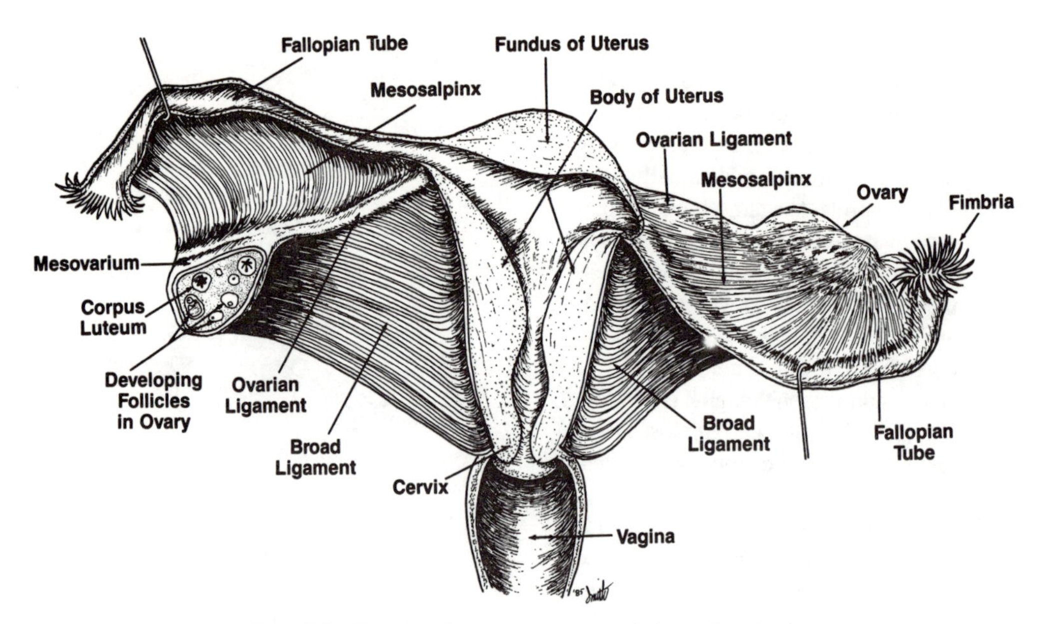

Fig. 15-7 Uterus and accessory structures (posterior view)

has a 2 cm duct which opens into the mucosal surface near the superior portion of the labia minora. A clear secretion is discharged during sexual arousal. Cysts and abscesses are not infrequent.

7. *The vagina* is a tube, about 3 inches long, lying between the bladder and urethra anteriorly, and the rectum posteriorly (Fig. 15-5). The cervix projects down about 1/2 inch, forming a space anteriorly and posteriorly (*the anterior and posterior fornix*). The vagina is an elastic organ, with an inner mucosa, a muscular wall and an outer connective-tissue layer.
8. *The uterus*, a hollow muscular organ, two by three inches long and about one inch thick, lies between the bladder and the rectum (Figs. 15-5, 15-7). The *body* is the largest portion. The smaller *cervix*, or neck, projects into the vagina. The round portion projecting up into the abdominal cavity is the *fundus*. The *fallopian tubes* are lateral extensions of the uterus. The *broad ligaments* attach the uterus to the pelvis on both sides. The *myometrium* is the thick smooth-muscle portion of the uterus, which undergoes profound growth during pregnancy. The inner lining is a soft, spongy layer of columnar epithelium, the *endometrium*, the surface of which is shed each month at the menses.
9. *The fallopian tubes (uterine tubes, oviducts)*. Each four-inch oviduct is suspended by a peritoneal fold, the *meso-salpinx* (Fig. 15-7). At the lateral end of each tube is a widened fringed portion (*fimbriae*) which receives the egg at ovulation. After ovulation, the fimbriae move close to the ovary, and the egg passes into the oviduct. Rhythmic contractions of the muscular layer and movement of cilia facilitate passage of the egg toward the uterus.
10. *Each ovary* is an almond-shaped nodular body, measuring 1 by 1½ inches, and held in place on the

lateral wall of the pelvis by the *ovarian ligament*. Microscopically, the ovary is made up of *follicles* (follicle: cup) in various stages of development or regression (Fig. 15-8). The mature form is the *Graafian follicle*, containing outer *theca cells* and inner *granulosa cells*. The latter surrounds a fluid-filled cavity containing the *egg (ovum)*. Also present in the ovary is a large *corpus luteum* (yellow-body), as well as connective-tissue and smooth-muscle cells, fibers and many blood vessels.

11. The *arterial supply to the pelvis* is from the *internal iliac arteries*, derived from the common iliacs.
12. The *pudendal nerves* arise from the sacral plexus, course near the spines of the ischia, and innervate

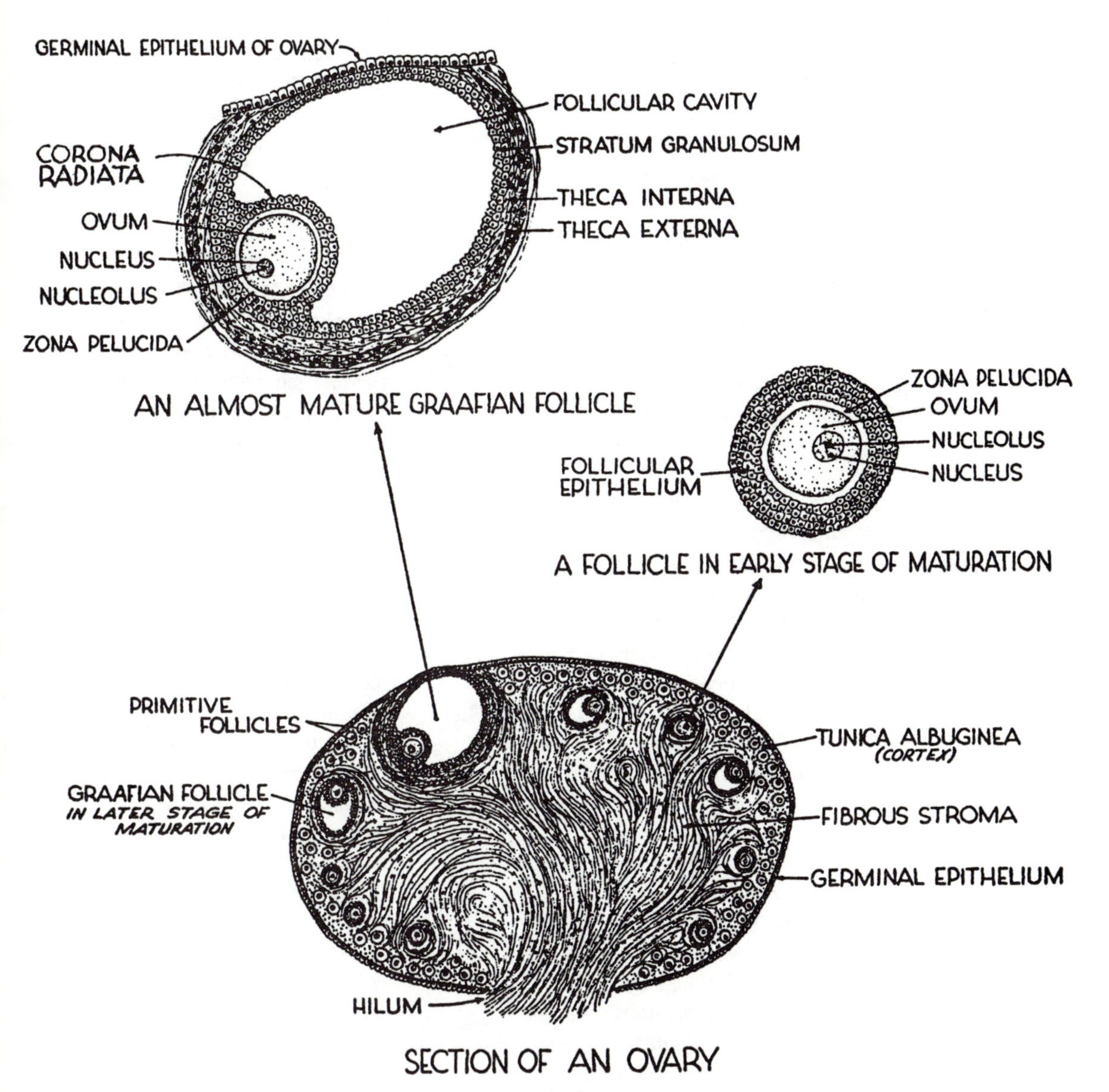

Fig. 15-8 Microscopic Anatomy of the Ovaries.

the perineum. Injections of lidocaine near the spines achieve good analgesia during childbirth.

B. *Physiology.*

1. *Hormonal control.*

a. *General.* A series of hormonal events takes place every 28 days. Known as the menstrual cycle, day number one begins with the first day of uterine bleeding (menses, menstruation). Cyclical hormonal changes occur in the pituitary, uterus, ovaries and vagina. As in the male, follicle-stimulating-hormone (FSH) and luteinizing hormone (LH) from the pituitary affect the gonads. *FSH* stimulates *growth of the follicle* (containing the egg) and the *secretion of female hormones* (collectively called *estrogens*) from the theca cells. The main estrogen is *estradiol*, responsible for *female secondary sex characteristics*, *growth of the maturing follicle*, *growth of the uterine lining (endometrium)* and negative feedback control of FSH. *LH has two main functions*: expulsion of the egg from the ovary each month (*ovulation*), and *formation of the corpus luteum* (from the old follicle). The corpus luteum secretes estrogens and another group of hormones, the progestins. The main progestin, *progesterone*, is responsible for the *secretory phase of the uterine cycle, glandular growth in the breast* and *negative feedback control of LH.*

b. *Puberty.* In the female, as in the male, almost no gonadal hormones are formed before age 9 or 10. As the hypothalamus matures, gonadotropin-releasing hormone stimulates production of FSH and LH. In response, the *ovaries produce estradiol, then progesterone.* Breastbuds and pubic hair appear about age 11. The breasts grow, and axillary hair appears (the maturing adrenal cortex is responsible for initial axillary and pubic hair in both sexes). The uterus and vagina enlarge. Uterine bleeding (the *menarche*) begins about two years after breast-bud development, and is often sporadic for several months. *Ovulation* takes place after the menarche. As puberty progresses, the body assumes more female proportions: the hips broaden, the forearms diverge more at the elbows, and scant body hair but much head hair is evident. The voice retains a high-pitched quality. Estradiol is not as anabolic as testosterone, and muscular development, bone size and general body growth is not as great as in the male. Estrogens cause the skin to have a smooth texture. Prepuberal characteristics such as voice, head hair-line, sparse body hair (compared to the male) and the distribution of body fat is retained and accentuated. Estradiol is responsible for the sexual drive (*libido*). In mammals, estrogens induce mating behavior, receptiveness of the female for the male, and nesting and maternal characteristics. As in the male, libido is modified by cerebral control. Libido increases at ovulation, and, sometimes, during the menses. After ages 40 to 50, a decrease in the responsiveness of the ovaries to FSH and LH, accompanied by irregular menstrual cycles, is the *menopause*. Although levels of estradiol and progesterone decrease, there is frequently little change in libido.

2. *Vaginal physiology*. The vagina has a dual function: copulation and birth. Mucous in the vagina during non-sexual times comes from uterine glands. During sexual arousal, Bartholin's glands are active, secreting increased mucus into the vaginal lumen. During orgasm, the muscular layer of the vagina contracts, moving semen into the cervix. Changes in the vaginal mucosa reflect cyclical endocrine changes. During ovulation, an increased number of superficial cells is seen on vaginal smear.
3. *Uterine physiology*. Changes in the endometrium over 28 days (average) constitute the uterine cycle (Fig. 15-9).
   a. *Menstrual phase: day 1 to 4*. Shedding of the superficial layer of the endometrium occurs (menstruation, menses).
   b. *Growth (proliferative) phase: day 4 to 14*. Under the influence of *FSH*, the ovarian follicle secretes *estradiol*, which promotes growth of the superficial layer of the endometrium.

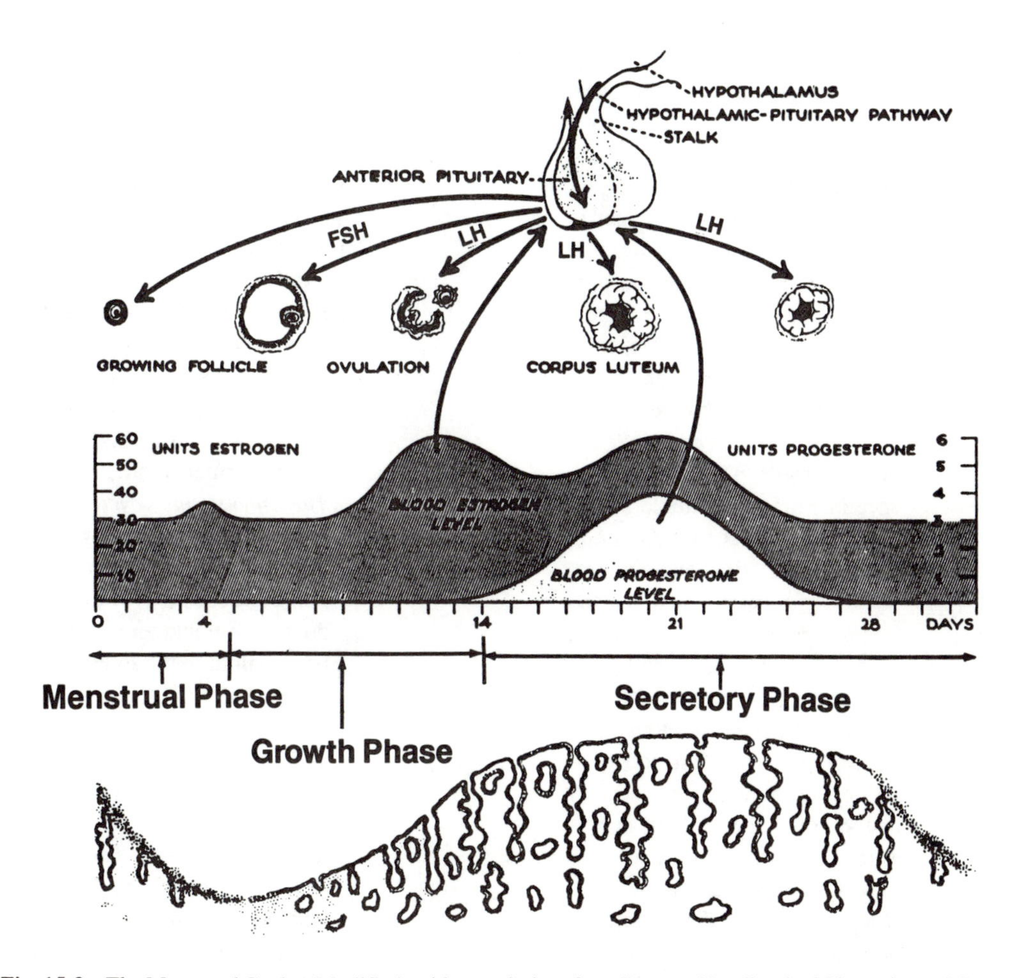

Fig. 15-9 The Menstrual Cycle. (Modified, with permission, from Benson, Handbook of Obstetrics and Gynecology, Lange, 1974).

Columnar cells, glands and vessels grow up and reconstitute the superficial layer.

c. *Secretory phase: day 14 to 28.* After ovulation (usually day 12 to 14), the corpus luteum, under the influence of LH, secretes both estradiol and progesterone. *Progesterone* causes endometrial glands to secrete nutrients, creating a stable environment for implantation of the fertilized egg. At ovulation, the cervix secretes a watery mucous, facilitating passage of sperm into the uterus. Cervical secretions then become more viscous, preventing passage of sperm. If implantation fails to occur, the corpus luteum shrinks, progesterone levels fall, and menstruation occurs.

4. *Ovarian physiology.* At birth, all of the eggs that will be ovulated in a lifetime are present. Primordial egg cells (oogonia) have become primary oocytes, containing the *diploid number* of chromosomes (22 pairs of autosomes, and two X chromosomes). Changes in the ovary over 28 days constitute the *ovarian cycle.* Under the influence of FSH, one follicle begins to enlarge to become a *Graafian follicle.* FSH stimulates the production of estradiol by theca and granulosa cells. On about the 14th day, LH causes ovulation, and the egg is expelled into the abdominal cavity. Although, on an average, ovulation occurs at mid-cycle, it may occur on any day of the month. The egg is attracted by the fimbria and moves into the fallopian tube. During movement down the fallopian tube, reduction of chromosome number takes place. The ovum now contains the *haploid number* of chromosomes (22 autosomes and an X chromosome) (Fig. 15-4). If fertilization does not occur, the egg is absorbed into the tubal or uterine mucosa. After ovulation, LH stimulates granulosa and theca cells to become lutein cells (*corpus luteum*), which secrete progesterone and estradiol.

C. *Mammary gland* (Fig. 15-10).

1. *Anatomy.* The adult female breast lies on the pectoralis major muscle, and is made up of many glandular lobules and a duct system interspersed in fatty tissue. Terminal ducts of the glands are arranged radially around the pigmented *nipple*, the secretory apparatus of the breast. The pigment extends outward for one or two centimeters as the *areola*, which contains smooth muscle fibers responsible for nipple erection.

2. *Physiology.*

a. *Puberty.* Under the influence of *estradiol*, puberty initiates a growth of glandular tissue plus an increase in fat. The nipples and areola enlarge and have a darker pigmentation.

b. *The menstrual cycle.* The breasts undergo monthly cyclical responses to hormones. *Estradiol* causes growth of the duct system and an accumulation of fluid in the interlobular tissue. Later in the cycle, *progesterone* stimulates growth and secretory activity of the glandular epithelium, causing an accumulation of fluid in the lobules and ducts. The result is tenderness of the breasts, which usually resolves at menstruation when the levels of estradiol and progesterone fall.

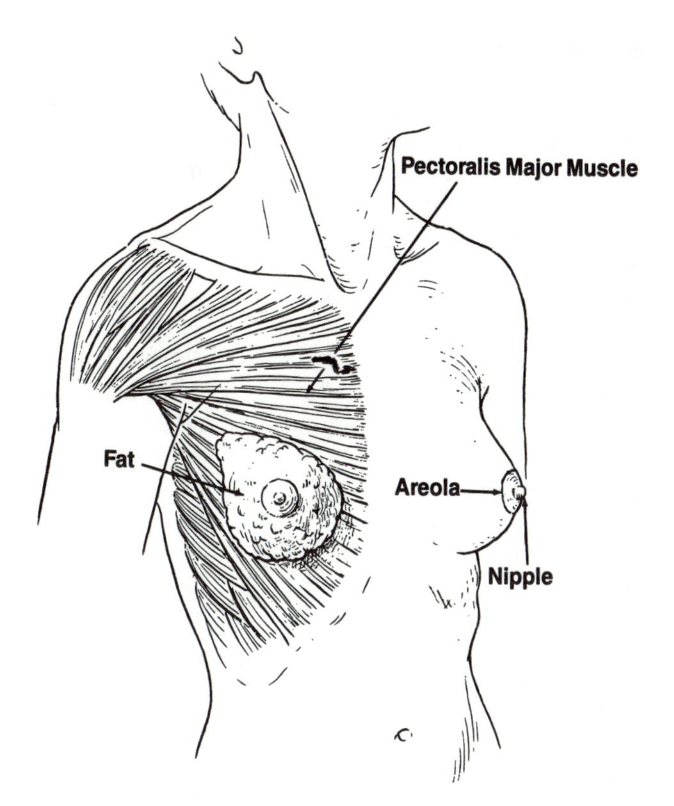

Fig. 15-10 The Breast.

D. *Female contraceptive methods*.

In contrast to the male, a multitude of contraceptive methods are available to the female. Removal of the uterus (hysterectomy) and tying the fallopian tubes (tubal ligation) should be considered permanent procedures. As with the vasectomy, even though the ends of the cut and tied fallopian tubes may later be rejoined, fertility is decreased. Current, common methods are the birth control pill, the intrauterine device, the diaphragm, spermicidal agents, postcoital douche, and the rhythm method. The most reliable methods are the pill, the condom plus a spermicidal agent, the IUD, and the diaphragm with a spermicidal agent, in decreasing order of effectiveness:

1. *The birth control pill*, an effective contraceptive method, contains an *estrogen* and a *progesterone*, in varying dosages. These hormones *block FSH and LH*, preventing ovulation. In some women, the pill contributes to venous thromboembolism. In others, it may cause weight gain.
2. The *condom* with a spermcidial agent has the added benefit of protection from venereal disease. A disadvantage is the reduction of sensation during coitus.
3. *The intra-uterine device (IUD)* is a plastic or combination plastic-copper device inserted through the cervix into the uterus. An inflammatory reaction of the uterine endometrium occurs, creating an environment in which the *fertilized egg does not implant.* The most serious disadvantage is infection.

4. *The diaphragm* is a rubber structure that may be coated with spermicidal jelly or cream and fits over the cervix to *prevent the entrace of sperm*. Disadvantages are a lack of spontaneity in sexual relations, the possibility of improper placement, and dislodgment during coitus.

III. *Pregnancy*.

A. *Coitus* (intercourse, mating, copulation) is the insertion of the erect penis into the vagina, followed by orgasm and ejaculation. In the female, orgasm is variable. Orgasm results from stimulation of the *hypothalamus*, as well as an area anterior and lateral to the hypothalamus, the limbic system. After ejaculation, sperm move into the cervix, uterus and fallopian tubes at the rate of about 3 to 6 mm per minute. The life of sperm and egg is about 2 days. Fertilization must take place at this time for pregnancy to occur.

B. *Fertilization* is the penetration of the egg by a sperm, restoring the diploid number (46) of chromosomes (Fig. 15-11). This usually occurs as the egg moves down the middle or outer third of the fallopian tube. The ovum contains one X chromosome. A sperm contains either an X or a Y. The male thus becomes the determiner of the sex of the baby. If a *male sperm ("Y")* reaches the egg, a male baby results; if a *female sperm ("X")* reaches the egg, a female baby is produced. After the head of the sperm enters the egg, the tail is lost and a barrier is set up, prohibiting the entrance of further sperm. The chromosomes of egg and sperm nuclei arrange themselves at the two poles of the fertilized egg and it begins to divide.

C. *Implantation*.

Cell division continues as the fertilized egg (zygote) moves through the

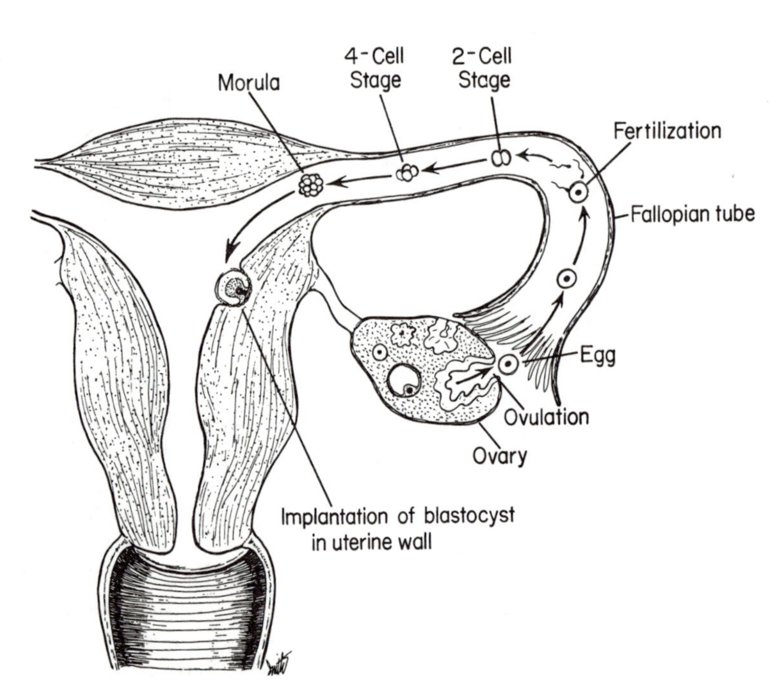

Fig. 15-11 Ovulation, Fertilization and Implantation.

fallopian tube. A cluster of cells, or *morula*, is formed. Further divisions create a hollow ball of cells (*blastocyst*) that *implants* in the uterine endometrium about *one week after fertilization*. The woman is now *pregnant*, and the implanted blastocyst is an *embryo*. Certain blastocyst cells (trophoblasts) form the placenta and the fetal membranes. Shortly after implantation (an hour or two), trophoblast cells begin to secrete *human chorionic gonadotropin (HCG)*, a hormone that maintains the corpus luteum in the early stages of pregnancy. *Progesterone* suppresses LH, blocking further ovulation and maintains the endometrium. Many fertilized eggs do not implant, or abort soon after implantation. Thus, a common physiological condition is non-implantation. *Twinning* takes place if the morula or blastocyst cleaves into two parts (or three, etc.). Both parts may develop into viable embryos, and are always the same sex. They are *genetically identical(monozygotic twins)*. When two sperm fertilize two different eggs ovulated at the same time, *fraternal(dizygotic) twins* are formed. In a sense they are not twins, but are brothers or sisters born at the same time. They may be of either sex, and have different genetic make-ups.

D. *Embryonic differentiation*.
Endometrial cells surrounding the blastocyst are *decidua cells*, and the area is the *decidua*. Trophoblast cells combine with decidual cells to form the *placenta*, an outer structure (*chorion*) and an inner membrane (*amnion*) that surrounds the embryo. *After the third month, the embryo is a fetus* (Fig. 15-12). The *amnion* (bag of waters) contains fluid that nourishes the fetus. Nutrient and waste exchange takes place between fetus and mother by *diffusion*. The two circulations are separate. However, most medicines and drugs used by the mother immediately affect the fetus.

E. *Placental hormones*.
Progesterone, maintained by *HCG*, causes endometrial cells to secrete large amounts of glycogen, proteins and other nutrients necessary for proper growth of the embryo. The level of HCG reaches a peak at about the 7th week, then declines to a low level that is maintained throughout pregnancy. The presence of HCG in blood and urine forms the basis for *pregnancy testing*. Detection of HCG is possible in blood about one week after implantation, and in urine a few days later. HCG has effects similar to *LH, FSH and TSH*. The placenta also secretes a hormone, *human placental lactogen (HPL)*, that regulates events in late pregnancy. The main estrogen in late pregnancy is *estriol* (reaching a peak at about the 7th month). Its secretion requires participation by the placenta, the fetal adrenal glands and fetal liver.

F. *Birth (parturition)*.
The exact stimulus for birth is unknown, but increased fetal activity plays a role. *Oxytocin stimulates contraction of the uterus*. The amnion ruptures and amniotic fluid is released, heralding impending birth. *Oxytocin also causes delivery of the placenta* after expulsion of the fetus.

G. *The breast in pregnancy*. The main physiological function of the mammary gland is to provide proper *nutrition* for the baby, as well as to protect the infant from infection during first few months of life by transferring *antibodies* from mother to baby. The breasts enlarge substantially after the second month of pregnancy because of increased amounts of estrogens and progesterone. These hormones are initially secreted by the *corpus luteum*

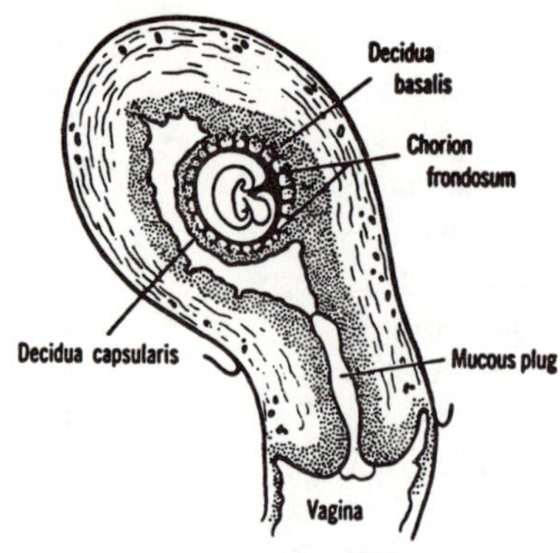

a. Embryo at One Month

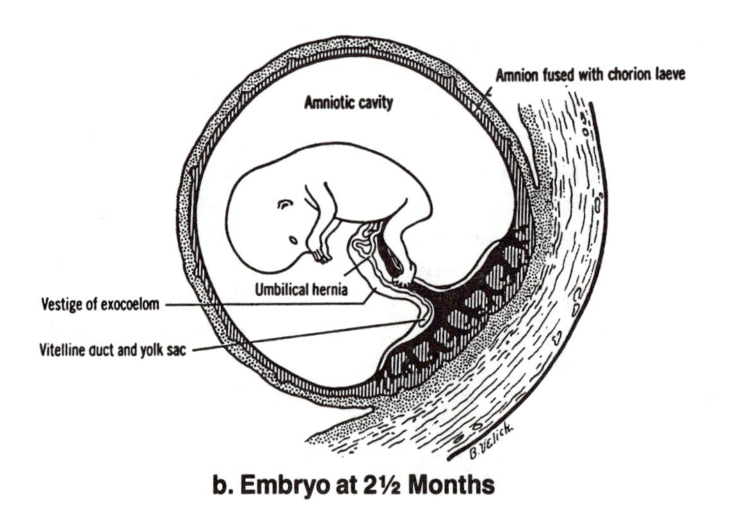

b. Embryo at 2½ Months

Fig. 15-12 Embryo and Fetal Membranes. (See ref., Fig. 15-4).

of pregnancy; after the eighth week they are secreted by the *placenta*.

1. *Delivery*. Before delivery, high estrogen and progesterone levels block prolactin. After delivery, these hormones fall. *Prolactin causes the production and secretion of milk from the lobules.* The maintenance of secretion requires periodic suckling and emptying of the lobules. The actual *ejection* (*"let-down"*) of milk from the nipple requires *suckling* and the *release of oxytocin* from the posterior pituitary. *Suckling also causes the secretion of prolactin.* If suckling continues, lactation will persist for months, even years. A yellow fluid, *colostrum*, is secreted during the last part of pregnancy and for the first day or two after delivery. It has a *high protein* content and *contains antibodies*. The output of antibodies from mother to baby is highest during the first week of life, insuring protection from infection for the 2 to 3 month period when the baby's immunological system is developing. Milk is

secreted one to three days after delivery. The cry of the infant, and in some cases emotional responses, may cause oxytocin release and lactation. (See Table 12-1).

2. *Summary of hormonal events of lactation.*

   a. *Estrogens* from the corpus luteum, and then the placenta, cause growth of the glandular ducts.

   b. *Progesterone* from the corpus luteum and then the placenta causes growth and secretory activity of glandular tissue.

   c. *Prolactin* from the anterior pituitary causes milk production and secretion from breast glands. Suckling stimulates prolactin secretion and milk production.

   d. *Oxytocin* release from the posterior pituitary is stimulated by suckling and other responses. It causes ejection of milk from the nipple.

IV. *Common disorders of the reproductive system*:

A. *An abortion* is the *termination of pregnancy before the fetus can survive, generally before 24 weeks* (or 500 grams). After this it is a *delivery*. An abortion may be spontaneous or induced. *Induced abortions* involve the use of drugs or surgical methods to evacuate the uterus. *A spontaneous abortion* (miscarriage) may be *threatened* (vaginal bleeding and cramping in the first 20 weeks of pregnancy), *incomplete* (passage of some products of conception), or *complete* (passage of all of the products of conception). Primary treatment for a threatened abortion is bedrest and sedation, for an incomplete abortion a dilatation and currettage (D and C), and nothing for a complete abortion.

B. *Bartholin cyst*. As mentioned, Bartholin's glands are located on each side of the vaginal opening. *Obstruction of a duct sometimes occurs from a bacterial infection*. The area is painful and swollen. Primary treatment: incision and drainage (I and D).

C. *Breast lumps*. Most are not cancerous, although the incidence increases with age.

1. *Fibrocystic disease* (mammary dysplasia, "lumpy breasts") is the most common disorder of the breast. About 50% of adult females have the condition. No treatment is necessary. However, the incidence of breast cancer is higher in these women.

2. *The fibroadenoma* is a benign lump in one breast, encountered in the 20 to 30 age group. Primary treatment is surgical removal of the lump, and laboratory analysis.

3. *Breast cancer* is the leading cause of cancer deaths in women, with a peak in the early menopausal age group. A common presentation is a *painless firm lump*, frequently felt in the upper outer quadrant. Diagnosis involves low-voltage soft-tissue x-rays (*mammograms*). Primary treatment consists of one of the following procedures: a "*lumpectomy*" (removal of the tumor), a *simple mastectomy* (removal of the breast only), a *radical mastectomy* (removal of the breast, pectoralis major and minor muscles, and axillary lymph nodes), an *extended radical mastectomy* (in addition to the above, removal of the internal mammary lymph nodes near the sternum) or a *modified radical mastectomy* (removal of the breast and axillary lymph nodes, preserving the pectoralis major muscle). Muscles and lymph nodes are removed because *metastases* have usually already taken place. These proce-

dures are often supplemented with *x-ray and chemotherapy treatments*.

4. *Anatomical and physiological problems after a mastectomy*.
   a. When the pectoralis major muscle is removed, some flexion and adduction of the arm is lost. The anterior part of the deltoid, as well as the coracobrachialis and long head of the biceps may be developed to help with flexion.
   b. Loss of lymphatic channels in the axilla causes obstruction of lymph-flow from the arm, and localized edema develops. Elevation of the arm and a special sleeve to provide compression are helpful.

D. *Cervical disorders*.

1. *Chronic cervicitis is a* smoldering inflammation of the cervix, sometimes with a turbid discharge. Primary treatment: an antibiotic and an aqueous vaginal cream of low pH.
2. *Cervical cancer*. Cervical cancer is the *third most common malignancy in women* (after breast and colon cancer) and is divided into two categories: noninvasive and invasive. Factors contributing to the development of cervical cancer are early coitus, poor penile hygiene, multiple sexual partners and a possible viral infection.
   a. *Noninvasive carcinoma, or carcinoma in situ* (in situ: local). Symptoms are usually absent. The only indication of malignancy is an *abnormal Pap smear* (Papanicolaou cytosmear, exfoliative cytology). The smear consists of a scraping of material from the cervix and posterior fornix. If malignant cells are present, *colposcopy* is performed. A colposcope (colpe: vagina) is an instrument with a magnifying lens and a bright light inserted into the vagina to examine the vagina and the cervix. If the cervix appears abnormal, a biopsy is done. Primary treatment includes either a therapeutic cone biopsy of the cervix, or removal of the uterus (hysterectomy).
   b. *Invasive carcinoma*. Early signs of invasive cancer are a *blood-tinged vaginal discharge and postcoital bleeding*. Intermenstrual spotting is frequent. Diagnosis is confirmed by a Pap smear and a biopsy. Primary treatment is based upon staging the lesion. Current therapy consists of supervoltage radiotherapy and the implantation of radium capsules in the pelvis around the uterus.

E. *Chromosomal abnormalities*.
When the zygote is formed, it contains *22 pairs of autosomes*, and either an *XX or XY sex chromosome pair*. Occasionally, one of the chromosomes breaks off or is lost, resulting in a total of 45 chromosomes. Sometimes, an extra chromosome is present, or 47 chromosomes. Other numbers are possible, but the above are the more common. A *karyotype* is constructed, in which the chromosome pairs are displayed and numbered in order of decreasing size from 1 to 22. A sample of amniotic fluid at about the 15th week of pregnancy (amniocentesis) permits displaying the karyotype of the fetus. Many chromosomal defects are incompatible with life. Of those that are, some degree of developmental abnormality, sterility, and mental retardation is common. The more important defects involve an extra chromosome (47 total chromosomes) at 21 (Trisomy 21) or at the sex chro-

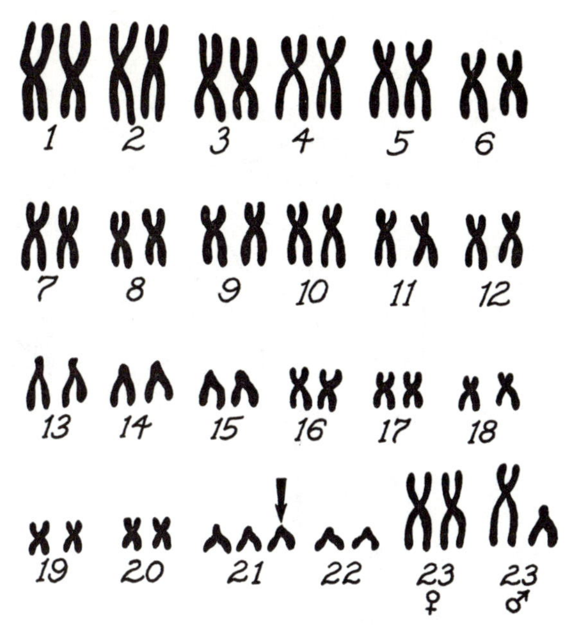

Fig. 15-13 Karyotype in Down's Syndrome (extra chromosome #21).

mosome pair (Fig. 15-13).

1. *Down's syndrome* (Trisomy 21, Mongolism) is the most common chromosomal abnormality. The occurance is high in babies of teenage mothers and women over age 35. The incidence is 1 per 1000 if the mother is age 25, and 1 per 100 if the mother is 40 or over. The syndrome is characterized by moderate to severe mental retardation and developmental abnormalities.
2. *Klinefelter's syndrome* (XXY, seminiferous tubule dysgenesis), is the most common sex chromosome abnormality. The person is a sterile eunuchoid-appearing (eunuch: a male castrated before puberty) male. Although some testosterone is secreted, the testes are small and the breasts are enlarged (gynecomastia). The legs are long and mild mental retardation is usually present.
3. "*Superfemale*" (XXX), the second most common sex chromosome abnormality, is a normal-appearing female. Some degree of mental retardation is common. However, these females usually have a normal menstrual history and are fertile. The offspring of XXX mothers are normal.

F. *Cryptorchidism (cryptorchism) is an undescended testicle*. The testes normally develop in the abdominal cavity. Shortly before birth they enter the inguinal canal and descend to the scrotum, carrying vessels, muscles and peritoneum. In cryptorchidism, a testicle remains in the abdominal cavity, or partially descends and lies in the inguinal canal. By the age of one, most undescended testes have descended with no treatment. Sometimes, surgical repair is necessary.

G. An *ectopic* (Gr: displaced) *pregnancy is one in which the blastocyst is implanted at a site other than the uterus. 90% occur in the fallopian tube* (tubal pregnancy) (Fig. 15-14). Implantation may occur in the peritoneum. As the embryo grows, *abdominal pain* is present. Menstrual periods

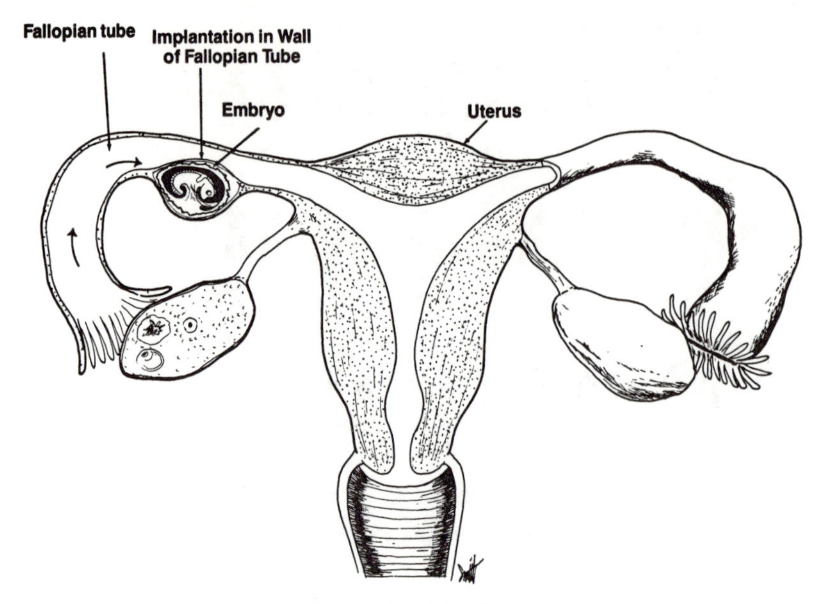

Fig. 15-14 Ectopic (Tubal) Pregnancy.

are irregular, with scanty flow or *spotting*, because insufficient gonadotropin is produced in the tubal placenta to maintain the uterine endometrium. Sometimes, a 4 or 5 inch mass can be felt lateral to the uterus. If rupture occurs, usually between the eighth and tenth week, abdominal pain is severe and may be accompanied by hemorrhage and shock.

1. *Useful tests*:
    a. *Ultrasound* (ultrasonography, echography) is the passage of sound waves into the body. A recording is made as they are reflected, or echoed, from various anatomical structures. Each tissue returns a different echo, which is converted into a series of pictures. The hazards of x-rays are thus avoided. Ultrasound may detect the presence of a tubal mass.
    b. *Culdocentesis* is the passage of a needle through the posterior fornix of the vagina into the cul-de-sac to obtain fluid. It may reveal blood in the peritoneal cavity.
    c. *Laparoscopy* is the insertion of a tube through the umbilicus into the abdominal cavity to visualize the tubes.
    d. *Culdoscopy* is the insertion of a tube through the posterior vaginal fornix for visualization.
2. Primary *treatment* for a tubal pregnancy is removal of the fallopian tube (*salpingectomy*) by an intra-abdominal operation (*laparotomy*). A ruptured tubal pregnancy resulting in hemorrhage and shock is treated by volume replacement with intravenous saline, typing and cross-matching for transfusions, and surgery.

H. *Endometriosis* is a disease of young women in which *endometrial tissue is present in non-uterine locations*, such as the ovaries and posterior cul-de-sac. Common symptoms are abnormal menstrual periods, pelvic pain, painful menstruation (dysmenorrhea) and painful intercourse (dyspareunia). Endometrial tissue may be transplanted from the uterus through the fallopian tubes into the peritoneal cavity at menstruation. It is thought that in some cases, a primitive embryonically implanted epithelium is transformed into endometrial tissue. Pregnancy is often curative. Primary treatment is the birth control pill: progesterone induces a pseudopregnancy which changes endometrial tissue. Sometimes, surgical excision of the implants is necessary.

I. *Infertility* is a relative *decrease in the ability to conceive. Sterility is a total loss of the ability.* In males, common causes are impotency (the inability to have an erection), abnormalities of sperm anatomy and motility, and a decrease in sperm number. A sperm count of 12.5 million or less per 5-day abstinent ejaculate usually indicates infertility. In females, common causes include a missed period (loss of ovulation), disorders of the fallopian tubes (often from a previous infection) and abnormal mucus secretion from the cervix, creating an environment hostile to sperm. In males, many cases of impotency are psychogenic and may respond to psychiatric consultation. The quality of sperm may be improved by avoiding alcohol, tobacco and caffeine. It may also be improved by adequate diet, exercise and sleep. Tight shorts should not be used, since they pull the testes close to the body, increasing the temperature and decreasing the sperm count. In females, clomiphene citrate is sometimes effective in inducing ovulation. Surgery may be beneficial to correct tubal scarring. The administration of an estrogen may restore normal cervical mucus.

J. *Prostate disorders.*
    1. *Prostatitis is a bacterial infection of the prostate*, usually caused by an extension of urethritis or a urinary tract infection. Perineal pain, fever, and a tender prostate on rectal exam are common signs. Primary treatment: an antibiotic.
    2. *Benign prostatic hypertrophy (BPH)* is a disorder of older males, and is thought to be caused by a relative decrease in the ratio of testosterone to estrogen. As testosterone declines, estrogen produced by the adrenal cortex seems to stimulate the central portion of the prostate, causing an *increase in cell mass*. The result is *compression of the prostatic urethra*, causing straining to void, dribbling, and sometimes urinary retention. (See Chapter 14) Primary treatment: relieving the retention by catheterization, and surgery to restore patency to the urethra. A common surgical procedure is a transurethral resection of the prostate (TURP).
    3. *Prostatic cancer* is the most common malignancy in males (other than skin cancer). It is slow growing, and is often discovered as an incidental finding. The prostate is hard and nodular. The enzyme *acid phosphatase* is plentiful in the prostate. Metastasis to bone commonly occurs, particularly in the thoracic and lumbar vertebrae and the sacrum. The enzyme *alkaline phosphatase* is high in bone, particularly as new bone is formed at the metastatic sites. Thus the blood levels of these enzymes *are often increased.* Symptoms include urinary retention if obstruction has taken place, and lower back pain if metastasis has occurred. Primary treatment depends on staging the cancer, but may include prostatectomy, transurethral resection, radiotherapy, and removal of the testes (orchiectomy), since cancer cells are stimulated by testosterone.

K. *Sexually transmitted diseases.*
The term "venereal disease", originally coined to include syphilis and gonorrhea (and later herpes), still car-

ries with it social condemnation and stigma. The phrase has been replaced with the more benign, "sexually transmitted disease", permitting the inclusion of vaginal infections, hepatitis B infection, non-gonoccocal urethritis, AIDS, and, possibly, body-lice. This discussion is limited to gonorrhea and herpes genitalis.

1. *Gonorrhea* is caused by the gram-negative bacterium Neisseria gonorrheae (gonoccocus). It is the number one public health problem in the United States (and has been for 30 years). It is becoming more resistant to antibiotics as time goes on, since mutant strains have developed. *The urethra of both sexes is infected*, producing a purulent *urethritis* several days after exposure. In the male, gonorrhea almost always causes painful urination (dysuria). In the female, signs and symptoms may be subtle, making her susceptible to the complications of an untreated infection. Primary treatment: an antibiotic. Strains resistant to antibiotics are common. Complications:
   a. *In the male*, untreated gonococci may migrate up the urethra to the prostate, causing *prostatitis*, or up the vas deferens to the epididymus, causing *epididymitis*. Primary treatment: an antibiotic.
   b. *In the female*, untreated gonorrhea may spread to *Bartholin's glands*, causing an *abscess or cyst*; or to the cervix, resulting in *cervicitis*. The most important complication is the movement of gonococci through the cervix and uterus into the fallopian tubes, where infection of the tubes (*salpingitis*) occurs. Involvement of the tubes and surrounding pelvic area is *pelvic inflammatory disease (PID)*. Signs and symptoms of acute PID include fever, bilateral pelvic pain (usually during the menses) and tenderness on movement of the cervix (stretching the broad ligament). The white blood count is high. Primary treatment: an antibiotic, sometimes intravenously. In advanced cases, removal of the uterus, fallopian tubes and ovaries (total abdominal hysterectomy and bilateral salpingectomy and oophorectomy) may be necessary.
2. *Herpes simplex is a DNA virus that causes painful blisters in and around the mouth, and on the genital area.* Type 1 usually infects the upper body, and type 2 the genital area. Type 2 is a common sexually transmitted disease. The primary infection lasts about one to four weeks. Recurrent lesions are less painful and debilitating, often emerge every month or two, and last for 7 to 10 days. They may be activated by fever, emotional stress, the menses, sunlight, infections and trauma. Genital lesions in women consist of painful vesicles and erosions on the labia, vagina or cervix. In males, the lesions are often located on the penis. Diagnosis is by viral culture, cytologic smear, or skin biopsy. *Primary treatment*: an analgesic, sitz-baths for vulvar lesions, cool compresses, and sometimes steroid therapy. The anti-viral ointment, acyclovir (Zovirax), is sometimes effective. Herpes is transmitted when it is active — that is, when the lesions are present and up to 7 days afterwards. In some people, the lesions recur. In others, recurrence takes place once or twice, and

never again.

L. *Uterine disorders.*

1. *A myoma (fibroid) is a benign tumor of the myometrium.* It is the most common disorder of the uterus. Seen in late reproductive years, the tumor is estrogen-dependent. Prolonged or abnormal menstrual bleeding is usually the first sign. Sometimes a hysterectomy is required.
2. *Polyps are small growths of the endometrium extending into the body of the uterus.* They are common in all age groups, but reach a peak about age 50. The main symptom is increased menstrual bleeding between periods (metrorrhagia), or post-menopausal bleeding. Primary treatment is removal of the polyps with a uterine curet (curettage).
3. *Dysfunctional uterine bleeding is abnormal bleeding throughout much of the 28-day cycle.* The mainstay of management, and often of diagnosis, is dilation and curettage.

M. *Vaginitis.* Common signs and symptoms are vaginal discharge, itching (pruritis) and irritation.

1. *Yeast vaginitis* (candidiasis, moniliasis) is a common *fungal infection,* and responds to an antifungal ointment such as miconazole (Monistat) or nystatin.
2. *Trichomonas vaginitis* (trichomoniasis) is caused by a motile, pear-shaped *protozoal parasite* that may infect the urinary tract of both sexes. It is thus a sexually transmitted organism. Primary treatment: metronidazole (Flagyl). The sexual partner may also require treatment.
3. *Gardnerella (Hemophilus) vaginalis* is a *bacterial infection* of the vagina, and responds to metronidazole.

-SIXTEEN-

REGIONAL ANATOMY

Regional, or topographical, anatomy is the study of regions of the body. This represents practical anatomy, because this is the way the physician and health professional actually approach the patient.

A. *Surface anatomy* (Fig. 16-1a: lips, cheeks; see also Fig. 8-1: eye; Fig. 8-10: ear; Fig. 11-1: nose). The *superficial temporal artery* is a pulse-taking artery palpable at the temporomandibular joint directly anterior to the tragus of the ear (Fig. 16-2).

B. *Internal structures.*
Fig. 16-3 shows a cross-section of scalp and skull. The mnemonic (memory device) "*SCALP*" is useful for remembering the various layers (Fig. 16-3).

II. *The neck.*

A. *Surface anatomy*. Most of the external part of the neck is formed by the two *trapezii*, the two *sternocleidomastoid* muscles, the *thyroid cartilage* ("Adam's apple") and the *supra and infra-hyoid muscles* anteriorly. Also anteriorly is the *hyoid bone*, palpable above the thyroid cartilage. The *tra-*

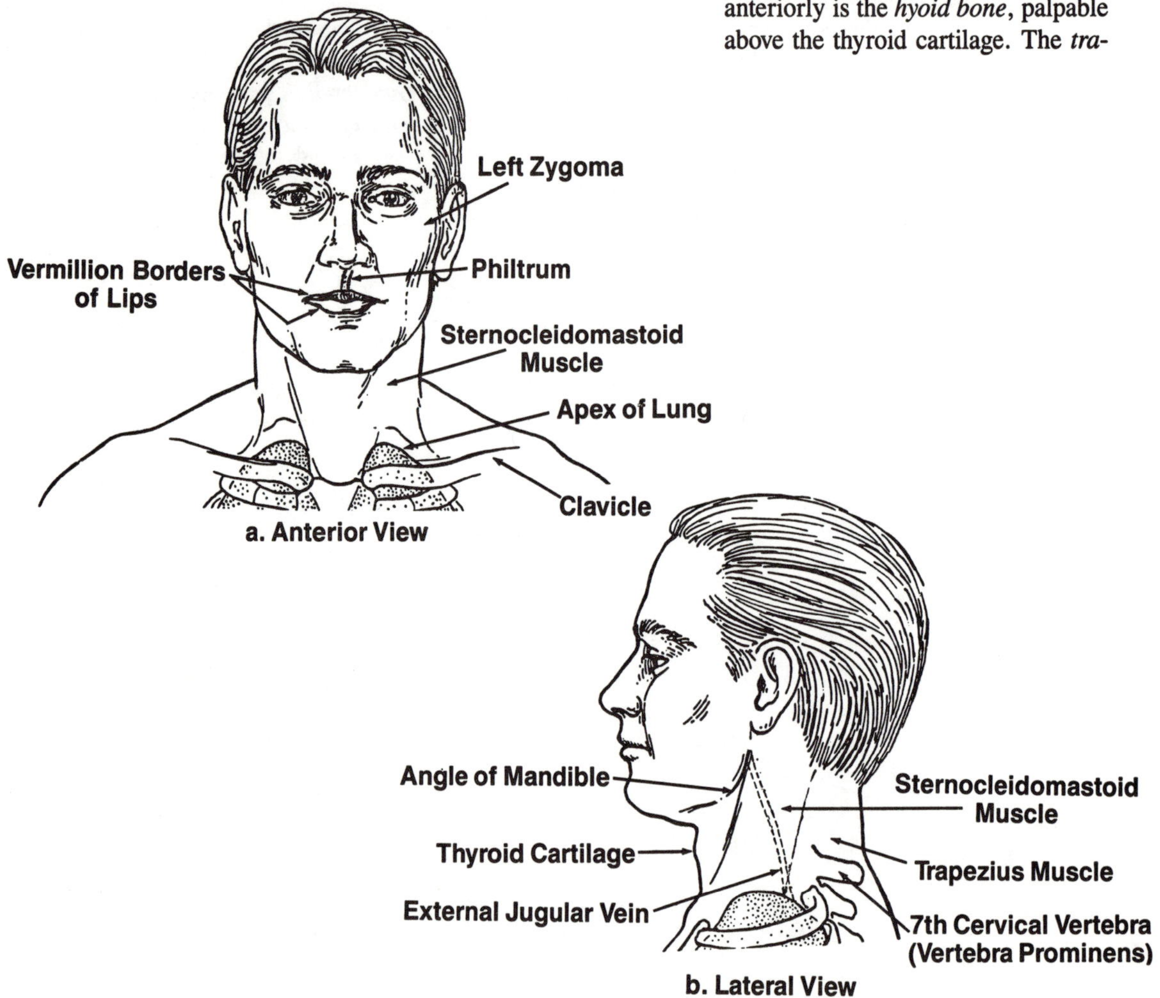

Fig. 16-1 Surface Anatomy of the Head and Neck.

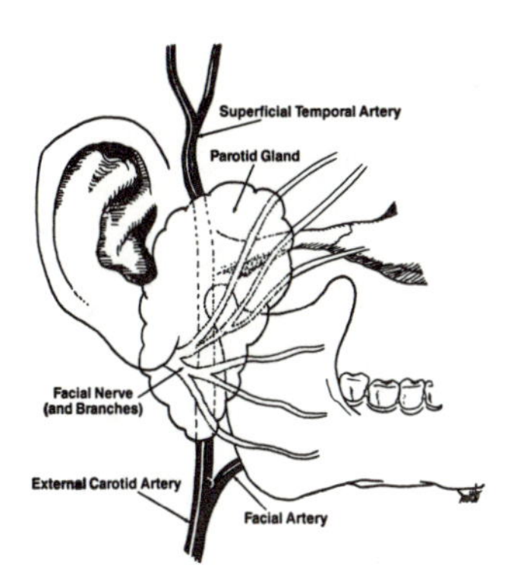

Fig. 16-2 Right Superficial Temporal Artery.

cheal rings lie below the cricoid cartilage. The *thyroid and cricoid cartilages* are palpable. The *thyroid gland* lies anteriorly on the trachea below the thyroid cartilage. It is *not* palpable unless enlarged (goiter). The *external jugular vein* crosses the sternocleidomastoid muscle. Posteriorly, the spine of the 7th cervical vertebra (*vertebra prominens*) is easily felt when the head is flexed (Fig. 16-1).

B. *Internal structures.*

1. *The vagus nerve, common carotid artery and internal jugular vein* lie deep to the sternocleidomastoid muscle in a thin fascial sheath (*carotid sheath*) (Fig. 16-4). The common carotid divides into the external and internal carotid arteries below the angle of the mandible.
2. *The supraclavicular fossa* is the region directly above each clavicle (Fig. 16-5). The *apex of the lung* lies in this area, and is thus subject to injury from a blow to the clavicle. Occasionally an enlarged lymph node is found here (sentinel node), raising the possibility of metastatic cancer.

III. *The thorax.*

A. *Surface anatomy:*

1. *Anterior and lateral.* Feel the *jugular notch* (suprasternal notch) at the top of the manubrium. Move your finger down the manubrium to a slight bump, the *sternal angle.* The second rib attaches here. Continue down the body of the sternum until the *xiphoid* process is felt at the bottom of the sternum. The *sternal borders* are the right and

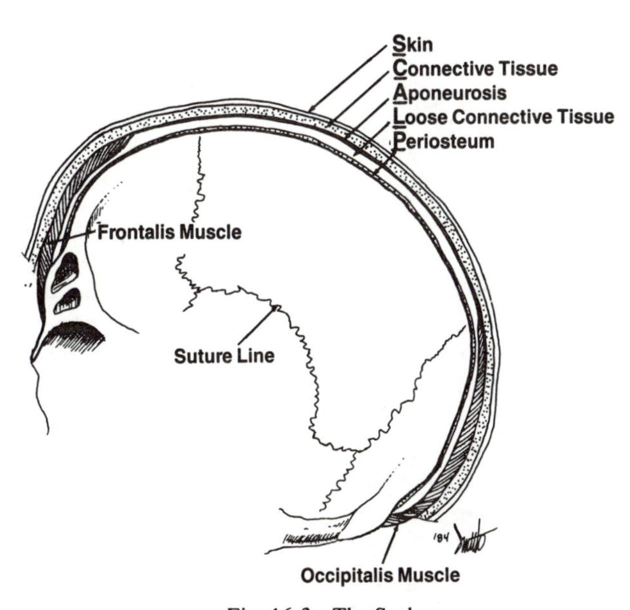

Fig. 16-3 The Scalp.

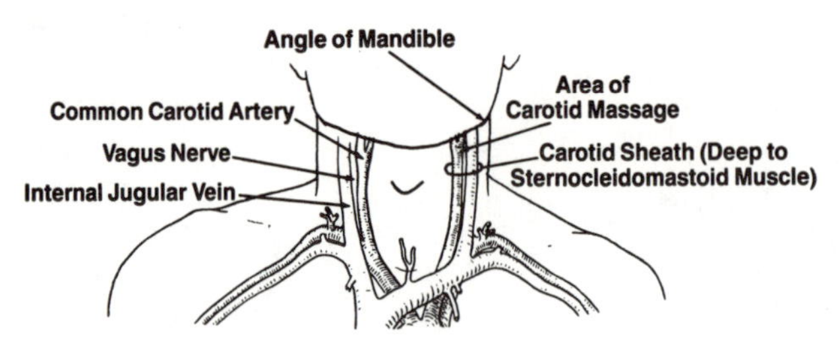

Fig. 16-4 Carotid Sheath.

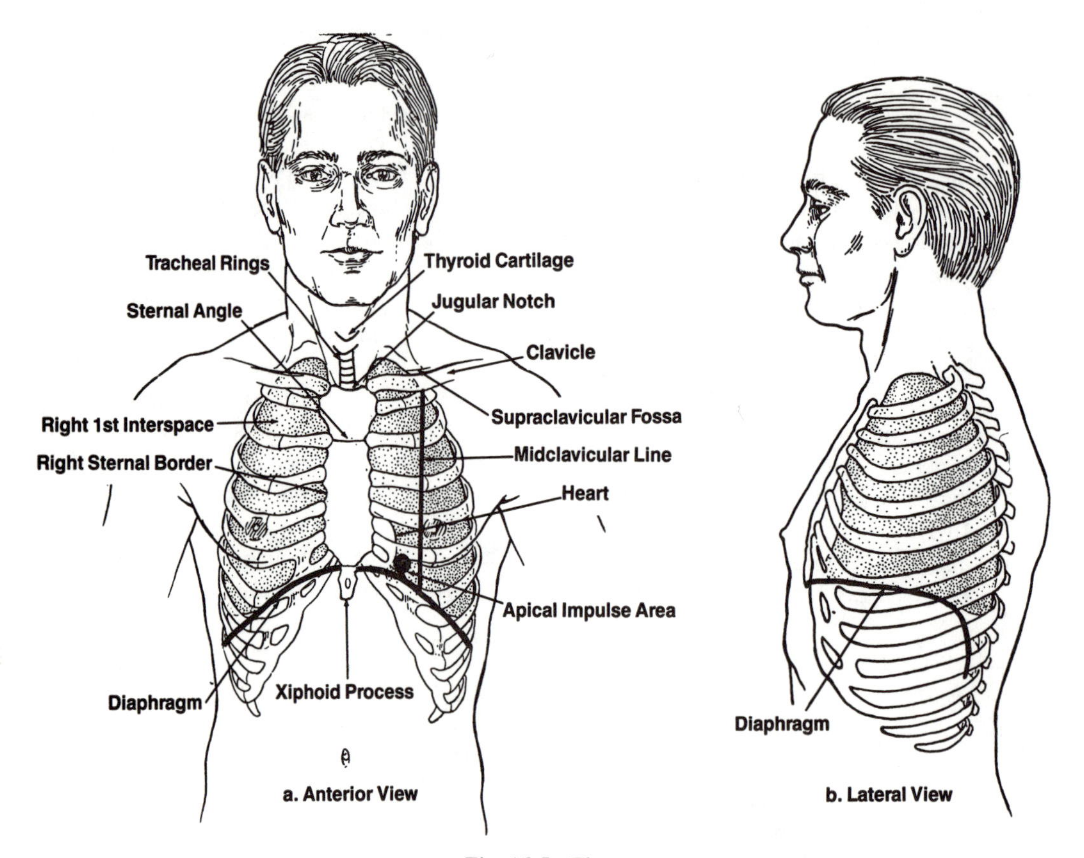

Fig. 16-5 Thorax.

left margins of the sternum, where the ribs attach (sterno-costal joints). The *pectoralis major muscle* forms the bulk of the musculature of the anterior thorax (Fig. 16-6). The *serratus anterior muscle* forms much of the lateral thorax. The ribs and interspaces are easily palpable. Since the first rib cannot be palpated, one uses the clavicle as the location of the first rib. The *diaphragm* separates the thorax from the abdominal cavity. It is a dome-shaped structure beginning at the xiphoid, extending laterally to about one cm below each nipple. It then dips to about ribs 7 or 8 and attaches posteriorly at about the level of the 10th thoracic vertebra and the 12th rib. During inspiration, the diaphragm moves downward and pushes the

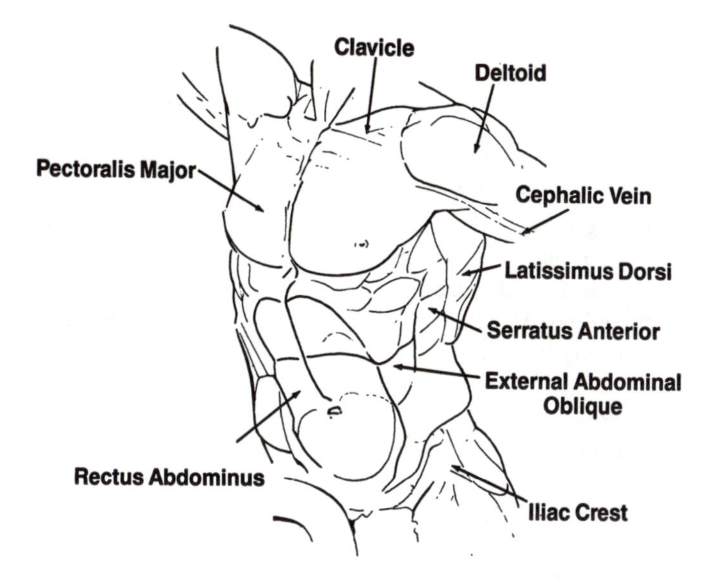

a. Anterior Aspect

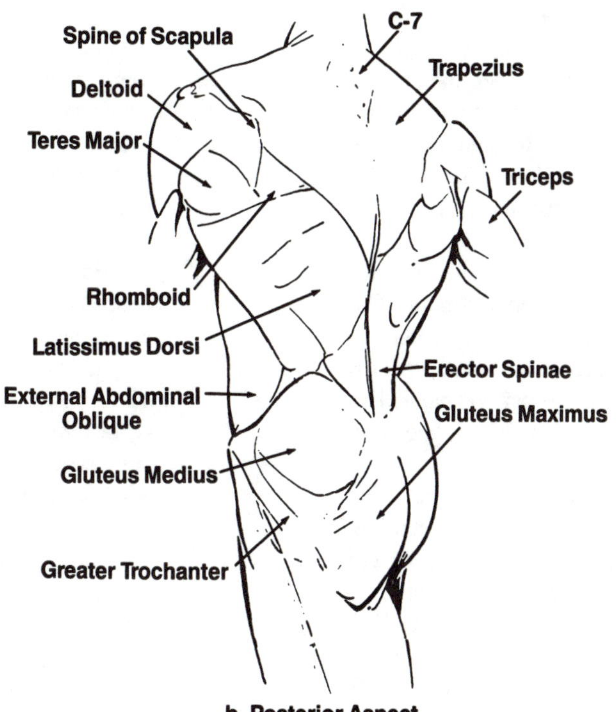

b. Posterior Aspect

Fig. 16-6 Torso. (Reproduced and modified, with permission, from Wolf-Heidegger, Atlas of Systematic Human Anatomy, S. Karger AG, Basel 1962).

upper abdominal structures down about 2 cm.

2. *Posterior* The upper back consists of the *trapezius* and *latissimus dorsi* muscles. Feel the spines of the thoracic vertebrae below the vertebra prominens. The *spines of the scapulae* are also easily palpable.
3. *Other features*. Many *heart mur-*

murs are heard with the stethoscope on the *left sternal border*. The *midclavicular line* is an imaginary line drawn from the midportion of the clavicle down towards the nipple (the midaxillary line is an imaginary vertical line bisecting each axilla). The left 5th interspace, 2 cm medial to the left midclavicular line, is where the *apical impulse* is usually located (Fig. 16-5a).

B. *Internal structures.*

1. *The mediastinum* is the region of the thorax posterior to the body of the sternum. It contains the heart, great vessels, arch of the aorta (plus the three arteries coming off of the arch) the thoracic aorta, the two brachiocephalic veins, the two vagus nerves, the thoracic duct, the esophagus, the trachea and the two main bronchi. The two phrenic nerves lie along the lateral margins of the pericardium. The azygous vein lies on the back body wall (Fig. 16-7).
2. *The pleural cavities* lie to the right and left of the mediastinum. The bases of both lungs run from the 7th rib anteriorly to about the 11th rib posteriorly. The parietal pleura lies about 3 cm lower than the lung bases (Fig. 1-5, 11-7).
3. *The infraclavicular region* is the area below the clavicle, containing the *subclavian artery and vein* (Fig. 16-8). The vein is a common site for catheter insertion. The catheter is inserted into the subclavian vein, then threaded into the brachiocephalic vein to the superior vena cava.
4. *The interspaces (intercostal spaces)*. Each space contains an *external and internal intercostal muscle* as well as an *intercostal vein, artery and nerve. Layers of tissue between skin and lung* are shown in *Fig. 16-9*. As mentioned, the lower level of the lungs reaches to about the 8th interspace on the sides and the 10th in back. Fluid (*pleural effusion*) in the pleural space in lung diseases may be removed by inserting a needle or catheter into the *7th interspace at the midaxillary line* (thoracentesis). The level of the 7th interspace at the midaxillary line is at about the level of the nipple. A chest tube may be inserted into the *2nd interspace at the midclavicular line*, or the 6th or *7th interspace at the midaxillary line*, in order to evacuate air in a *pneumothorax* or blood in a *hemothorax*. The needle or tube is inserted *directly above the rib*, since the intercostal vessels and nerve lie along the lower margin.

IV. *The upper extremity.*

A. *Surface anatomy* (Fig. 16-10).

1. Muscles of the *shoulder* area include the *deltoid*, and contributions by the latissimus dorsi posteriorly and the pectoralis major anteriorly. The jutting bony prominence at the lateral shoulder is the *acromion* process of the scapula. Palpate the *clavicle*.
2. The *arm* (brachium) is composed of the *biceps* anteriorly and the *triceps* posteriorly. Both the deltoid and triceps are common intramuscular injection sites.
3. At the *elbow* feel the *lateral and medial epicondyles* of the humerus, as well as the *olecranon* process of the ulna between them. The ulnar nerve lies superficially at the medial epicondyle. Compression results in tingling of the 4th and 5th fingers (the "funnybone"). The anterior part of the elbow is the *antecubital fossa*. *Superficial veins* are easily seen here. Palpate the *biceps tendon* and the *brachial artery* medially, the major blood-pressure taking artery.

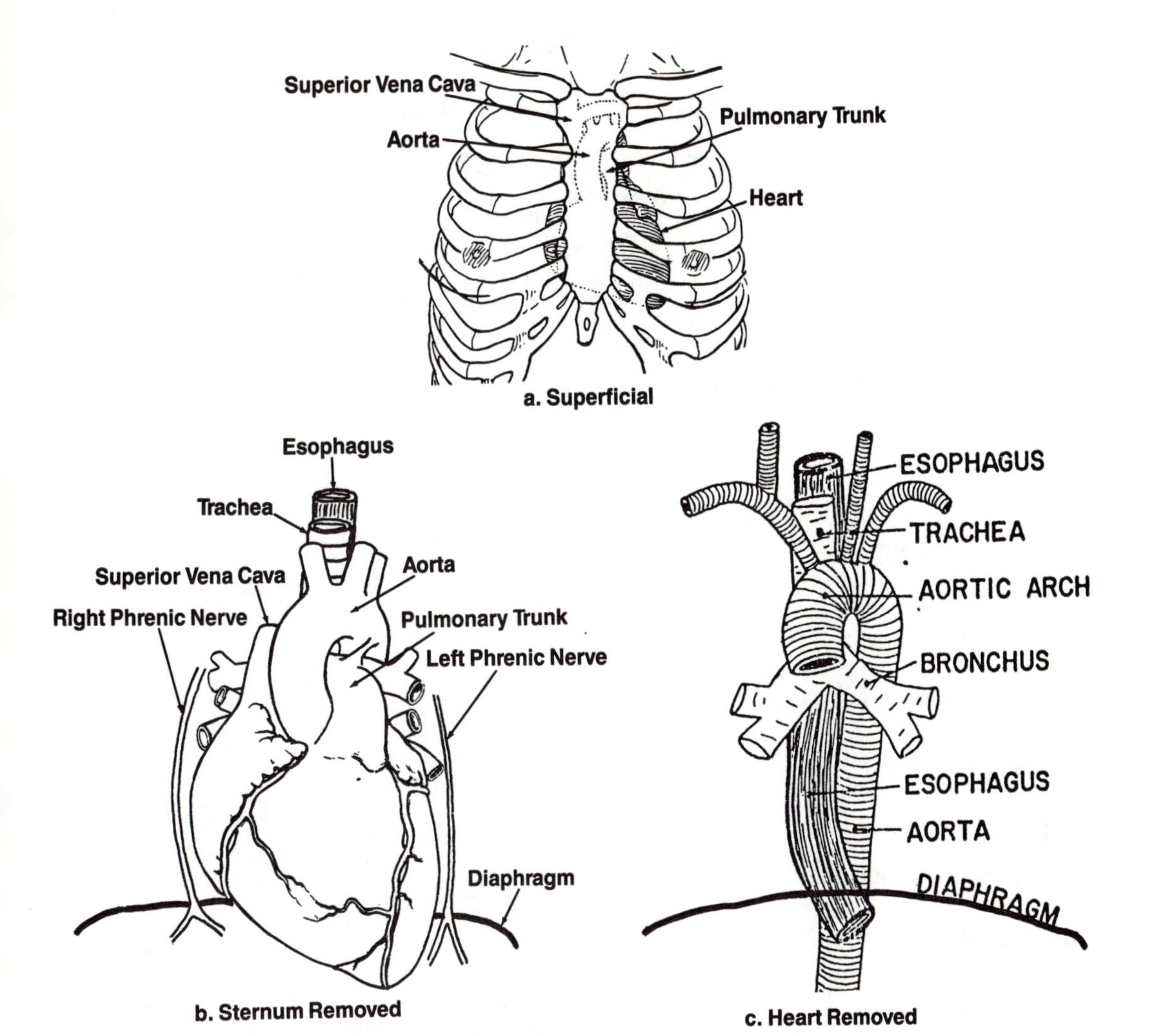

Fig. 16-7 Mediastinum.

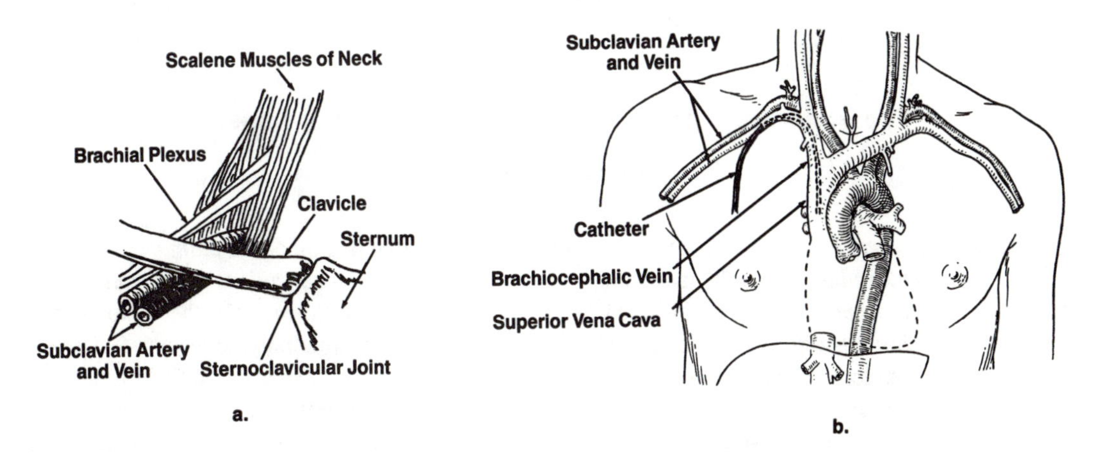

Fig. 16-8 Infraclavicular Region (with Subclavian Catheter in Place).

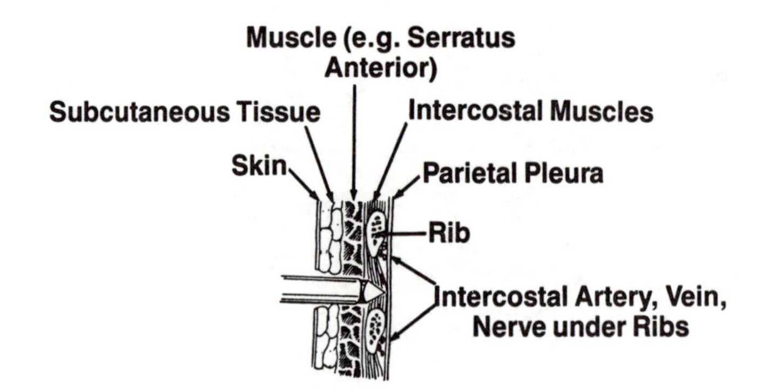

Fig. 16-9 An Intercostal Space. (Reproduced, with permission, from Suratt, Manual of medical Procedures, Mosby, 1982).

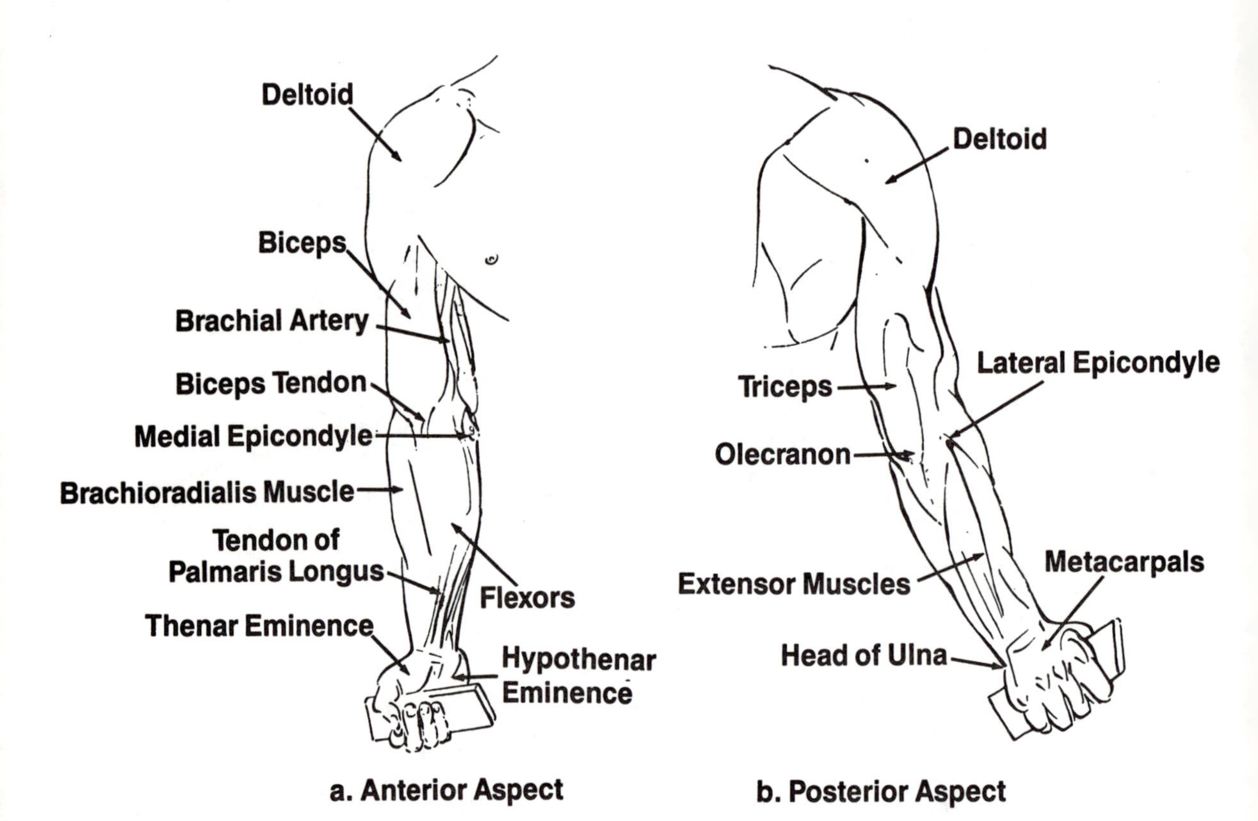

Fig. 16-10 (See ref., Fig. 16-6).

4. The *forearm* consists of the large *brachioradialis muscle* laterally, the *forearm flexors* anteriorly and the *forearm extensors* posteriorly. The tendons of these muscles are easily seen at the wrist.
5. *The wrist* is the area of the radiocarpal joint. The flexed hand reveals the tendon of the *palmaris longus*, as well as *tendons of the hand and finger flexors*. On the flexor surface, medially, one can feel the pisiform bone. On the lateral side is the superficial *radial artery*, a major pulse-taking artery. The dorsal aspect of the wrist contains the

tendons of the hand and finger extensors. The *head of the ulna* is palpable on the extensor side of the medial wrist.

6. *The hand* is composed primarily of *metacarpals*. Dorsally lies a venous plexus and *extensor tendons*. The *superficial veins* on the hand dorsum are common sites for *intravenous infusions* (Fig. 16-11). Also on the hand dorsum, with the thumb extended, are the tendons of the extensor pollicis longus and brevis (with an indentation between them). This is the *anatomical "snuffbox"* (Fig. 16-12). Tenderness here after a fall on the outstretched hand, may indicate a navicular fracture. Anteriorly lie flexor tendons and two muscular areas of the palmar hand, the *thenar eminence* at the thumb and the *hypothenar eminence* at the little finger (Fig. 16-10a).

B. *Internal structures*.

1. *The axilla* (armpit), bounded by the *pectoralis major muscle* anteriorly and the *latissimus dorsi* muscle posteriorly, contains fat and lymph nodes (Fig. 16-13). The *axillary artery, vein and brachial plexus* lie deep. Enlarged *lymph nodes* in this region may indicate breast cancer that has metastasized.

2. *The antecubital fossa*, bounded primarily by the brachioradialis and forearm flexors, separates the arm and forearm. The fossa contains the *tendon of the biceps*, the *median nerve*, the *brachial artery* and the *basilic, median cubital and cephalic veins* (Fig. 16-14). The ulnar nerve lies outside the fossa. The medial and lateral antebrachial cutaneous nerves lie superficially and are sometimes injured during venipuncture by inexperienced

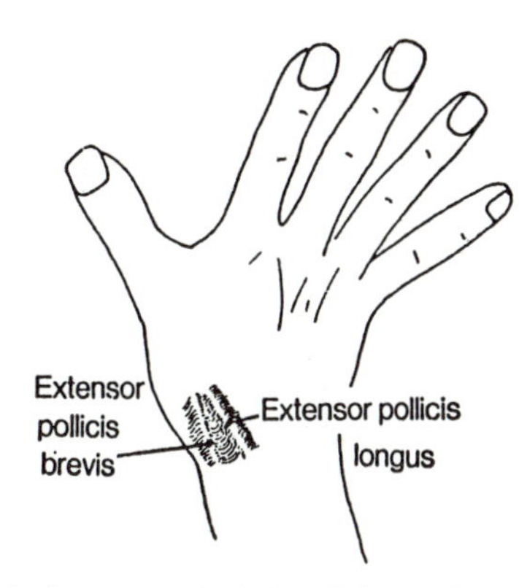

Fig. 16-12 Anatomical Snuff Box. (Reproduced, with permission, from Goldberg, Clinical Anatomy Made Ridiculously Simple, MedMaster, 1984).

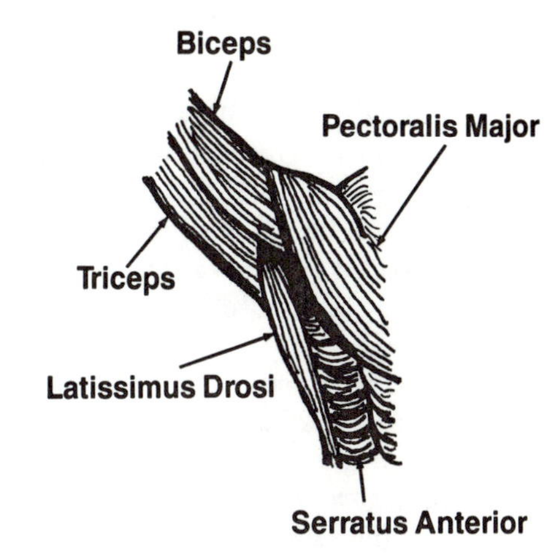

Fig. 16-13 Right Axilla.

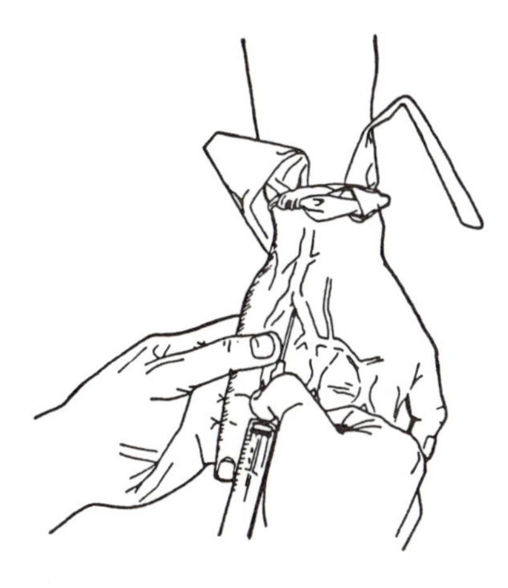

Fig. 16-11 Hand Veins. (Reproduced, with permission, from Abridged Textbook of Advanced Cardiac Life Support, American Heart Association, 1983).

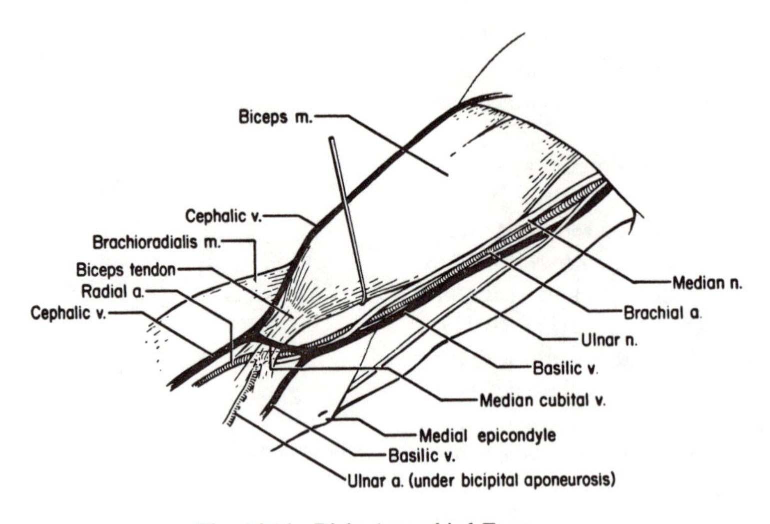

Fig. 16-14 Right Antecubital Fossa.

hands. The basilic and cephalic are the most common veins for drawing blood (Fig. 16-15). Occasionally, intravenous infusions are started here, and a cardiac catheter may be inserted.

V. *The abdomen.*

A. *Surface anatomy.* The boundaries of the abdominal cavity are formed by the *diaphragm superiorly,* and the *inguinal ligament and pelvic basin inferiorly* (Fig. 16-16). The anterior portion of the abdomen consists of the *rectus abdomini muscles* medially, the *abdominal oblique muscles* laterally, and posteriorly, the lower portion of the *latissimus dorsi* and the *erector spinae muscles* (Fig. 16-6). The *spleen* lies deep to the lower *left anterior ribs,* the *liver* lies deep to the lower *right anterior ribs,* and the *kidneys* lie deep to the *lower posterior ribs.* Thus, injuries to the ribs in these areas may cause *intra-abdominal* bleeding if the liver or spleen are lacerated, and *retroperitoneal* bleeding in the case of kidney damage (Figs. 16-17, 16-18).

1. *Abdominal regions.* Fig. 16-19 demonstrates the division of the abdomen into *quadrants.* Disorders originating in intra-abdominal, retroperitoneal, mediastinal and pleural structures may cause

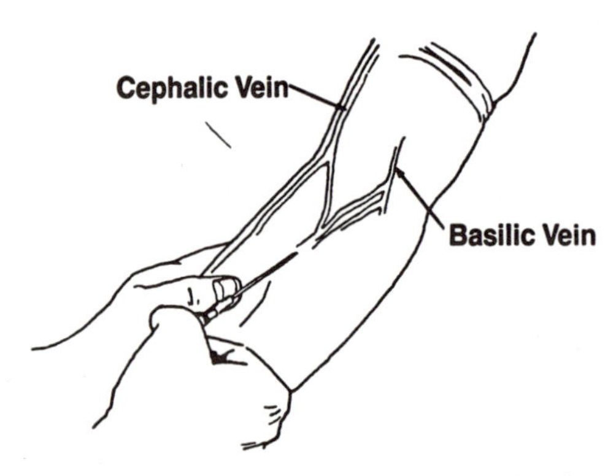

Fig. 16-15 Right Antecubital Region. (See ref., Fig. 16-11).

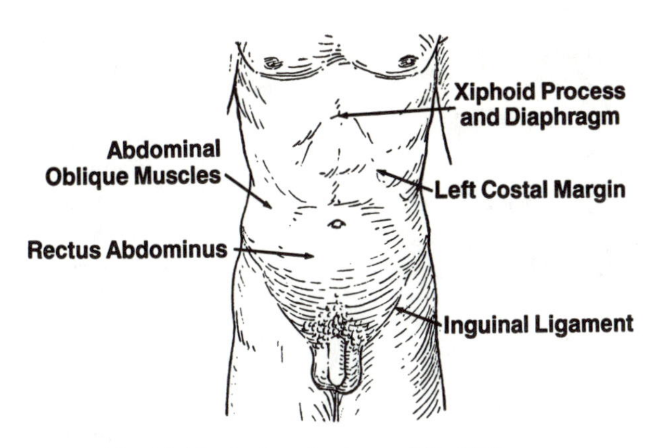

Fig. 16-16 Surface Anatomy of the Abdomen.

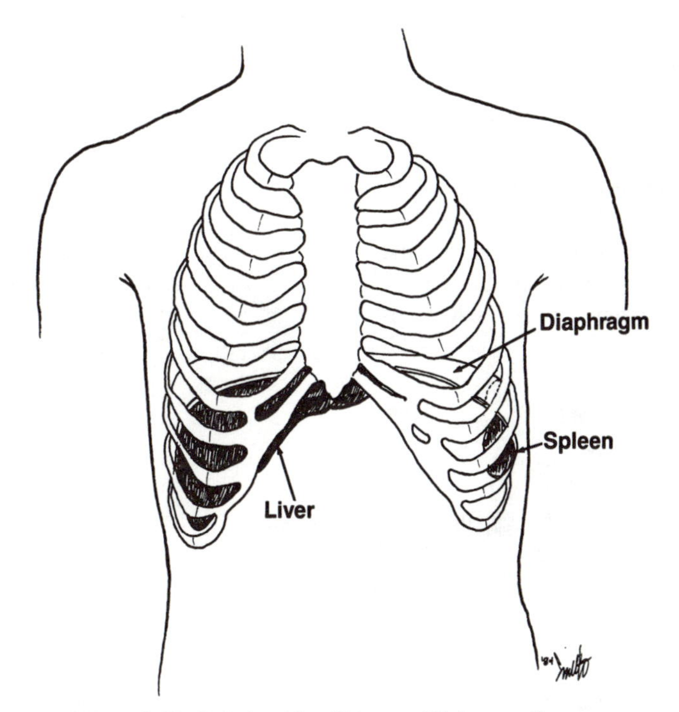

Fig. 16-17 Relationship of Liver and Spleen to rib cage.

pain in these areas:

a. *Right upper quadrant (RUQ)* pain may be felt from the liver, gall-bladder and lower right lung. Gall-bladder pain often radiates to the back and right shoulder.

b. *Left upper quadrant (LUQ)* pain may be felt from the stomach, spleen, pancreas and lower left lung. Sometimes, pancreatic pain radiates to the back.

c. *Right lower quadrant (RLQ)* pain may be felt from the right ovary, fallopian tube, appendix, right kidney and ureter. Disorders of the kidney and ureter may cause flank and back pain.

d. *Left lower quadrant (LLQ)* pain may be felt from the left ovary,

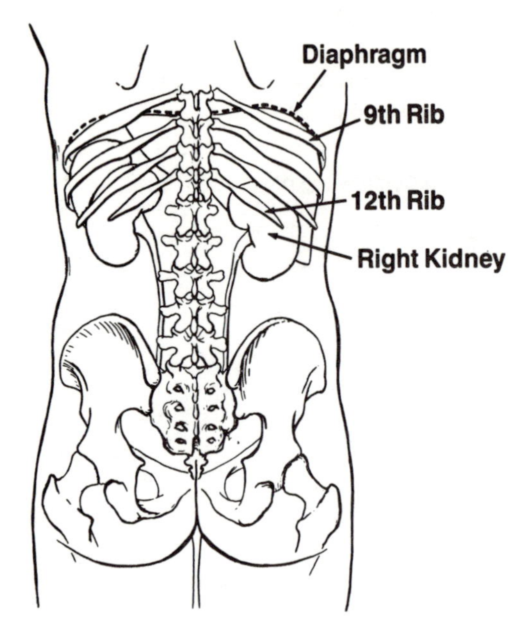

Fig. 16-18 Relationship of kidneys to posterior ribs.

fallopian tube, sigmoid colon, left kidney and ureter.

e. The *periumbilical region* is the area surrounding the umbilicus. Pain from the stomach, pancreas or appendix may be felt here.

f. The *hypogastric, or suprapubic, region* is the area above the pubis. Pain from the bladder, fallopian tubes or uterus may be felt here.

g. The *epigastric, or substernal, region* is the area below the xiphoid. Pain from the distal stomach, duodenum, esophagus or heart may radiate here.

2. The *inguinal region* is the area *superior to the inguinal ligament*. The ligament runs from the anterior superior iliac spine to the pubic bone. Superior and inferior to the ligament are weak areas in the abdominal wall because of vessels and the absence of muscle. Hernias may occur here. *A hernia, or rupture, is the protrusion of a structure through a weak point*. In the abdomen, this usually involves the *parietal peritoneum,* but a part of intestine may accompany it. A her-

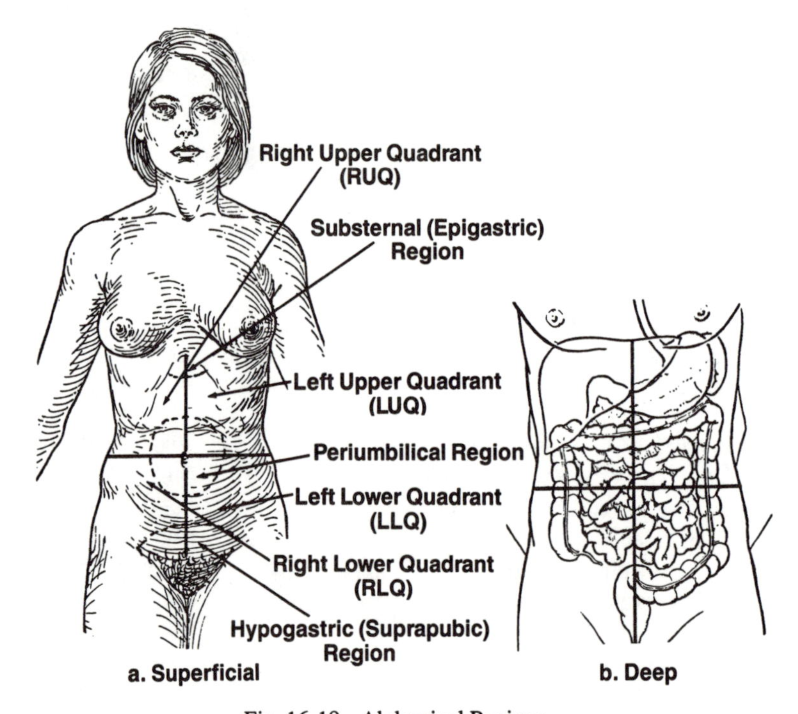

Fig. 16-19 Abdominal Regions.

nia above the inguinal ligament is an *inguinal hernia*; below the ligament, a *femoral hernia* (Fig. 16-20).

3. *Other features.*
   a. In thin people, *pulsations of the abdominal aorta* may be felt with the palm of the hand on the umbilicus.
   b. *Palpation of the liver, spleen and gall-bladder* are sometimes possible when these organs are enlarged from disease. It is important to note how far below the costal margins these organs may be felt (e.g., "the liver is palpable 2 cm below the right costal margin").
   c. *McBurney's point* is a point midway along an imaginary line drawn from the anterior superior iliac spine to the umbilicus. This is the location of the appendix.

B. *Internal structures.*
   1. The *intraperitoneal* organs are described in chapter 13 (see Figs. 13-1 and 13-13).
   2. The *layers between skin and peritoneum* are shown in Fig. 16-21.
   3. *The pelvis* is the bony basin created by the two *hip bones (os coxae), sacrum and coccyx*. The upper part (false pelvis) contains the lower abdominal contents. The true pelvis is the lower portion of the basin and houses the sigmoid colon, rectum, bladder and reproductive organs (Fig. 16-22).
   4. *Retroperitoneal structures* are shown in Fig. 13-9, chapter 13.

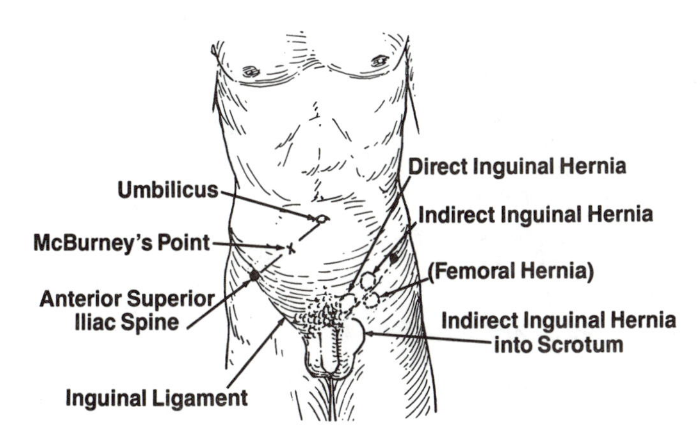

Fig. 16-20 The Inguinal Region and Hernias.

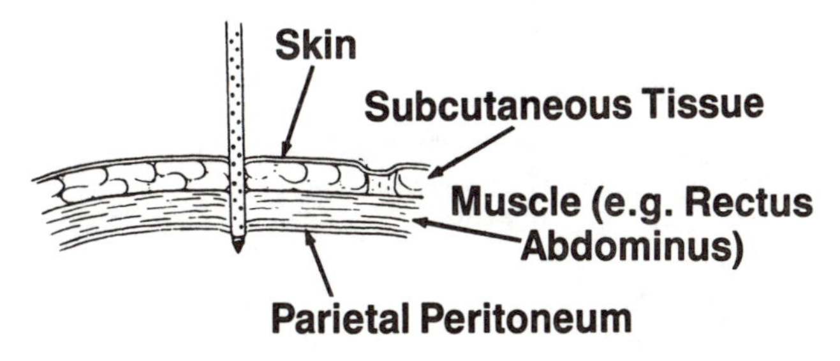

Fig. 16-21 Abdominal Layers. (See ref., Fig. 16-9).

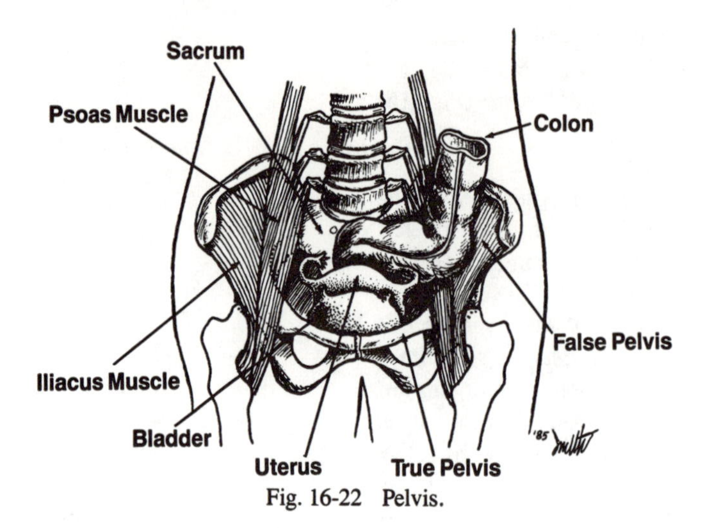

Fig. 16-22 Pelvis.

VI. *The lower extremity.*

A. *Surface anatomy.* Palpate the *iliac crest* and *pubic bone* anteriorly (Fig. 16-23). The *femoral artery* is felt inferior to the inguinal ligament. Posteriorly, the buttock consists of the *gluteus maximus* muscle and, slightly superior, the *gluteus medius*. Both

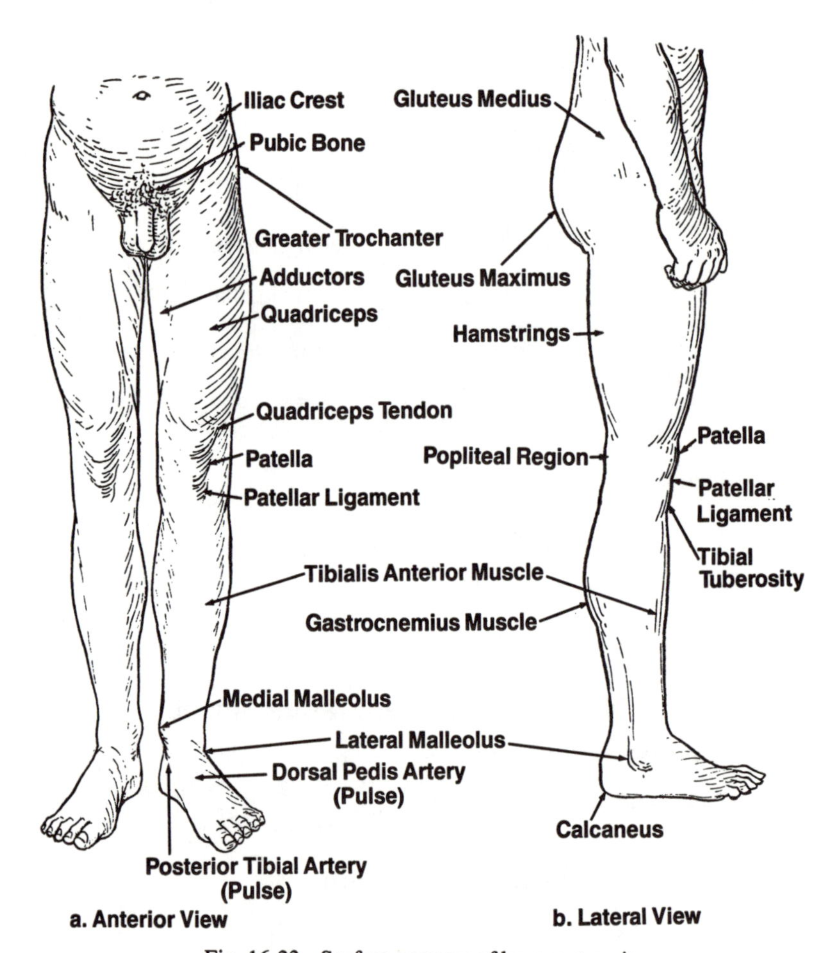

Fig. 16-23 Surface anatomy of lower extremity.

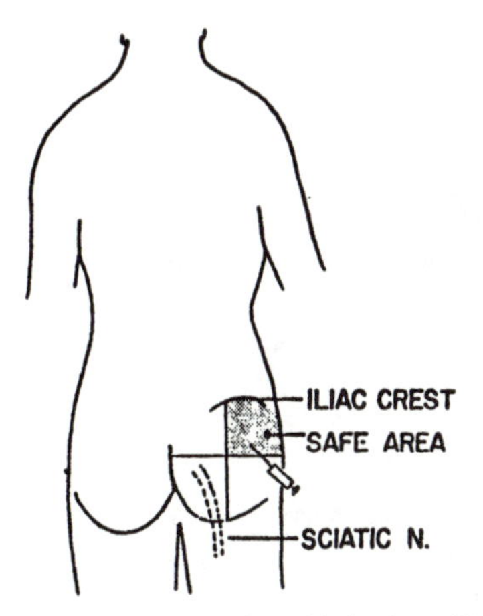

Fig. 16-24 Safe area for gluteal injection. (See ref., Fig. 16-12).

muscles are intramuscular injection sites (Fig. 16-24). The gluteus medius is preferred, because in thin and older people one might injure the sciatic nerve. Palpate the *sacrum* and *os coccyx*. At the inferior area of the gluteus maximus, palpate the *ischial tuberosity*. Note the superficial *greater saphenous vein* running along the medial aspect of the thigh and leg. The *greater trochanter* of the humerus is palpable at the lateral aspect of the thigh. The bulk of the anterior thigh consists of the *quadriceps*. The posterior thigh consists of the *hamstrings*. The *adductor muscles* lie on the medial aspect of the thigh. Feel the *tendons of the hamstrings* in back of the knee. At the knee, palpate the *patella* (kneecap). Above the patella is the *quadriceps tendon*. Below is the *patellar ligament*. Feel the patellar ligament as it attaches to the *tibial tuberosity*. The *epicondyles of the femur and tibia* are palpable on both sides of the knee. The *head of the fibula* is palpable on the lateral aspect of the knee. Follow the tibia down. Slightly lateral, one may palpate the *tibialis anterior muscle*. Posteriorly, the *calf* is composed of the *gastrocnemius* and *soleus muscles*. The distal tibia is palpable at the ankle as the

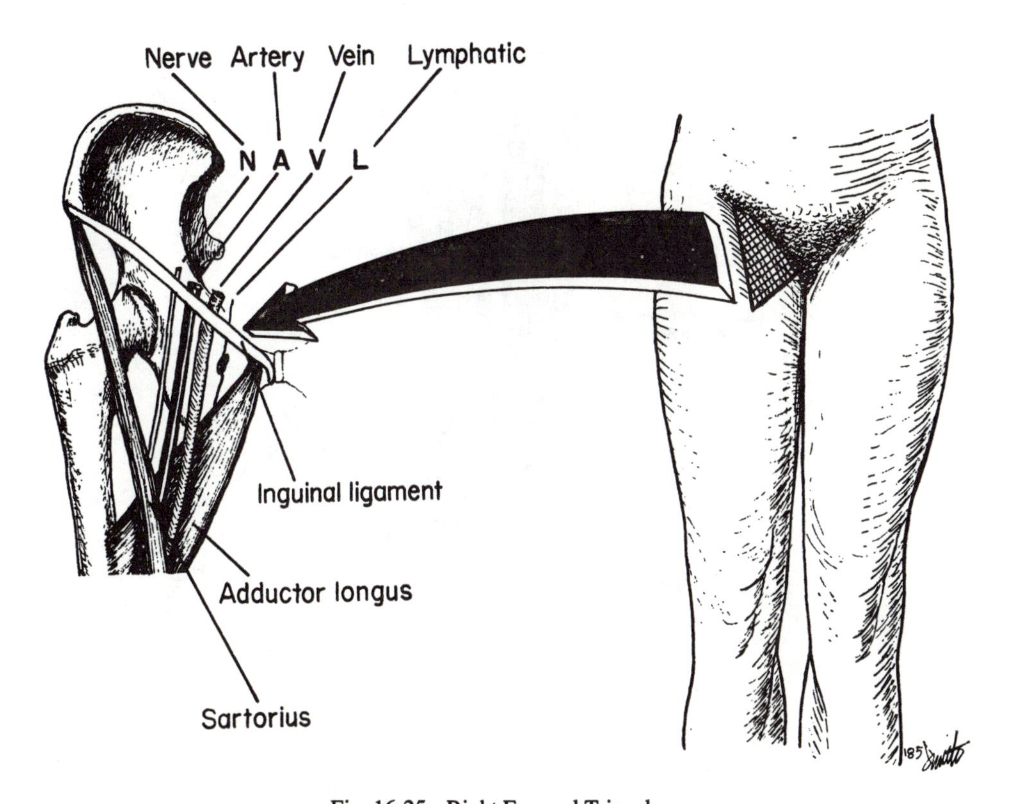

Fig. 16-25 Right Femoral Triangle.

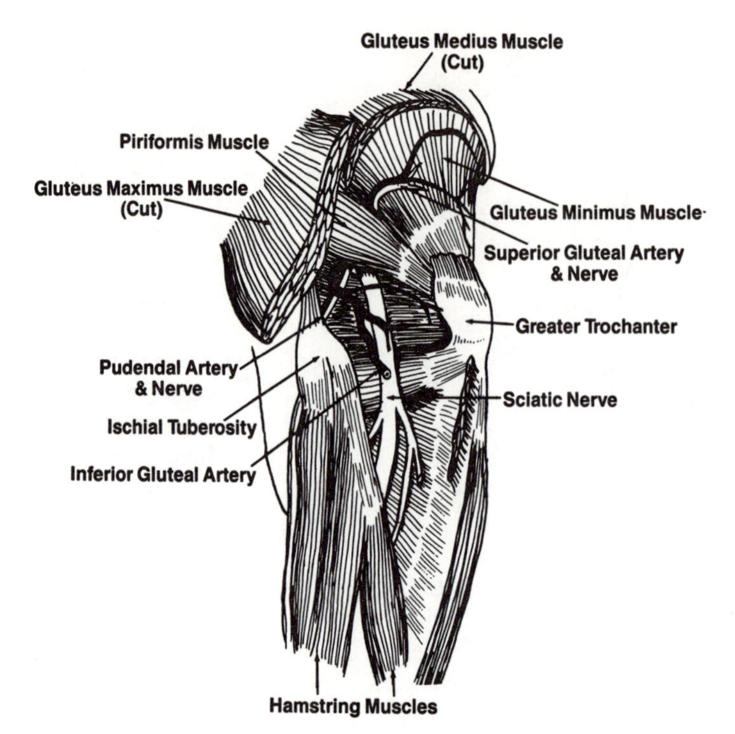

Fig. 16-26 Right Gluteal Region.

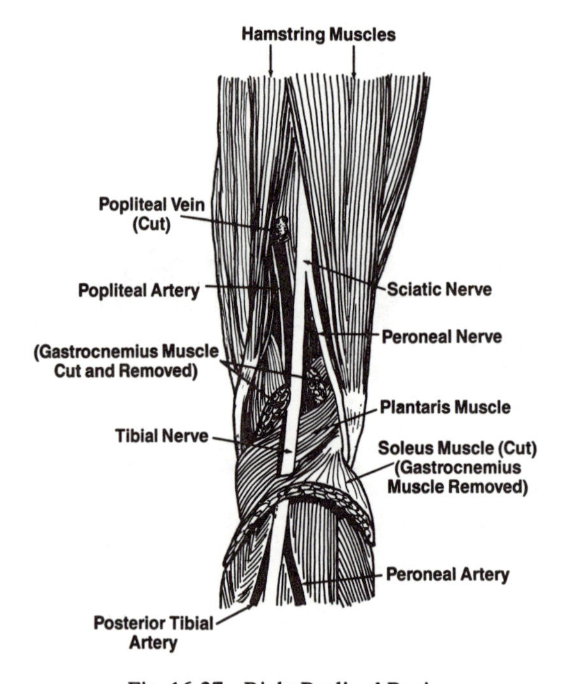

Fig. 16-27 Right Popliteal Region.

medial malleolus, and the distal fibula as the *lateral malleolus*. The top of the foot is the dorsal aspect. The *heel is the calcaneus. Important arterial pulses* are the *posterior tibial*, behind the medial malleolus, and the *dorsalis pedis* on the foot dorsum.

B. *Internal structures*.

1. The *femoral region, or triangle*, is the area inferior to the inguinal ligament (Fig. 16-25). The triangle is bounded by the sartorius and the adductor longus as shown. From lateral to medial are the femoral nerve, artery, vein and lymphatics *("NAVL")*. The *femoral artery* is important in cardiopulmonary resuscitation, because it is the largest palpable artery for checking the *pulse* and drawing *arterial blood gases*. The femoral artery passes through the adductor muscles posteriorly (adductor canal) and emerges as the popliteal artery. A *femoral hernia* is the least common of the groin hernias, and occurs primarily in females.
2. The *gluteal region* lies deep to the gluteus maximus muscle, and contains the external rotators of the thigh. The *sciatic nerve* emerges below an external rotator, the piriformis muscle, and descends to the thigh and leg (Fig. 16-26). The *superior and inferior gluteal arteries*, as well as the *pudendal artery and nerve*, lie here.
3. The *popliteal* fossa is the area in back of the knee. It contains the *popliteal artery and vein*, and the *tibial and peroneal nerves* (Fig. 16-27).

INDEX